Modeling Dynamic Systems

Editors

Matthias Ruth
Bruce Hannon

Springer
New York
Berlin
Heidelberg
Barcelona
Budapest
Hong Kong
London
Milan
Paris
Santa Clara
Singapore
Tokyo

Bruce Hannon

Matthias Ruth

Modeling Dynamic Biological Systems

With a Foreword by Simon A. Levin

With 255 Illustrations and a CD-ROM

Springer

Bruce Hannon
Department of Geography
220 Davenport Hall, MC 150
University of Illinois
Urbana, IL 61801
USA

Matthias Ruth
Center for Energy and
Environmental Studies
and the Department of Geography
Boston University
675 Commonwealth Avenue
Boston, MA 02215
USA

Cover photo: ©TSM/Craig Tuttle, 1997

Library of Congress Cataloging in Publication Data
Hannon, Bruce M.
Modeling dynamic biological systems / by Bruce Hannon and Matthias Ruth.
p. cm. — (Modeling dynamic systems)
Includes bibliographical reference and index.
ISBN 0-387-94850-3 (hardcover : alk. paper)
1. Biological systems—Computer simulation. 2. Biological systems—Mathematical models. I. Ruth, Matthias. II. Title. III. Series.
QH324.2.H36 1997
574′.01′13—dc20 96-38281

The CD-ROM contains the run time version of the STELLA II software which is fully compatible with later versions of the software. All these versions of the software are collectively referred to in the text as STELLA.

Printed on acid-free paper.

Acquiring editor: Robert C. Garber
Production coordinated by Impressions Book and Journal Services, Inc., Madison, WI, and managed by Terry Kornak; manufacturing supervised by Jacqui Ashri.
Typeset by Impressions Book and Journal Services, Inc., Madison, WI.
Printed and bound by Maple-Vail, York, PA.
Printed in the United States of America.

9 8 7 6 5 4 3 2 1

ISBN 0-387-94850-3 Springer-Verlag New York Berlin Heidelberg SPIN 10545133

Foreword

A mathematical model is a caricature, a deliberate oversimplification of reality. As such, its weaknesses may be transparent, its limitations obvious. This is not so, perhaps, with the invisible model a decision maker or researcher must inevitably use. But some form of model must be implicit in informed decision making, lest those decisions become random acts of whimsy, and some form of model similarly guides any experimental design. For all the limitations of any model, the likelihood is that the explicit consideration of assumptions is a step in the direction of better understanding and better decision making. One goal of this book is to develop that thesis, and to provide the reader the tools to become a better decision maker or better scientist. A more general goal is simply to make the power of dynamical simulation models available to the widest possible audience of researchers as aids to exploring the dynamics of ecosystems.

The use of dynamic mathematical models in ecology is not new; indeed, it has a rich and glorious history. In the early part of the 20th century, the brilliant mathematician, Vito Volterra, challenged by his son-in-law, Umberto D'Ancona, to explain the oscillations of the Adriatic fisheries, formulated a simple but now-classic pair of differential equations to show how the interaction between predator and prey could drive sustained oscillations. More sophisticated models, incorporating historical effects via Volterra's own specialty, integral equations, were explored; but it was the simplicity of the ordinary differential equation models that captured the attention of later generations. Indeed, the power of simple models as tools for understanding is also a central theme of this book.

Volterra was not alone in laying out the foundations of today's mathematical ecology. Alfred Lotka, an actuary and part-time genius, developed similar equations, as did the Russian V. I. Kostitzin. Generations of mathematicians explored the complexities of these apparently simple equations; and even today, new and esoteric discoveries about bifurcations and chaos are being made. But this research has largely been the domain of mathematicians, and at times has made little contact with biological fact or application. It is time, in the words of the authors, for a "democratization of modeling," driven by their conviction that modeling is too important and too much fun to be left to mathematicians. It is also the case that modeling in the hands of the uninformed can be a dangerous thing. Hence, as modeling becomes more widely used, it is essential for those who will use the output as well as those who will do the modeling to learn the complexities and limitations to the greatest extent possible; the touchstone of democracy is an informed electorate.

This book could not have been imagined 20 years ago, or consummated even 10 years ago. The advent of computers indeed introduced a new dimension, the ability to go beyond analytical treatments and to use simulation as an experimental tool for much more complex models. High speed computation has led to increases in what can be done at a dizzying rate. But the initial result of this revolution was to replace one kind of theoretical elite—the mathematician—with another—the computer nerd. The power of the modeling was no more in the hands of the proletariat than it was when the keys to the kingdom were in the intricacies of dynamical systems theory. Hannon and Ruth have set out to change that. Using STELLA as a platform, they have produced the modeling cookbook for those who thought they hated modeling. Their goal, admirably achieved, was to show that some forms of modeling are no harder to operate than a VCR, and that there are intellectual and practical rewards waiting for those willing to venture forth.

The student of this text will be well rewarded, learning about modeling in the only way one really can . . . by putting one's modeling muddy boots on and slogging through exercises. Along the way, the hardy traveler will enjoy a wonderful tour of a wide range of applications in ecology, while learning both techniques and pitfalls. Modeling is just one tool available to the researcher, to be used in concert with observation, experimentation, and hypothesis construction. Simulation such as that taught in this book is just one part of the process, to be combined with analysis and thought. But it is a powerful tool for exploring the consequences of our assumptions, and Hannon and Ruth have provided an introduction that opens this world to a far greater part of the research community than could have been included a decade ago. Word processing seemed forbidding when it was first made available but now is a skill performed more easily by most Americans than arithmetic or spelling. Modeling too can be made easily available, at least in its simplest forms. Word processing is no substitute for thoughtful composition, and STELLA similarly has limits. For the advanced researcher, or one who wants to become one, more sophisticated approaches beckon. Nonetheless, we are nearing the day when every researcher will be able to use simple tools such as STELLA to construct representations of their hypotheses, and to use these to explore broad classes of scenarios unavailable through experimentation. Read on and join the revolution.

Simon A. Levin
Princeton University

Series Preface

The world consists of many complex systems, ranging from our own bodies to ecosystems to economic systems. Despite their diversity, complex systems have many structural and functional features in common that can be effectively modeled using powerful, user-friendly software. As a result, virtually anyone can explore the nature of complex systems and their dynamical behavior under a range of assumptions and conditions. This ability to model dynamic systems is already having a powerful influence on teaching and studying complexity.

The books in this series will promote this revolution in "systems thinking" by integrating skills of numeracy and techniques of dynamic modeling into a variety of disciplines. The unifying theme across the series will be the power and simplicity of the model-building process, and all books are designed to engage readers in developing their own models for exploration of the dynamics of systems that are of interest to them.

Modeling Dynamic Systems does not endorse any particular modeling paradigm or software. Rather, the volumes in the series will emphasize simplicity of learning, expressive power, and the speed of execution as priorities that will facilitate deeper system understanding.

Matthias Ruth and Bruce Hannon

Preface

The problems of understanding complex system behavior and the challenge of developing easy-to-use models are apparent in the fields of biology and ecology. In real-world ecosystems, many parameters need to be assessed. This requires tools that enhance the collection and organization of data, interdisciplinary model development, transparency of models, and visualization of the results. Neither purely mathematical nor purely experimental approaches will suffice to help us better understand the world we live in and shape so intensively.

Until recently, we needed significant preparation in mathematics and computer programming to develop such models. Because of this hurdle, many have failed to give serious consideration to preparing and manipulating computer models of dynamic events in the world around them. This book, and the methods on which it is built, will empower us to model and analyze the dynamics characteristic of human–environment interactions.

Without computer models we are often left to choose between two strategies. First, we may resort to theoretical models that describe the world around us. Mathematics offers powerful tools for such descriptions, adhering to logic and providing a common language by sharing similar symbols and tools for analysis. Mathematical models are appealing in social and natural science, where cause-and-effect relationships are confusing. These models, however, run the risk of becoming detached from reality, sacrificing realism for analytical tractability. As a result, these models are only accessible to the trained scientist, leaving others to believe or not believe the model results.

Second, we may manipulate real systems in order to understand cause and effect. One could modify the system experimentally (e.g., introduce a pesticide, some CO_2, etc.) and observe the effects. If no significant effects are noted, one is free to assume the action has no effect and to increase the level of the system change. This is an exceedingly common approach. It is an elaboration of the way an auto mechanic repairs an engine, by trial and error. But social and ecological systems are not auto engines. Errors in tampering with these systems can have substantial costs, both in the short and long term. Despite growing evidence, the trial-and-error approach remains the meter of the day. We trust that, just like the auto mechanic, we will be clever enough to clear up the problems created by the introduced change. Hand–eye tinkering is the American way. We let our tendency toward optimism mask the new problems.

However, the level of intervention in social and ecological systems has become so great that the adverse effects cannot be ignored. As our optimism about repair begins to crumble, we take on the attitude of patience toward the inevitable—unassignable cancer risk, global warming, fossil fuel depletion—the list is long. We are pessimistic about our ability to identify and influence cause-and-effect relationships. We need to understand the interactions of the components of dynamic systems in order to guide our actions. We need to add synthetic thinking to the reductionist approach. Otherwise, we will continue to be overwhelmed by details, failing to see the forest for the trees.

There is something useful that we can do to turn from this path. We can experiment using computer models. Models give us predictions of the short- and long-term outcomes of proposed actions. To do this we can effectively combine mathematical models with experimentation. By building on the strengths of each, we will gain insight that exceeds the knowledge derived from choosing one method over the other. Experimenting with computer models will open a new world in our understanding of dynamic systems. The consequences of discovering adverse effects in a computer model are no more than ruffled pride.

Computer modeling has been with us for nearly 40 years. Why, then, are we so enthusiastic about its use now? The answer comes from innovations in software and powerful, affordable hardware available to every individual. Almost anyone can now begin to simulate real world phenomena on their own, in terms that are easily explainable to others. Computer models are no longer confined to the computer laboratory. They have moved into every classroom, and we believe they can and should move into the personal repertoire of every educated citizen.

The ecologist Garrett Hardin and the physicist Heinz Pagels have noted that an understanding of system function, as a specific skill, needs to be and can become an integral part of general education. It requires the recognition (easily demonstrable with exceedingly simple computer models) that the human mind is not capable of handling very complex dynamic models by itself. Just as we need help in seeing bacteria and distant stars, we need help modeling dynamic systems. We *do* solve the crucial dynamic modeling problem of ducking stones thrown at us or of safely crossing busy streets. We learned to solve these problems by being shown the logical outcome of mistakes or through survivable accidents of judgment. We experiment with the real world as children and get hit by hurled stones, or we let adults play out their mental model of the consequences for us and we believe them. These actions are the result of experimental and predictive models, and they begin to occur at an early age. In the complex social, economic, and ecological world, however, we cannot rely on the completely mental model for individual or especially for group action, and often we cannot afford to experiment with the system in which we live. We must learn to simulate, to experiment, and to predict with complex models.

In this book, we have selected the modeling software STELLA® and MADONNA, which combine together the strengths of an iconographic programming style and the speed and versatility of desktop computers. Programs such as STELLA and MADONNA are changing the way in which we think. They enable each of us to focus and clarify the mental model we have of a particular phenomenon, to augment it, elaborate it, and then to do something we cannot otherwise do: to run it, to let it yield the inevitable dynamic consequences hidden in our assumptions and the structure of the model. STELLA, MADONNA, and the Macintosh, as well as the new, easy-to-use, Windows-based personal computers, are not the ultimate tools in this process of mind extension. However, their relative ease of use make the path to freer and more powerful intellectual inquiry accessible to every student.

These are the arguments for this book on dynamic modeling of biological systems. We consider such modeling as the most important task before us. To help students learn to extend the reach of their minds in this unfamiliar yet very powerful way is the most important thing we can do.

Bruce Hannon
Urbana

Matthias Ruth
Boston

Contents

Part 1

Introduction

1

Modeling Dynamic Biological Systems

> Data can become information if we know the processes involved. Information can become knowledge if we see the system that is operating. But knowledge only becomes wisdom when we can see how any system must change, and can deal with that reality.
>
> Peter Allen,
> Coherence, Chaos and Evolution in the Social Context,
> *Futures* 26:597, 1994.

1.1 Process and Art of Model Building

Biologists, from those who study the mechanisms of the nerve cell to those who study ecosystems, are in one way or another inescapably involved in dynamic modeling. This book is dedicated to biologists, with an understanding of at least some of the problems they face. This book is about the process and art of modeling. We define the process of model building as an unending one—one with rewards typically proportional to the effort extended. The art of modeling is implied throughout this book, first by virtue of our continuing reference to the style of the modeling approach, second by the plethora of modeling examples from nearly every field in biology, and finally by regular reference to the use of analogy.

Modeling style is important. Throughout our life we have learned to develop models in our minds of the processes that we face every day. We do solve an amazing class of dynamic problems, such as hitting baseballs and driving cars, by acquiring through trial and error the skills that are necessary to put the various components of a dynamic system together in our mind, draw the necessary conclusions, and react acordingly. However, the more complex the system, the less we are able to sufficiently grasp in our mind its workings and to prepare our actions. We simply cannot hold the many aspects of a dynamic process in mind at once. We need to be able to capture our knowledge, and possibly that of others, in a consistent and transparent way so that we can better understand, and act in, a changing world. There are by now general rules and computer programs available that have been found useful in letting the modelers quickly get down to the business of capturing their experience in those little chips inside the electonic box. Such knowledge capturing is essential to both learning and understanding.

But just as we needed microscopes and telescopes to extend the reach of the eye, we need dynamic simulation to extend the reach of our minds. In this process, the computer becomes a facilitator, but it does not substitute for our ability to develop an understanding of complex dynamic systems; rather, it requires the process and skill of modeling that we called for earlier. The computer is a means by which we can enlarge our reach into as yet unexplored territory, and we need to accustom ourselves to the possibilities it opens for us. In this sense, the computer is not unlike other great technologies that required we familiarize ourselves with them and come to understand their powers and limitations.

We have long been accustomed to machinery which easily out-performs us in *physical* ways. *That* causes us no distress. On the contrary, we are only too pleased to have devices which regularly propel us at great speed across the ground—a good five times as fast as the swiftest human athlete—or that can dig holes or demolish unwanted structures at rates which would put teams of dozens of men to shame. We are even more delighted to have machines that can enable us physically to do things we have never been able to do before: they can lift us into the sky and deposit us at the other side of an ocean in a matter of hours. These achievements do not worry our pride. But to be able to *think*—that has been a very human prerogative.[1]

Dynamic modeling is a process of extending our knowledge, and the computer is only the means towards this end. The history of dynamic modeling is traced back to World War II and the immense technical effort mounted by the scientists involved in the development of automated computation. Dynamic modeling was done before then, but to escape the use by the skilled mathematicians we had to wait for the advent of machine computation, the development by Forrester of the basic computer logic, and the eventual emergence of this development to the personal computer in the form of STELLA. STELLA is a graphical programming process that evokes the most easily accessible form of symbolic understanding by humans, the use of icons. As you will find throughout the text, the classification of variables is quite simple and the resulting icons associated with them are quite appropriate for capturing all the parts that influence a system's behavior. An experienced STELLA modeler literally sees and understands the dynamic process through the arrangement of these icons. This is part of what we mean by the modeling art. Now the art can be practiced by anyone with knowledge of basic mathematics and the ability to use a personal computer.

The only way to achieve the facile use of the icons of dynamic modeling is to use them repeartedly in many different applications. Thus our book is a carefully arranged set of models of biological processes that build on each other to various degrees and make use of the same modeling tools in various contexts. We have tried to arrange the models from simple to more complex and by scale, from small to large, from simple growth models of a cell to a rather involved set of interacting spatial ecosystems. As you work through these models, you will not only become fluent in the use of the

[1]R. Penrose, *The Emperor's New Mind,* New York: Penguin Books, 1989.

modeling language STELLA, but you will also develop a new way of thinking about dynamic systems. Practice is the foundation of the modeling art.

Throughout the examples, we try to show how the principal idea of one model can be used again in a different application. The basic growth models are elements of large models. The law of mass action in chemistry is used in epidemic and ecosystem modeling. The play of analogy is dangerous in that it can be misleading, but the loss is never more than a blush of embarrassment and a little electricity. Much new science seems to ride on analogy: Its use is the final piece in the construct of the modeler's art.

The goals of dynamic modeling are to explain and, with enough effort and luck, to predict. The dynamic events occurring in the real world are multifaceted, interrelated, and difficult, perhaps impossible, to understand. To reduce our worry and to sate our curiosity about such events, we pose and then try to answer specific questions about the dynamic processes that seem to comprise these events. Of necessity, we abstract from most of the details and attempt to concentrate on some portion of the larger picture, a particular set of features of the real world. The resulting models are true abstractions of reality. They force us to face the results of the structural and dynamic assumptions we have used in our abstractions.

The modeling process is necessarily complicated and it is unending. Well-posed questions lead us develop a model; that model leads us to more questions. If we are good at modeling, we approach reality. If we are expert at modeling and we perisist at it, we approach reality asymptotically. We observe what we call real events from the world; we abstract a version of these events, highlighting our view of the important elements; then we build and run a dynamic model; we draw conclusions from the results of the model; and finally, we compare our explanation of the events with the reality of such events. If we are really good, we make and test predicitons with the model. According to John L. Casti,[2] good models are the simplest that explain most of the data from an operating system and yet do not explain it all, leaving some room for the model or theory to grow. Good models must have elements that directly correspond to objects in the real world. All models are necessarily simple constructs of reality; some are just too simple, some just too complex.

The biological modeling process is certainly a knowledge-capturing one, and yet it is exceedingly difficult to do well. Nature seems ineffable. The ultimate difficulty stems from the complexity of the living system, both its structure and its dynamics, and the impossibility of removing ourselves fully to observational status. Even an expert in biochemistry, in medicine, or in ecosystem analysis cannot make much headway against such complexity alone. A team of such people, expert in the various aspects of a problem, may be the answer. As a team, researchers can more easily see

---

[2]John L. Casti, *Alternate Realities: Mathematical Models of Nature and Man*, New York, John Wiley and Sons, 1989.

anthropomorphism in each other. We suspect that such team modeling will become much more common in the near future. The elements that make such a team successful are, of course, the possession of real and pertinent expertise, compatible personalities, a common concept of the questions to be answered, and a common modeling language. Because the level of modeling expertise is likely to vary among even the best set of experts, a simple modeling language is needed—one that can be understood by each of the team in a very short time. The programming language STELLA fills that requirement.[3] Our book contains the guidelines for the most likely successful process of team or individual modeling, and it is based on that language. Many generations of students—at universities, in government, and in corporations—have contributed to the process described herein.

1.2 Static, Comparative Static, and Dynamic Models

Three general types of models can be defined. The first type are models that represent a particular phenomenon at a point of time, a static model. For example, a map of the world might show the location and size of a city or the location of infection with a particular disease, each in a given year. The second type are comparative static models that compare some phenomena at different points in time. This is like using a series of snapshots to make inferences about the system's path from one point in time to another without modeling that process itself.

Other models describe and analyze the very processes by which a particular phenomenon is created. We may develop a mathematical model describing the change in the rate of migration to or from a city, or the change in the rate of the spreading of a disease. Similarly, we may develop a model that represents the change of these rates *over time*. This latter type are dynamic models that attempt to capture the change in real or simulated time.

With the advent of easy-to-use computers and software, we can all build on the mathematical descriptions of a system and carry them further. The world is not a static or comparative static process, and so the models treating it in that way will become obsolete, and are perhaps even misleading. We can now investigate in great detail and with great precision the system's behavior over time, its movement toward or away from equilibrium positions, rather than restrict the analysis to an equilibrium point itself.

An understanding of the dynamics and changing interrelationships of systems, such as social, biological, and physical systems, is of particular importance in a world in which we face increasing complexity. In a variety of disciplines scientists ask questions that involve complex and changing interrelationships among systems. How do mutation and natural selection affect

---

[3]The producers of STELLA, High Performance Systems, Inc., offer an identical product that is called Ithink and marketed to the business community.

the distribution of genetic information in a population? How does a vaccination program affect the spread of a disease? All good modeling processes begin (and end) with a good set of questions. These questions keep the modeler focused and away from the miasma of random exploration.

Models help us understand the dynamics of real-world processes by mimicking with the computer the actual but simplified forces that are assumed to result in a system's behavior. For example, it may be assumed that the number of people migrating from one country to another is directly proportional to the population living in each country and decreases the further these countries are apart. In a simple version of this migration model, we may abstract away from a variety of factors that impede or stimulate migration, besides those directly related to the different population sizes and distance. Such an abstraction may leave us with a sufficiently good predictor of the known migration rates, or it may not. If our answers do not compare sufficiently well with reality, we re-examine the abstractions, reduce the simplifying assumptions, and retest the model for its new predictions. The results will not only be a better model of the system under investigation but, most importantly, a better undersanding of our conception of that system, showing us whether we were indeed able to identify and properly represent the essential features of that system.

We cannot overstress the fact that one should keep the model simple, even simpler than one knows the cause-and-effect relationship to be, and only grudgingly complexify the model when it does not produce the real effects. After all, it is not the goal to develop models that capture all facets of real-life systems. Such models would be useless because they would be as complicated as the systems we wanted to understand in the first place. The real quest of dynamic modeling is to "discover" the hopefully few rather simple underlying principles that together bring the observed complexity. This is our meaning of simplicity.

Each element of the model is specified by initial conditions, and the computer works out the system's responses according to the specified relations among the model elements. The initial conditions may derive from actual measurement, such as the number of people living on an island, or estimates, such as estimates of the number of voles living in a specific garden. The estimates, in turn, could be derived from empirical information or even reasonable guesses by a modeling team. Models built on such uncertain parameters may still be of great value, providing a picture of a particular processes, rather than exact information. Documentations of the parameters and assumptions, always necessary at each step in the modeling process, is important when the modeler's judgment is used.

In the end, models can be no better than the modelers. Hence the elegant statement by Botkin is very appropriate,

> by operating the model the computer faithfully and faultlessly demonstrates the implications of our assumptions and information. It forces us to see the implications, true or false, wise or foolish, of the assumptions we have made. It is not so much

that we want to believe everything that the computer tells us, but that we want a tool to confront us with the implications of what we think we know.[4]

1.3 Analogies, Anomalies, and Reality

For many years, physicists have known of the analogous relations between the principle variables in the basic equations of hydraulics, electricity, and mechanical systems. Force, springs, dampers, inertia, velocity, and displacement have their counterparts in voltage, current, resistance, inductance, and capacitance, and again, pressure, mass flow, frictional loss, and vorticity. Coulomb apparently was convinced that the attractive force between charged particles was of the same form as the gravitational attraction between planetary bodies, put forth by Newton centuries before. These analogies are more than curiosities. They show a common world view of such important phenomena. As scientists developed each of these disciplines in turn, they recognized the heritage of hard-won successes in describing parts of the real world.

Not only were these analogies useful in physics and engineering, but it is well known that the great economist Walras produced his equations of the economy from the principles of hydraulics. Before him, the medical doctor and physiocratic economist Quesnay divined his input–output tables of the French economy from an analogy with the circulation of blood in the body. Analogy, carried to the right level of detail, is the cornerstone, if not the foundation, of the creative enterprise.

Analogies abound among economics, biology, and chemistry. For example, the most common production functional form is

$$\text{Rate of Production} = Q = A\, L^{\alpha}\, K^{\beta},$$

where A, α, and β are constants; L is the input of labor; and K is the input of capital services in a process that produces, say, widgets at rate Q. But to a chemist, this is the law of mass action at work. Q is the production rate of some product made from the reaction of L and K, two chemicals that combine to produce the product. The constants α and β reflect temperature or perhaps pressure effects on the reaction rate. Or if you were an epidemiologist, you would say that Q is the rate at which people are getting sick, L is the size of the healthy population, K is the size of the sick population, and *A* is the contact coefficient. If you are a metapopulaton theorist in ecology, L would be the number of patches occupied by an inferior species, K the number of superior species, A the colonization rate of the superior species, and Q the rate of conversion of the inferior patches to superior ones.

---

[4]D. Botkin, Life and Death in a Forest: The Computer as an Aid to Understanding, in: C. Hall and J. Day (eds.), *Ecosystem Modeling in Theory and Practice: An Introduction with Case Studies*, New York: John Wiley and Sons, 1977, p. 217.

Some ecologists use a variant analogy to the economic production function described earlier. They say that Q, an insect birth rate, for example, equals a maximum growth rate for the insect (A) times a series of factors, each of whose value varies from 0 to 1, where 1 is the factor value associated with the maximum possible growth rate. These are usually graphically based, experimentally derived factors that naturally show diminishing returns. Examples of such factors are temperature and humidity. Under the exact condition of optimal temperature for the growth rate for this insect, the temperature factor would be 1.0.

This process is actually identical to the production function use in economics. In both cases the factors are assumed to be completely independent. Capital and labor can be substituted for one another without concern about the availabilty of the other. But actually, it clearly takes labor to make more capital to substitute for the displaced labor. So the factors are not actually independent, although they are commonly assumed to be. Neither are temperature and humidity independent, despite the assumptions in the insect model, for example. For a single firm, the independent factor assumption is not a terribly bad one, but to make this assumption for the economy as a whole is absurd. Such assumptions are usually a matter of expediency of model building. Be careful how you use them.

Analogies can also help you spot anomalies. Lightman and Gingerich[5] point to the *retrorecognition* phenomenon, where anomalies in one theory are only recognized when they are explained later by a superseding theory. For a variety of reasons, we scientists are essentially blind to those facts not explained by the dominant theory. By the use of appropriate analogy, *de rigor* for the 19th century likes of Lord Kelvin and J. C. Maxwell, we should be able to turn up anomalies in our current explanations of the way things work. A rule of thumb for anomaly finding is to push your model to its reasonable limits.

Our discussion on the use of analogy is intended to raise your optimism about the idea of dynamic modeling. But not everyone is optimistic. We might be able to accurately simulate some very complex biological system, but can we actually learn more about these systems from such models? Can we learn to make wiser decisions from our modeling exercises? We think the answer is yes but our view has its dissenters.[6] The counterargument is based on the idea that human problem solving is very context dependent, while most computer models are not. This sounds to us more like a complaint than a fatal criticism, that is, we should be able to model in a context-dependent way when that is needed. We should be able to model in such a way that the guiding rules in the model shift as the context shifts. We further

---

[5]A. Lightman and O. Gingerich, When Do Anomalies Begin?, *Science* 255:690–695, 1991.

[6]For an excellent summary of this argument, see P. Denning, Modeling Reality, *American Scientist* 78:495–498, 1990.

argue that the terrific complexity of biological, ecological, and sociological systems can mask the possibility of simple underlying rules. These rules, when used together in a model, might cause the system behavior to appear exceedingly complex. The quest is to find the underlying rules. Such a quest pushes us well beyond simulation. It is what we mean by the term *dynamic modeling.*

1.4 Model Components

The most important elements of a system are the state variables. State variables are indicators of the current status of the system. They are the variables on which all the other calculations in the model are based. State variables come in two flavors: conserved and nonconserved. A conserved state variable represents an accumulation or stock of something—water, people, materials, or information. These stocks are created and destroyed by the results of the control variables in the system. But nonconserved state variables, such as price and temperature, are also possible. Clearly, the temperature of a hot body sitting in a cool room will determine the rate that the body cools. The changing price of a natural resource will signal the changes in its rate of optimal depletion. To maintain simplicity in the model, strive to minimize the number of its state variables.

System elements that represent the action or change in a state variable are called *flows* or *control variables.* As a model unfolds in time, control variables update the state variables at the end of each time step. Examples for control variables are the rate of flood water inflowing to a reservoir, the rate of water release from that reservoir, and the rate of water evaporation from its surface, all acting to change the water contained or "conserved" in the state variable, the reservoir.

The remaining set of variables in any model might be classed as convertors or transforming variables. They take in parameters, or perhaps the results of calculations elsewhere in the model, and transform these inputs still further. These results are relayed on to other such transforming variables or to that special class of transforming variables, the control variables. These interactions in a model are often classed in terms of feedback—the flow of information from a state variable through a chain of transforming variables, and ultimately back to the control variables, change that state variable, and so continue in an ever-changing loop, perhaps finally reaching a steady state, or maybe race off to infinity, to zero, or to chaos. The nature of such circulations of information is negative or postive. Negative feedback tends to force state variables toward goals set up either implicitly or explicitly in the model. Positive feedback tends to do the opposite: It allows variables to re-enforce differences rather than to minimize these differences. Positive feedback, as we shall see in the text, has some surprising results. Negative feedback is the basic idea of the controlled dynamic system, and ultimately it is the confederate of the causally oriented theorist.

Variation in feedback processes can be brought about by nonlinear relationships. Such nonlinear relationships are present if a control variable does not depend on other variables in a linear fashion but changes, for example, with the square root of some other variable. As a result of nonlinear feedback processes, systems can exhibit complex dynamic behavior. Consequently, system modelers must pay special attention to nonlinearities, particularly lagged effects, that describe the relationships among model components.

Throughout this book we encounter a variety of nonlinear feedback processes that give rise to complex system behavior. Let us develop a simple model to illustrate the concepts of state variables, flows, and feedback processes. We will then return to discuss some principles of modeling that will help you to structure the model building process in a set of steps.

1.5 Modeling in STELLA

To explore modeling with STELLA, we will develop a basic model of the dynamics of a fish population. Assume you are the sole owner of a pond that is stocked with 200 fish that all reproduce at a fixed rate of 5% per year. For simplicity, assume also that none of the fish die. How many fish will you own after 20 years?

In building the model, we will utilize all four of the graphical tools for programming in STELLA. The appendix has a Quick Help Guide to the Software, should you need one. The appendix also describes how to install the STELLA software and models of the book. Follow these instructions before you proceed. Then double-click on the STELLA icon to open it.

On opening STELLA, you will be faced with the High-Level-Mapping Layer, which we will not need now. To get to the Diagram Layer, click on the downward-pointing arrow in the upper left-hand corner of the frame:

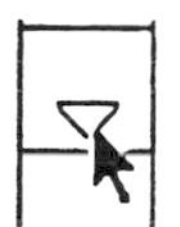

The Diagram Layer displays the following symbols, *building blocks,* for stocks, flows, converters, and connectors (information arrows):

Click on the globe to access the modeling mode:

In the modeling mode you can specify your model's initial conditions and functional relationships. The following symbol indicates that we are now in the modeling mode:

We begin with the first tool, a stock (rectangle). In our example model, the stock will represent the number of fish in your pond. Click on the rectangle with your mouse, drag it to the center of the screen, and click again. Type in FISH. This is what you get:

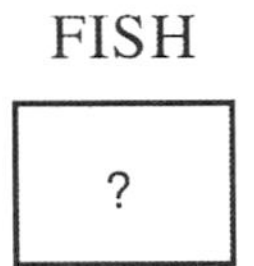

This is the first state variable in our model. Here we indicate and document a state or condition of the system. In STELLA, this stock is known as a *reservoir*. In our model, this stock represents the number of fish of the species we are studying that populate the pond. If we assume that the pond is one square kilometer large, the value of the state variable FISH is also its *density,* which will be updated and stored in the computer's memory at every step of time (DT) throughout the duration of the model. The fish population is a stock, something that can be contained and conserved in the *reservoir;* density is not a stock, it is not conserved. Nonetheless, both of these variable are state variables. So, because we are studying a species of fish in a *specific area* (one square kilometer), the population size and density are represented by the same rectangle.

Inside the rectangle is a question mark. This is the remind us that we need an initial or starting value for all state variables. Double-click on the rectangle. A dialogue box will appear, which happens often in STELLA. The box is asking for an initial value. Add the initial value you choose, such as 200, using the keyboard or the mouse and the dialogue keypad. When you have finished, click OK to close the dialogue box. The question mark will have disappeared.

We must decide next what factors control (i.e., add to or subtract from) the number of fish in the population. Since we assumed that the fish in your pond never die, we have one control variable: REPRODUCTION. We use the *flow* tool (the right-pointing arrow, second from the left) to represent the control variable, so named because it controls the states (variables). Click on the flow symbol; then click on a point about 2 inches to the left of the rectangle (stock) and drag the arrow to POPULATION, until the stock becomes dashed, and release. Label the circle REPRODUCTION. This is what you will have (Fig. 1.1):

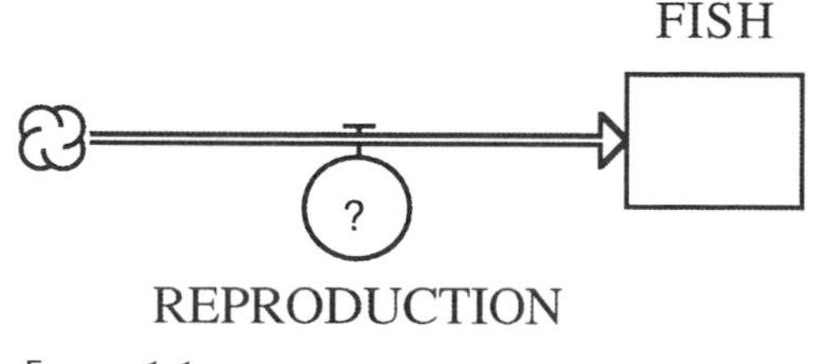

FIGURE 1.1

Here, the arrow points only into the stock, which indicates an inflow. But, you can get the arrow to point both ways if you want it to. You do this by double-clicking on the circle in the flow symbol and choosing Biflow. The Biflow enables you to add to the stock if the flow generates a positive number and to subtract from the stock if the flow is negative. In our model, of course, the flow REPRODUCTION is always positive and newly born fish go only *into* the population. Our control variable REPRODUCTION is a uniflow—new fish per annum.

Next we need to know how the fish in our species reproduce; not the biological details, just how to accurately estimate the number of new fish per annum. One way to do this is to look up the birth rate for the fish species in our pond. Say we find that the birth rate = 5 new fish per 100 adults each year, which can be represented as a transforming variable. A transforming variable is expressed as a *converter*, the circle that is second from the right in the STELLA toolbox. So far REPRODUCTION RATE is a constant; later we will allow the reproduction rate to vary. The same clicking and dragging technique that got the stock on the screen will bring up the circle. Open the converter and enter the number 0.05 (5/100). Down the side of the dialogue box is an impressive list of "built-in" functions that we can use for more complex models.

At the right of the STELLA toolbox is the *connector* (information arrow). We use the connector to pass on information (about the state, control, or transforming variable) to a circle, to a control or transforming variable. In this case, we want to pass on information about the REPRODUCTION RATE to REPRODUCTION. Once you draw the information arrow from the transforming variable REPRODUCTION RATE to the control and from the stock FISH to the control, open the control by double-clicking on it. Recognize that REPRODUCTION RATE and FISH are two required inputs for the specification of REPRODUCTION. Note also that STELLA asks you to specify the control: REPRODUCTION = . . . "Place right hand side of equation here."

Click on REPRODUCTION, then on the multiplication sign in the dialog box, and then on FISH to generate the equation:

$$\text{REPRODUCTION} = \text{REPRODUCTION RATE} * \text{FISH} \quad (1)$$

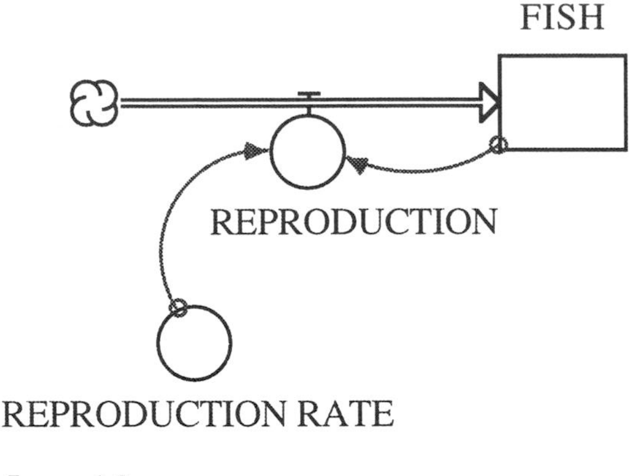

FIGURE 1.2

Click on OK and the question mark in the control REPRODUCTION disappears. Your STELLA diagram should now look like Figure 1.2.

Next, we set the temporal (time) parameters of the model. These are DT (the time step over which the stock variables are updated) and the total time length of a model run. Go to the RUN pull-down menu on the menu bar and select Time Specs . . . A dialogue box will appear in which you can specify, among other things, the length of the simulation, the DT, and the units of time. We arbitrarily choose DT = 1, length of time = 20, and units of time = years.

To display the results of our model, click on the graph icon and drag it to the diagram. If we wanted to, we could display these results in a table by choosing the table icon instead. The STELLA icons for graphs and tables are, respectively,

When you create a new graph pad it will open automatically. To open a pad that had been created previously, just double-click on it to display the list of stocks, flows, and parameters for our model. Each one can be plotted. Select FISH to be plotted and, with the >> arrow, add it to the list of selected items. Then set the scale from 0 to 600 and check OK. You can set the scale by clicking once on the variable whose scale you wish to set and then on the ↕ arrow next to it. Now you can select the minimum on the graph, and the maximum value defines the highest point on the graph. Rerunning the model under alternative parameter settings will lead to graphs that are plotted over different ranges. Sometimes these are a bit difficult to compare with previous runs because the scaling has changed.

Would you like to see the results of our model so far? We can run the model by selecting RUN from the pull-down menu. We get Figure 1.3.

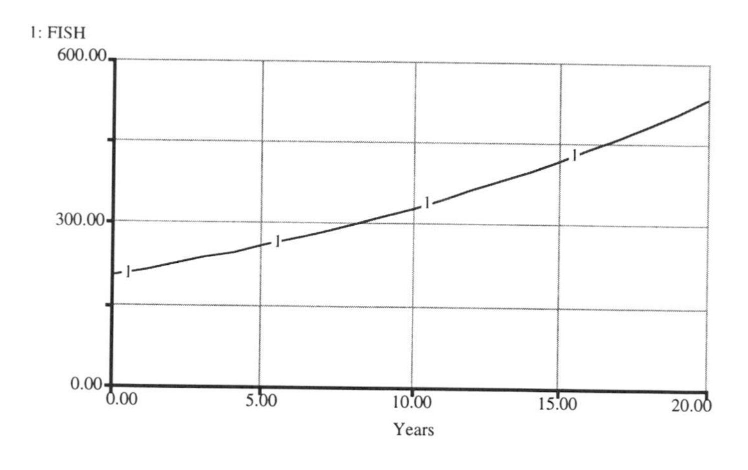

FIGURE 1.3

We see a graph of exponential growth of the fish population in your pond. This is what we should have expected. It is important to state beforehand what results you expect from running a model. Such speculation builds your insight into system behavior and helps you anticipate (and correct) programming errors. When the results do not meet your expectations, something is wrong and you must fix it. The error may be either in your STELLA program or your understanding of the system that you wish to model, or both.

What do we really have here? How does STELLA determine the time path of our state variable? Actually, it is not very difficult. At the beginning of each time period, starting with time = 0 years (the initial period), STELLA looks at all the components for the required calculations. The values of the state variables will probably form the basis for these calculations. Only the variable REPRODUCTION depends on the state variable FISH. To estimate the value of REPRODUCTION after the first time period, STELLA multiplies 0.05 by the value FISH (@ time = 0) or 200 (provided by the information arrows) to arrive at 10. From time = 1 to time = 2, the next DT, STELLA repeats the process and continues through the length of the model. When you plot your model results in a table, you find that, for our simple fish model, STELLA calculates fractions of fish from time = 1 onward. This problem is easy to solve, for example, by having STELLA round the calculated number of fish, with a built-in function that can do that, or just by reinterpreting the population size as thousands of fish.

This process of calculating stocks form flows highlights the important role played by the state variable. The computer carries that information, and only that information, from one DT to the next, which is why it is defined as the variable that represents the *condition* of the system.

You can drill down in the STELLA model to see the parameters and equations that you have specified and how STELLA makes use of them. Click on the downward-pointing arrow at the right of your STELLA diagram.

The equations and parameters of your models are listed here. Note how the fish population in time period *t* is calculated from the population one small time step DT earlier and from all the flows that occurred during a DT.

The model of the fish population dynamics is simple. So simple, in fact, we could have solved it with pencil and paper, using analytic or symbolic techniques. The model is also linear and unrealistic. So let us add a dimension of reality and explore some of STELLA's flexibility. This may be justified by the observation that as populations get large, mechanisms set in that influence the rate of reproduction.

To account for feedback between the size of the fish population and its rate of reproduction, an information arrow is needed to connect FISH with REPRODUCTION RATE. The connection will cause a question mark to appear in the symbol for REPRODUCTION RATE. The previous specification is no longer correct; it now requires FISH as an input (Fig. 1.4).

Open REPRODUCTION RATE. Click on the required input FISH. The relationship between REPRODUCTION RATE and FISH must be specified in mathematical terms, or at least, we must make an educated guess about it. Our educated guess about the relationship between two variables can be expressed by plotting a graph that reflects an anticipated effect the variable (REPRODUCTION) will have on another (FISH). The feature we will use is called a *graphical function.*

To use a graph to delineate the extended relationship between REPRODUCTION RATE and FISH, we click on Become Graph. Set the limits on

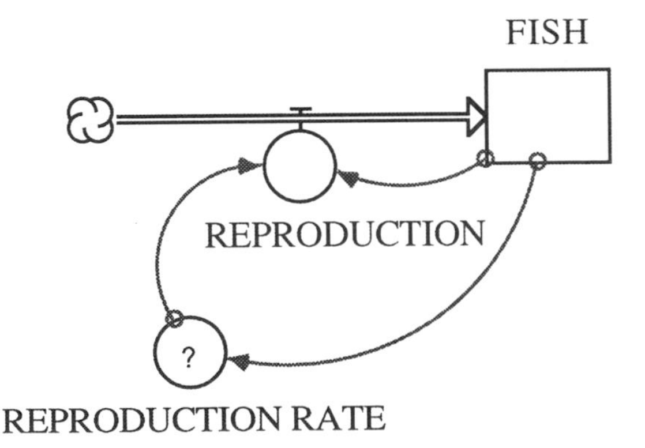

FIGURE 1.4

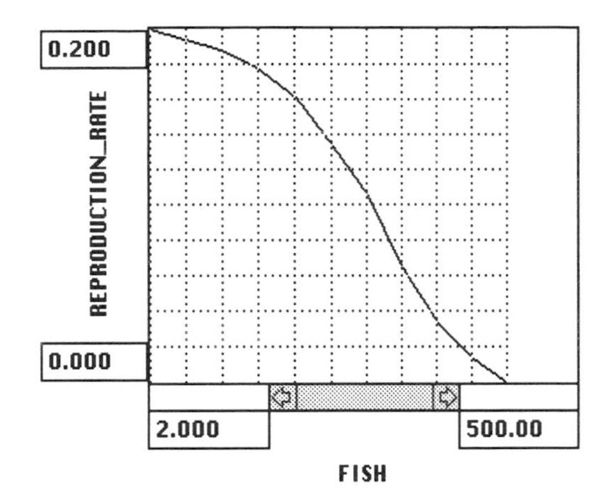

FIGURE 1.5

the FISH at 2 and 500; set the corresponding limits on the REPRODUCTION RATE at 0 and 0.20, to represent a change in the birth rate when the population is between 0 and 500. Here we are using arbitrary numbers for a made-up model. Finally, use the mouse arrow to draw a curve from the maximum birth rate and population of 2 to the point of zero birth rate and population of 500.

Suppose a census of the fish population was taken at three points in time. The curve we just drew goes through all three points (Fig. 1.5). We can assume that if a census had been taken at other times, it would show a gradual transition through all the points. This sketch is good enough for now. Click on OK.

Before we run the model again, let us speculate what our results will be. Think of the graph for FISH through time. Generally, it should rise, but not in a straight line. At first the rise should be steep: The initial population is only 200, so the initial birth rate should be very high. Later it will slow down. Then, the population should level off at 500, when the populations density would be so great that new births tend to cease. Run the model (Fig. 1.6). We were right!

This problem has no analytic solution, only a numerical one. We can continue to study the sensitivity of the answer to changes in the graph and the size of DT. We are not limited to a DT of 1. Generally speaking, a smaller DT leads to more accurate numerical calculation for updating state variables and, therefore, a more accurate answer. But note that the interpretation of the REPRODUCTION RATE changes as the DT changes. Choose Time Specs from the RUN menu to change DT. Change DT to reflect ever-smaller periods until the change in the critical variable is within measuring tolerances. Start with a DT = 1 and reduce it to 0.5, 0.25, 0.125 . . . for subsequent runs, each time cutting it into half of its previous value. We may also change the numerical technique used to solve the model equations. Euler's method is chosen as a

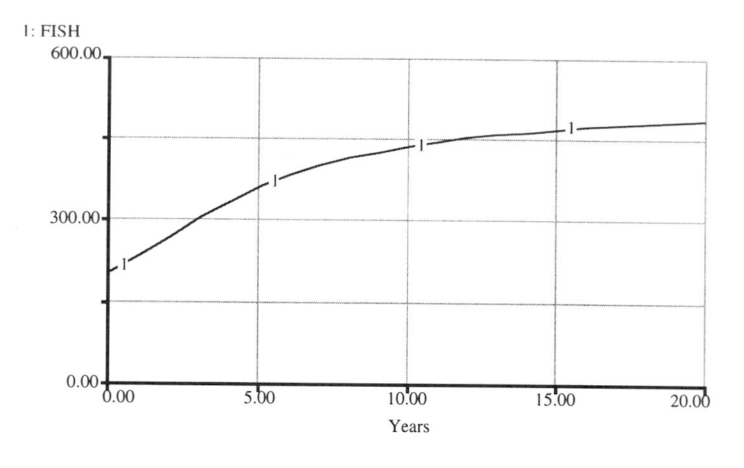

FIGURE 1.6

default. Two other methods, Runge-Kutta-2 and Runge-Kutta-4, are available that update state variables in different ways. We will discuss these methods later.

Start with a simple model and keep it simple, especially at first. Whenever possible, compare your results against measured values. Complicate your model only when your results do not predict the available experimental data with sufficient accuracy, or when your model does not yet include all the features of the real system that you wish to capture. For example, as the owner of a pond, you may want to extract fish for sale. What are the fish population dynamics if you wish to extract fish at a constant rate of 3% per year? To find the answer to this question, define an outflow form the stock FISH. Click on the converter, then click onto the stock to have the converter connected to the stock, and then drag the flow from the stock to the right. Now fish disappear from the stock into a *cloud*. We are not explicitly modeling where they go. What you should have developed thus far as your STELLA model is illustrated in Figure 1.7.

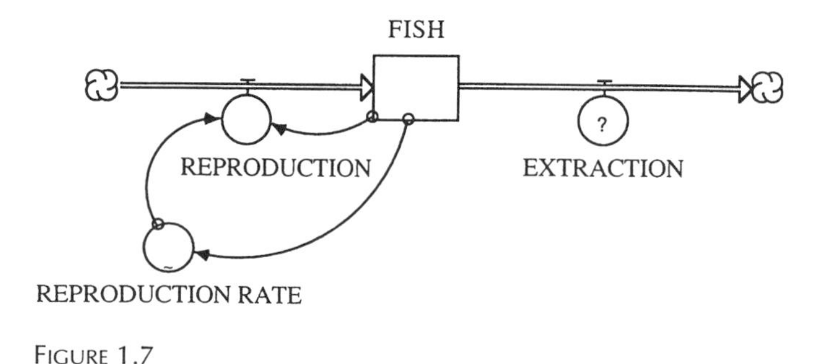

FIGURE 1.7

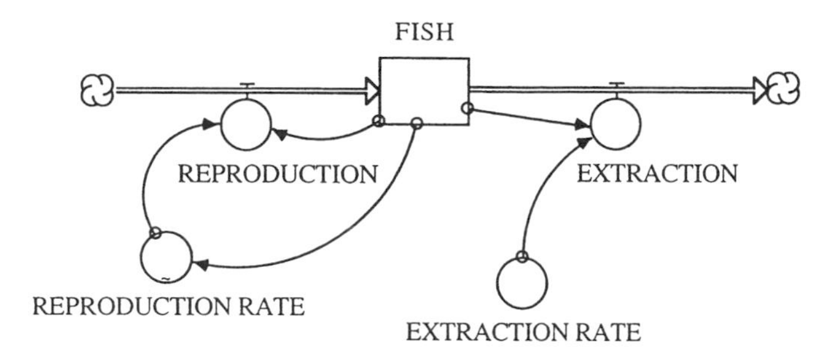

FIGURE 1.8

Next, define a new transforming variable called EXTRACTION RATE and set is to 0.03. Specify the outflow as

$$\text{EXTRACTION} = \text{EXTRACTION RATE} * \text{FISH} \tag{2}$$

after making the appropriate connections with information arrows. Your model should look like Figure 1.8.

Run your model again. The time profile of the fish population in your pond is seen in Figure 1.9.

You can easily expand this model, for example, to make the decision on EXTRACTION endogenous to your model, or to introduce unforeseen outbreaks of diseases in your pond or other problems that may occur in a real-world setting. When your model becomes increasingly complicated, try to keep your STELLA diagram as organized as possible, so it clearly shows the interrelationships among the model parts. A strong point of STELLA is its ability to demonstrate complicated models visually. Use the hand symbol to move model parts around the diagram; use the paintbrush to change the

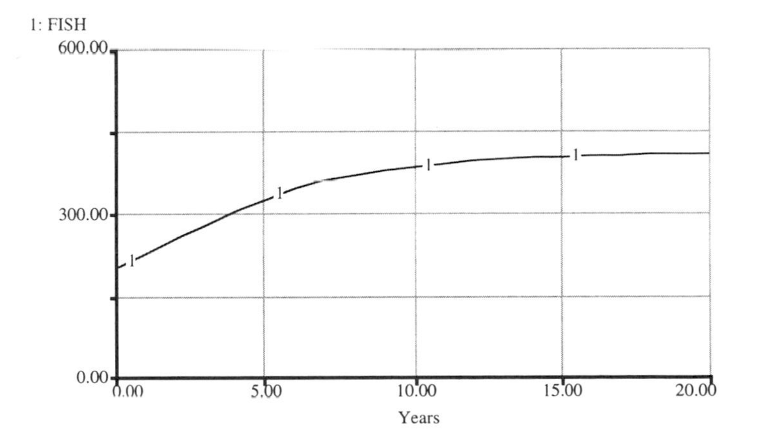

FIGURE 1.9

color of icons. The dynamite will blast off any unnecessary parts of the model.

Be careful when you blast away information arrows. Move the dynamite to the place at which the information arrow is attached and click on that spot. If you click, for example, on the translation variable itself, it will disappear, together with the information arrow, and you may have to recreate it.

As the model grows, it will contain an increasing number of submodels, or modules. You may want to protect some of these modules from being changed. To do this, click on the Sector symbol (to the left of the A in the next set of pictures) and drag it over that module or modules you want to protect. To run the individual sectors go to Sector Specs . . . in the Run pull-down menu and select the ones that you wish to run. The values of the variables in the other sectors remain unaffected.

With each additional feature, the diagram of the model increases in size. In larger models that have highly interdependent components, we need to make a large number of connections. With an increasing number of connections, or information arrows, the readability of the model can be seriously reduced. Use "Ghosts" of icons to avoid crossing arrows and increase the transparency of the model structure. You can create a ghost, for example, of a translation variable by first clicking on the ghost icon:

Once you clicked on the ghost icon, move it to the variable that you want to duplicate. Click on the symbol you want to duplicate. The ghost icon then changes its appearance into that of the symbol you clicked on. You can now place this duplicate of the original anywhere in the diagram and connect it with information arrows to the relevant parts of the model. The ghost you created is the same icon as its original but printed slightly lighter in color than the original.

By annotating the model, you can remind yourself and inform others of the assumptions underlying your model and its submodels. This is important in any model, but especially in larger and more complicated models. To do this, click on the Text symbol (the letter A) and drag it into the diagram. Then, type in your annotation.

The tools we mentioned here are likely to prove very useful when you develop more complicated models and when you want to share your models and their results with others. STELLA contains very helpful tools, which we hope you will use extensively. Space limitations preclude us from describing all of STELLA's features. The appendix provides a brief overview of STELLA features.

Make thorough use of your model, running it over again and always checking your expectations against its results. Change the initial conditions and try running the model to its extremes. At some point you will want to perform a formal sensitivity analysis. Later, we will discuss STELLA's excellent sensitivity analysis procedures and how you can use the MADONNA software included with this book to speed up sensitivity analyses and to complement your modeling skills with additional modeling tools.

1.6 Principles of Modeling

Although our title of this section may seem somewhat ostentatious, we surely have learned something general about the modeling process after many years of trying. So here is our set of 10 steps for the modeling process. We expect you to come back to this list once in a while as you proceed in your modeling efforts, and to challenge and refine these principles. A good set of principles should be useful to the novice and aid in speeding the process of learning to become an effective modeler.

1. Define the problem and the goals of the model. Set the questions you want the model to answer. The power of a good set of specific questions is hard to overstate. Good questions focus the mind on some aspect of the entire system in which your subsystem of interest is embedded. Appropriate generalization will come with time. Spend a lot of time defining the question(s) to be answered by your model.
2. Select the state variables, those variables that are to be the indicators of the status of the system through time. Designate the condition for nonnegativity of the state variables. Some state variables are conserved, some are not. Identify those in your model. Keep the number of state variables as small as possible. Purposely avoid complexity in the beginning. Minimize the number of state variables. Record the units of the state variables.
3. Designate the control variables, those flow controls that will change the state variables. Note which state variables are donors and which are recipients with regard to each of the control variables. Note whether lagged effects should be included either in the controls or in the variables that compose the controls. Be sure to set these flows as biflows if appropriate. Also, note the units of the control variables. Are there any useful analogies to apply here? Keep it simple at the start. Try to capture only the essential features. Put in one type of control as a representative of a class of similar controls. Add the others as needed, in step 10.

4. Select the parameters for the control variables. Note the units of these parameters and control variables. Ask yourself: Of what are these controls and their parameters a function? Do you expect some of these variables to be *lagged* or *delayed* functions of some of the other variables? Complexify begrudgingly.
5. Check your model for compliance with any appropriate physical, economic, or other laws, for example, the conservation of mass, energy, and value and any continuity requirements. Also, check for consistency of units. Look for the possibilities of division by zero, negative volumes or prices, etc. Use conditional statements if necessary to avoid these violations. Fully document your parameters, initial values, the units of all variables, assumptions, and equations before going on.
6. Chose the time and space horizon over which you intend to examine the dynamic behavior of the model. Choose the length of each time interval for which state variables are being updated by reference to the space over which the dynamics occur, and mainly by reference to the fastest rate of change you expect in your model. Then choose the numerical computation procedure by which flows are calculated. Set up a graph showing the most important variables and guess their variation before running the model.
7. Run the model. Are your results reasonable? Are your questions answered? Choose alternative lengths of each time interval for which state variables are updated. Choose alternative integration techniques. Explain any differences.
8. Do a sensitvity analysis of the parameters and initial values in the model. Try out these small changes singly and collectively within their reasonable extremes and see if the results in the graph still make sense. Revise the model to repair errors and anomalies.
9. Compare the results to experimental data. This may mean shutting off parts of your model to mimic a lab experiment, for example.
10. Revise the parameters, perhaps even the model structure, to reflect greater complexity and to meet exceptions to the experimental results, repeating steps 1 through 10. Do the results of this model suggest a new set of questions? They should.

1.7 Why Model?

Now that we introduced you to modeling, the software, and general principles of modeling, it is time to step back and ask ourselves again an important question: Why, and for what purposes, do we develop models? Dynamic modeling has four possible general uses:

- First, you can experiment with models. A good model of a system enables you to compare your result to those available from the real system and to change the model components to see how these changes affect the rest of the system. You can experiment, form and run scenarios, and bypass the inherent risk aversion to making changes in a real system.

- Second, a good model enables prediction of the future course of a dynamic system. Some modelers want only to explain what is going on; others aspire to a higher and more difficult (and dangerous) calling: forecasting. A good model will highlight gaps in what we know about the system we are studying. A good model will indicate the normal fluctuations in a complex system. Such variations are sometimes the cause of great alarm and much un-needed change. Conversely, observation of variation that is unexpected from real-world experience could signal the need for action. A good model should be able to indicate the results of these corrective actions.
- Third, a good model is a good thought-organizing device. Sometimes most of the value in modeling comes from a deeper understanding of the variables involved in a system that people are routinely struggling to control. Modeling requires that you assemble the group of experts on the various parts of the system to be understood. Each group member gains a better understanding of the system and of the skills and knowledge of his or her colleagues. Good modeling stimulates further questions about the system behavior and, in the long run, the applicability to other systems of any newly discovered principles.
- Fourth, a good model is a growing storage device for data and ideas that the enterprise has struggled long and hard to find and learn. Most often such data and insights are left to gather dust in printed form or to reside quietly in some distant computer. An ever-developing model should capture the knowledge of the enterprise, those lessons learned through the years about how the system actually works.

1.8 Model Confirmation

When do you know that you developed a "good" model? Giving an answer to this question often is rather difficult. By definition, all models abstract away from some aspects of reality that the modeler perceives less relevant than others. As a result, the model is a product of the modeler's perceptions. Consequently, one model is likely not the same as the models developed for the same system by other modelers who have their own, individual perceptions. Plurality of, and competition among, models is therefore required to improve our collective understanding of real-world processes. The more open and flexible the modeling approach, and the more people engaged in the specific modeling enterprise, the greater the chance of important discovery.

By enclosing a selected number of system components in the model and determining the model-system's behavior over time solely in response to the forces inside the model, the model becomes closed. Real systems, in contrast, are not closed but open, allowing for new, even unprecedented development in response to highly infrequent but dramatic changes in their environment. It is therefore not possible to completely verify a model by comparing model results to the behavior of the real system. There may have been extenuating circumstances that led the real system to behave differently from the model. The model itself may not have been incorrect, but just *incomplete* with regard to those circumstances. Such circumstances will always be present, precisely due to the fact that it is the modeler's goal to capture only the essentials of the real system and to abstract away from other factors. Consequently, complete verification of a model can only be done with regard to the consistency, or logical accuracy, of its internal structure.

Generally, there are two ways to test a model. First, one can withhold some of the basic data that were used to set up the model to determine the model parameters, data that represent the real world to the extent that it can be measured. Then the model can be used to predict the data used. For example, one might develop a model to predict the future population of the United States based on actual data from, say, 1900 to 1960 and then use the resulting model to predict the (known) population data from 1970 to 1990. If the prediction is a success, the later data can be incorporated into the model, and prediction for the next 20 or 30 years can be made with some reasonable degree of confidence. Another way to test the goodness of a model is to predict the condition in some heretofore unmeasured arena and then proceed to measure these variables in the field. For example, suppose that a model is being deviced to predict the location of an endangered species. The model is built and calibrated on the known habitats and then applied to the rest of the likely geography to qualify these places as likely locations.

If it is not possible, by definition, to verify a model by comparing its results to the performance of the real system, how can we know that we really captured the essentials of that system in the model? We know that our model is not unique—there are always other ways to construct a model. The guide for selection is always to first try to choose the simplest form. We may compare the model results to reality, not to verify but to *confirm*, and if we are unable to reproduce at least the trends observed in the real system, we know something is missing or wrong, and we are forced to revise our model or to check the accurracy of the data that went into specifying the model parameters and initial conditions.

Ironically, things can also become more problematic if the model results coincide well with our observations of the real system. The problem, for example, lies in the possibility that errors in the model cancel each other. Such mis-specifications are difficult to detect, and this is why we will start in many chapters of this book with a theory of model behavior rather than with observations of real systems. Combining theory, observations, and, indeed, intuition in a disciplined way in dynamic models is especially important when we make use of easy-to-learn and easy-to-use software packages. These devices allow us to develop models that can get ahead of ourselves. At each point in the model construction process, it is important to be able to justify the assumptions that are made.

Once the model is built on a theoretical base and observations, or "reasonable" initial conditions and parameter values, we let it yield the consequences of the forces built into the model. The choice of observations versus "reasonable" values is basically a choice of providing a predictive or descriptive model. For a description of the role of feedback mechanisms for system development, it is frequently sufficient to concentrate on those forces and the appropriate parameter range rather than precise numerical values.

To confirm our model results, we may compare them to appropriate data. The greater the number of instances in which model results and reality coincide under a variety of different scenarios, the more probable it is that the model captures the essential features of the real system it attempts to portray. We can increase our confidence in the model further by comparing its results to other models for the same or similar systems. The latter approach is of particular interest if the model was descriptive rather than predicitve. Finally, we hope that we develop, through practice in modeling, experience and new understanding of system dynamics that enable us to more easily detect problems in model specifications. This is a learning processes and we hope our book makes a significant contribution.

1.9 Extending the Modeling Approach

The models developed in this book are all built with the graphical programming language STELLA. In contrast to the majority of computer languages available today, STELLA enables you to spend the majority of your time and effort on understanding and investigating the features of a dynamic system, rather than writing a program that must follow some complicated, unintuitive syntax. With its easy-to-learn and easy-to-use approach to modeling, STELLA provides us with a number of features that enhance the development of modeling skills and collaboration of modelers. First, there is the knowledge-capturing aspect of STELLA. The way we employ STELLA leads not only to a program that captures the essential features of a system but, more importantly, results in a process that involves the assembly of experts who contribute their specialized knowledge to a model of a system. Experts can see the way in which their knowledge is incorporated into the model because they can pick up the fundamental aspects of STELLA very quickly. They can see how their particular part of the whole is performing and judge what changes may be needed. They can see how their part of the model interacts with others and how other specialists formulate their own contribution. As a result, these experts are more likely confident that the whole model is able and accurate than if they had entrusted a programmer with their insight into the system's workings. Once engaged in this modeling process, experts will sing its praises to other scientists. This is the knowledge-capturing aspect of the modeling process.

These same experts are likely to take STELLA into the depths of their own discipline. But, most importantly, they will return to their original models and repair them to meet broader challenges than first intended. Thus models grow along with the expertise and understanding of the experts. This is the expert-capturing part of our process. It is not only based on informed consensus but it has the possibility of continuing growth of the central model.

So we avoid the cult of the central modeler on the mainframe computer, who claims to have understood and captured the meaning of the experts. We avoid producing a group of scientists who, unsure of how well their knowledge has been caputured, and so, in their conservative default position, deny any utility or even connection to the model.

The process is not risk free and it requires delicate organization. The scientists who do participate in the modeling endeavor will reveal an approach to their field. There, no doubt, will be errors and omissions in that approach, and thus they may worry about criticism. Therefore, the rules of interaction must gently recognize their courage. The process does promise to make young scientists wiser and older scientists younger.

A general strategy for modeling with experts is to build a STELLA model of the phenomena that all expect will be needed to answer the questions posed by them at the outset. That model should cover a space and have a time step that is comensurate with the detail needed for the questions. The whole space and time needed for the model may exceed reasonable use of the desktop computer. Eventually, any modeling enterprise may become so large that the program STELLA is too cumbersome to use. However, translaters exist for the final STELLA model, converting its equations into either C+ or FORTRAN code. Translated models can be duplicated and put into parallel modes, adjoining cells on a landscape, for example, and run simultaneously on large computers. For models of intermediate size, a program like MADONNA[7] might well be used. It translates text files of STELLA equations and compiles them for much faster running. It allows sensitivity testing and the fitting of the principal state variable results to known data or known conditions by varying specified parameters within specified tolerances.

These processes are now being combined into a seamless one, running from desktop to supercomputer and back, with little special effort on the part of the body of experts. For example, in spatial ecological modeling we use STELLA to capture the expertise of a variety of life science professionals. We

[7] The latest versions of the software, documentation, and general description are available from the following internet site: http://www.kagi.com/authors/madonna/default.html.

then electronically translate that generic model into C+ or FORTRAN and apply it to a series of connected cells, for example, as many as 120,000 in a model of the Sage Grouse.[8] The next step is to electronically initialize these now cellularized models with a specific Geographic Information System of parameters and initial conditions maps. We then run the cellularized combine on a large parallel processing computer or a large network of paralleled workstations. In this way, the knowledge-capturing features of STELLA can be seamlessly connected to the world's most powerful computers.

The cellular or parallel approach to building dynamic spatial models with STELLA and running those models on ever more powerful computers is receiving increasing attention in landscape ecology and environmental management. An alternative, but closely related, approach has been chosen by Ruth and Pieper[9] in their model of the spatial dynamics of sea level rise. The model consists of a relatively small set of interconnected cells, describing the physical processes of erosion and sediment transport. Each cell of the model is initialized with site-specific data. These cells are then moved across the landscape to create a mosaic of the entire area to be covered by the model. In its use of an iconographic programming language, its visual elements for data representation, and its representation of system dynamics, the model is closely related to pictorial simulation models[10] and cellular automata models.[11] The approach is flexible, computationally efficient, and typically does not require parallel-processing capabilities. Though slightly awkward, it is also possible to use STELLA to carry out object-oriented models.

It is the intention of this book to teach you how to model, not just how to use models. STELLA was chosen towards this end because it is a very powerful tool for building dynamic models. The software also comes with an optional Authoring version that enables you to develop models for use by others who are uninterested in the underlying structure of the model. However, because model development and understanding are the purpose of this book, we do not discuss the Authoring version.

The basics of the STELLA programming language are outlined briefly in the next section. To easily follow that introduction, install STELLA on your disk. The installation procedure is explained in the appendix. Then open the STELLA program by double-clicking on the STELLA program icon on your disk.

---

[8]J. Westervelt and B. Hannon, A Large-Scale, Dynamic Spatial Model of the Sage Grouse in a Desert Steppe Ecosystem, Mimeo, Department of Geography, University of Illinois, Urbana, Illinois, 1993.

[9]M. Ruth, and F. Pieper, Modeling Spatial Dynamics of Sea Level Rise in a Coastal Area, *System Dynamics Review*, 10:375–389, 1994.

[10]A.S. Câmara, F.C. Ferreira, J.E. Fialho, and E. Nobre, Pictorial Simulation Applied to Water Quality Modeling, *Water Science and Technology*, 24:275–281, 1991.

[11]T. Toffoli, and N. Margolus, *Cellular Automata: A New Environment for Modeling*, Cambridge, MA: MIT Press, 1987.

2

Exploring Dynamic Biological Systems

> Art may be said . . . to overcome, and advance nature, as in these Mechanicall disciplines.
>
> Wilkins, *Mathematical Magick,* 1648.

2.1 Simple Population Dynamics

In this chapter we return to the concepts and ideas presented in Chapter 1, and explore in more detail the dynamics of seemingly simple dynamic population models. In the process of developing and exploring these models, you will learn more about the features of the STELLA and MADONNA software packages. The findings of this exploration should sensitize your perception of dynamic processes and help you develop your dynamic modeling skills.

Let us begin with a simple model of a population N in a given ecosystem with carrying capacity K. The initial size of the population is $N(t = 0) = 10$, and the carrying capacity is $K = 100 =$ constant. For population sizes below the carrying capacity, N will increase. Above the carrying capacity, N will decrease. The maximum rate of increase of N is $R = 0.1$, measured in individuals per individual in N per time period. A convenient specification for the change in the population size is the logistic function

$$\Delta N = R * N * \left(1 - \frac{N}{K}\right) \tag{1}$$

To set up the STELLA model for our investigation of the dynamics of this population, use the reservoir icon for the stock N, the flow symbol for ΔN, and converters for the transforming variables R and K. Specify the flow ΔN as a biflow by clicking on the biflow option in the dialogue box in which you specify the flow:

ΔN
○ UNIFLOW ◉ BIFLOW
☐ Unit conversion

Your STELLA model should now look like Figure 2.1.

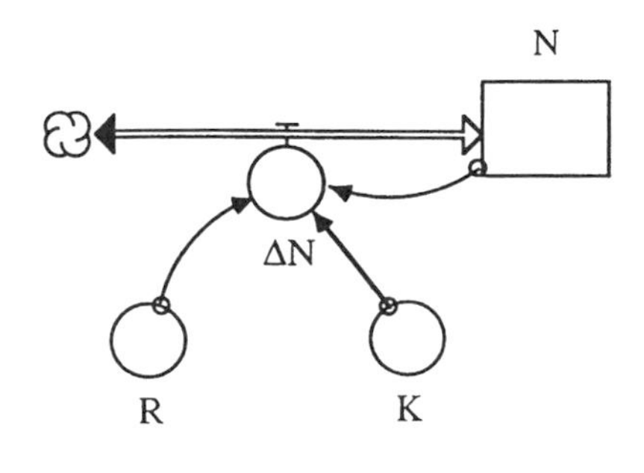

FIGURE 2.1

Set up a graph to plot N and K over time, and specify in the Time Specs menu a DT of 1 and the length of the model run to extend from time period 0 to 120, and specify the units of time as months. Before you run the model, make an educated guess of the population size N over time. Will N reach the carrying capacity? Will the approach be exponential? Will N overshoot K? What you should get when you run your model is illustrated in Figure 2.2.

The population size N asymptotically approaches K, and this approach is at first fairly rapid—as long as N is far below K—but the increase slows down as K is approached. The ratio of N/K approaches 1 as N increases,

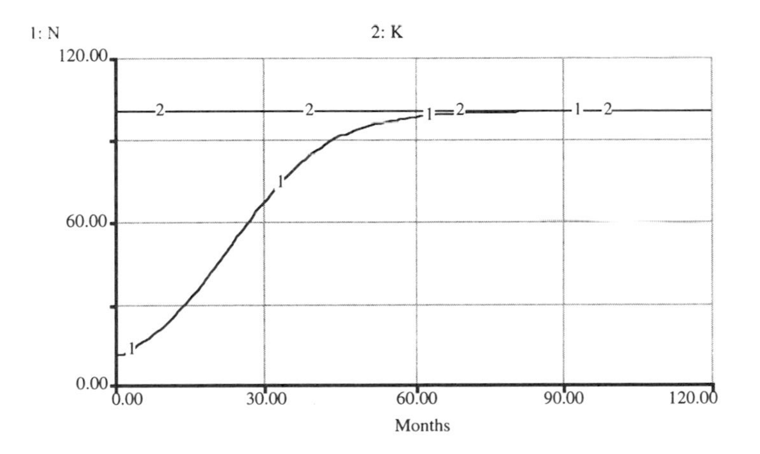

FIGURE 2.2

and thus (1 – N/K) becomes ever smaller. As a result, ΔN approaches zero, but it never quite gets there. Experiment with alternative values for R, K, and initial population sizes, and observe the resulting population dynamics. Always make an educated guess about the results before you run the model.

SIMPLE POPULATION DYNAMICS

```
N(t) = N(t - dt) + (ΔN) * dt
INIT N = 10
INFLOWS:
ΔN = R*N*(1-N/K)

K = 100
R = .1
```

2.2 Simple Population Dynamics with Varying Carrying Capacity

Let us now explore the dynamics of this system by making small changes to the parameter values. We have already modeled in the previous chapter the case in which the rate of natural increase changes as a function of the population size. Another parameter that may not be constant over time is the carrying capacity K. For example, there may be seasonal fluctuations in the physical environment that affect the resource base on which our population feeds. For simplicity, we assume that these seasonal fluctuations occur along a sinewave around the carrying capacity of 100. Double-click on the converter for K, then type "100 +" and scroll down in the list of built-in functions to find SINWAVE to add SINWAVE to the value 100. The built-in function SINWAVE requires an amplitude and period for its specification. We set those arbitrarily to 10 and 12, respectively. You should now have

$$K = 100 + \text{SINWAVE}(10,12) \tag{1}$$

which yields a carrying capacity that fluctuates between 90 and 110 over the course of a 12-month period. The STELLA diagram should look as before, but the results are different because of the change in the specification of ΔN (Fig. 2.3).

How will this change in the carrying capacity over time affect the population size N? Because the carrying capacity has only little influence on N as

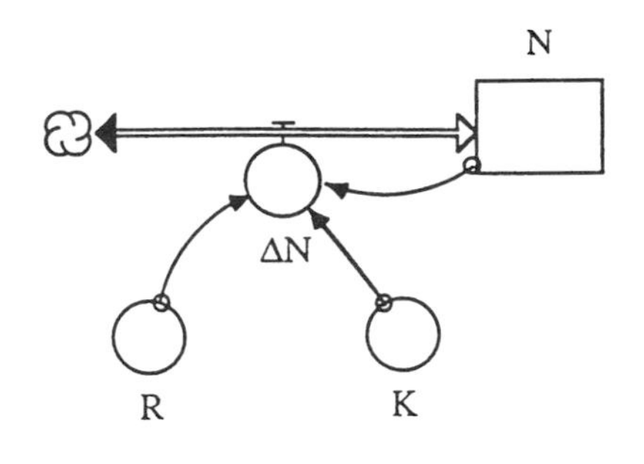

FIGURE 2.3

long as N is small, we would expect the change in K not to alter the early sigmoidal growth phase of N. However, as N gets larger, K has an increasing influence on the subsequent changes in N. When you run the model, you should find that this is indeed the case. Look closely at the graph and recognize, however, that the changes in N and K are not exactly in sync with each other. Rather, an increase in K is not instantaneously matched by an increase in N. Can you explain why (Fig. 2.4)?

Again, explore the dynamics of the system by successively running the model for alternative specifications of R, K, and initial population sizes. For example, enlarge the range over which K fluctuates over the course of a year. Alternatively, abandon the assumption that K fluctuates along a sinewave and make it a random variable. You can do so with the built-in

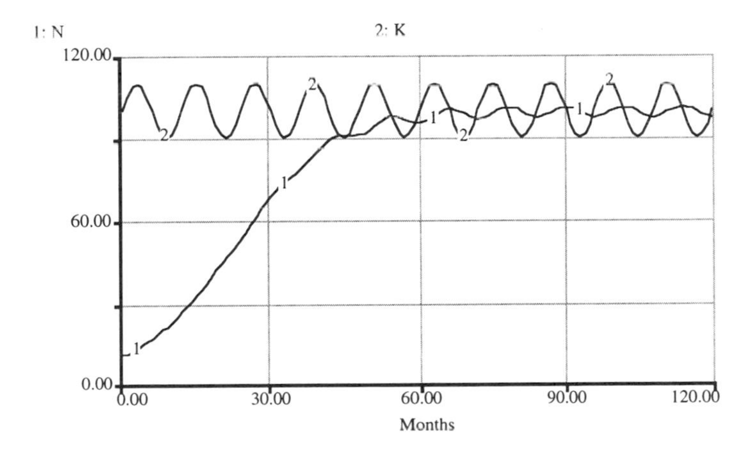

FIGURE 2.4

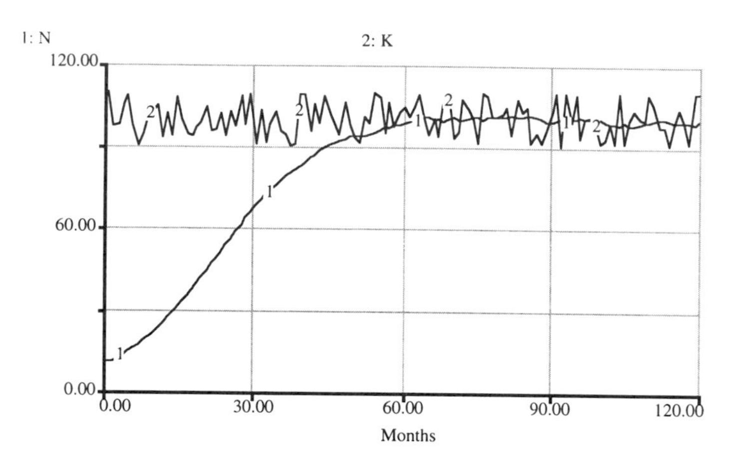

FIGURE 2.5

function RANDOM, which requires that you specify upper and lower bounds. For example, specify

$$K = 100 + \text{RANDOM}(-10,10) \qquad (2)$$

and K will fluctuate randomly between the values 90 and 110. The results of one model run with K specified as in equation (2) are shown in Figure 2.5. The time paths of this system are different from run to run because of the random number.

SIMPLE POPULATION MODEL
WITH VARYING CARRYING CAPACITY

```
N(t) = N(t - dt) + (ΔN) * dt
INIT N = 10
INFLOWS:
ΔN = R*N*(1-N/K)

K = 100 + SINWAVE(10,12)
R = .1
```

2.3 Sensitivity and Error Analysis with STELLA and MADONNA

Let us reflect for a moment on the models that we developed so far. We have hypothesized about the workings of a dynamic system—the influences on births and deaths in a population, and possible fluctuations in the maximum number of individuals of that population that can be sustained in a given environment. We have not concerned ourselves with real data describing real populations in a real environment. Rather, we were interested in the general features of such systems.

The modeling approach that we chose here is distinct from a data-driven, statistical approach. Statistical, or as they are sometimes called, empirical models are a kind of disembodied representation of some well-studied phenomenon. They have no connection to reality other than the purely mathematical. The systematical alternative, the kind we have used in this and similar books,[1] strives to represent as much as possible the reality of the dynamic phenomenon. Some refer to this form of modeling as the *mechanical approach,* but this term seems to us too wooden and likely to leave the impression that we think nature is just another mechanical process, rather like the engine of an auto. We do not think so.

In systematical modeling, we build into the representation of the phenomenon that which we know actually exists, such as the birth and death processes of populations. Our systematical alternative, therefore, starts with an advantage over the purely statistical or empirical modeling schema. This advantage allows the systematical model to be used in more related applications than the empirical model—the systematical model is more transferable to new applications. But the empirical model does have one advantage: In the process of its evaluating on the data gathered about the phenomenon, the mean and standard deviations of the coefficients are found. The corresponding parameters and initial values in our models are not so elaborated. Such values are at first usually found or derived from the pertinent literature, often without a given variation.

Once the systematical model has performed to meet a general sanity test, the parameters and initial values need to be flexed to determine the sensitivity of the model results with regard to the choice of parameters and initial conditions. This process is time consuming and is usually allocated to the drudgery part of modeling. But it is essential. Just how effective is a model that responds with dramatic difference when one of its parameters is changed slightly? The point is not whether sensitivity analysis needs to be done but how can it be done efficiently? Our view is that STELLA is a very efficient tool for building the structure of the systematical model and for performing sensitivity analyses.

To conduct a sensitivity analysis, for example, on the parameter R of our population model, choose Sensi Specs . . . (Sensitivity Specification) from the Run pull-down menu, and choose the parameter R—by clicking on it and selecting it—as the one for which you want to perform a sensitivity analysis. Type in the dialogue box "# of Runs" 5 to generate five sensitivity

---

[1]B. Hannon and M. Ruth, *Dynamic Modeling,* New York: Springer-Verlag, 1994.
M. Ruth and B. Hannon, *Modeling Dynamic Economic Systems,* New York: Springer-Verlag, 1997.

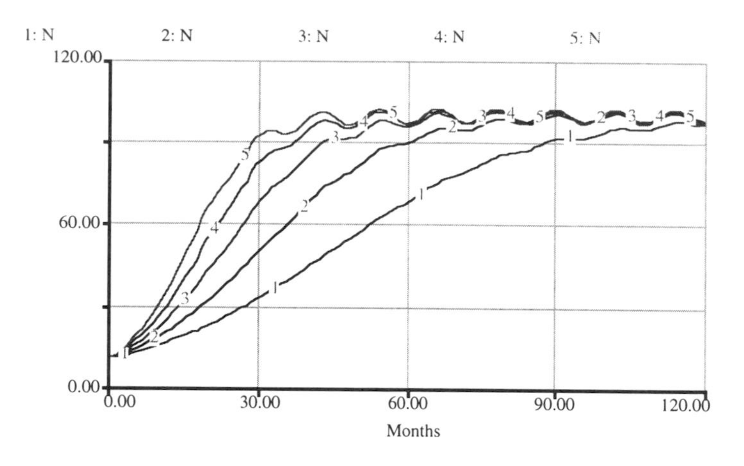

FIGURE 2.6

runs. Then provide start and end values for R. If you chose Incremental as the variation type, STELLA calculates the other values from the start and end values that you specified such that there are equal incremental changes in A from run to run. Plot the five resulting curves for N in the same graph by choosing the Graph option in STELLA's Sensitivity Specs menu. Run the model with the S-Run command and observe the resulting graph. In Figure 2.6 you will find the results for R varying from run to run incrementally between 0.05 and 0.15, and a carrying capacity that is specified as

$$K = 100 + \text{SINWAVE}(10,12) \tag{1}$$

If you wish to let them vary from run to run along a normal distribution with known mean and standard deviation, choose the Distribution option instead of Incremental. When you specify Seed as a positive integer, you ensure the ability to replicate a particular random number sequence in subsequent sensitivity runs. Specify 0 as the seed and a "random" seed will be selected. If you do not wish to make use of the normal distribution of the random numbers used in the sensitivity analysis, click on the bell curve button.

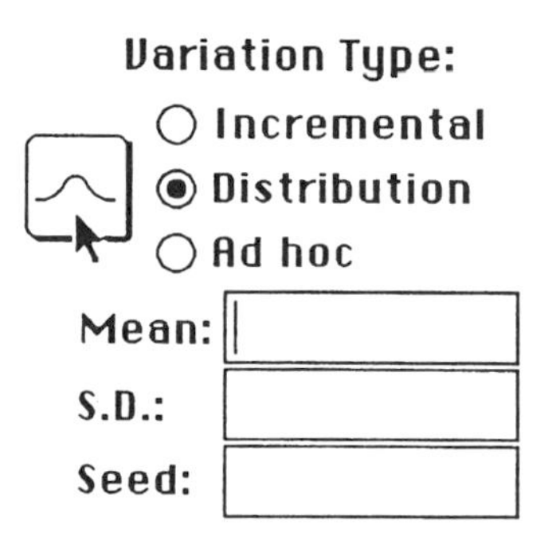

This button will change its appearance, and you now need to specify a minimum, maximum, and a seed for your sensitivity analysis.

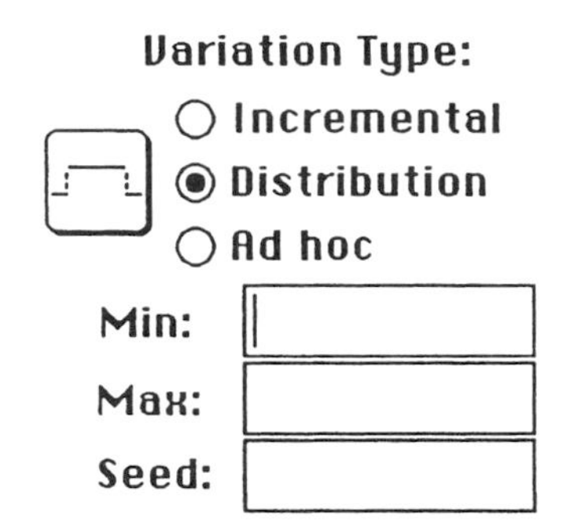

A final choice for the specification of sensitivity runs in STELLA is not to change parameters in incremental intervals or along distributions. You can specify ad hoc values for each of the consecutive runs.

The sensitivity procedures in STELLA enable us to investigate the influence of the choice of parameter values and initial conditions on model runs. As our models become larger, these procedures are significantly slowed down. Also, we may want to investigate the sensitivity of model results with regard to the choice of DT or the numerical integration method. To do these sensitivity analyses efficiently, they should be done on the compiled form of the model. The model will run much faster in the compiled form, but it is harder to change the structural connections. We see the physical process of modeling building in two parts: (1) building and testing the structural form of the model, and (2) compiling the resulting equations and performing sensitivity analysis on the parameters and initial values. This second step is the main reason we have enclosed MADONNA with our book.[2]

To run sensitivity analyses of the model that you already developed in STELLA, navigate to the STELLA equations by clicking on the downward-pointing triangle on the right-hand side of the STELLA diagram.

Then save these equations as text, open MADONNA, and then open the equations in MADONNA where they are automatically compiled or translated

[2]For a quick reference guide to MADONNA for the Macintosh, see the appendix. MADONNA for Windows can be downloaded at http://www.kagi.com/authors/madonna/default.html as it becomes available.

into a form more related to the basic language of the computer. Here the equations are harder to change (structurally), but the parameters and initial values can be easily changed and the results of such changes found quickly.

Once you have opened the equations in MADONNA, change the STOP-TIME to 120 and the DT to 1 to conform with the model runs that we already had in the previous sections of this chapter. Next, choose New Window in the Graph pull-down menu and select INIT N as the variable you wish to plot. Then, go to Multiple Runs . . . in the Model menu and choose N as the parameter in which to perform a sensitivity analysis. Then choose to perform 51 sensitivity runs on initial values of N ranging from 10 to 210. Click on OK and observe the speed at which MADONNA performs these runs. The results in Figure 2.7 are derived for the simple population model of the previous chapter with a carrying capacity randomly varying between 90 and 110. Can you explain why the graph is not symmetrical around N = 100?

MADONNA provides a number of additional features that are not found in STELLA, including alternative integration techniques, the calculation of means and standard deviations for multiple runs, and more. Explore these features and return to MADONNA as you proceed through this book and as you develop your own dynamic models of biological systems.

MADONNA and STELLA provide various methods for sensitivity analysis of alternative parameter values and initial conditions. Still a strategy must be developed for changing these values. First, you must select that variable that is most important to the problem—the total population, for example, in a long and elaborate population model. Then you must select a reasonable range for each of the parameters and initial values. Next you try out these ranges in the model, one at a time, to see if a change in any of the parameters or initial values has any significant effect on the main

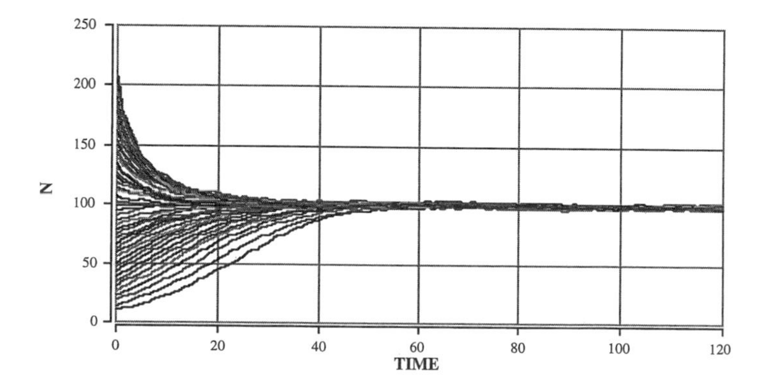

FIGURE 2.7

variable of interest. Some may show considerable leverage—a small change begets a major change in the main variable, others will show little effect. From this you could tentatively conclude that more must be known about these sensitive parameters or initial values, or the model needs to be restructured to eliminate them. But the situation is really more complex. It is possible that several of these parameters will act together in some particular way to produce rather sharp changes in the main variable, while when acting alone they have no effect. This means that one needs to vary all the parameters and initial values simultaneously in a great plethora of combinations with the hope of finding the critical combinations. This is sometimes called the *Monte Carlo technique,* and it can be done in STELLA and MADONNA.

You can easily specify a whole series of parameter variations in either of these programs and make the hundreds of runs needed to reasonably explore the combinations for their collected sensitivity. Examination of the results can lead you to those parameter groups that can cause trouble. Again, you might consider eliminating this combination by a structural change in the model or invest more effort in narrowing the real range of these parameters through extended research.

Another rather interesting approach is to change each of the parameters to a sine or cosine function, varying the mean value of the parameter, similar to what we have done in the preceeding section of this chapter. Each parameter is assigned its own frequency, and the model is run with all parameters and initial values varying in this way. A spectral analysis can be performed on the variations in the main variable, with the hope of finding certain critical frequencies, leading you directly to the parametrical culprits.

Generally though, big computer-based models create a demand for big computers as they are needed to sift through the parameter and initial value specification problem. There is no easy way around this situation. The problem is actually larger than the parameter problem discussed here. There are many sources of modeling error. Gertner[3] wisely advocates the use of Error Budgets as a way of pinning down the critical areas of error sources. He and his colleagues have developed the methods of breaking down the source of error in several categories [Input Measurement, Sampling, Components of the model (sets of equations), Grouping and Computational]. They are able to isolate the sources of variation in the main variable. With such information the model can be effectively revised or the

[3]G. Gertner, X. Cao, and H. Zhu, A quality assessment of a Weibull based growth projection system, *Forest Ecology and Management* 71:235–250, 1991; G. Gertner, B. Guan, Using an error budget to evaluate the importance of component models within a large-scale simulation model, *Proc. Conf. on Math. Modeling of Forest Ecosystems,* Frankfurt am Main: Germany, J.D. Verlag, 1991, pp. 62–74.

data collection effort intelligently redirected. For very large spatial dynamic models with thousands of cells, the testing problem is very great, seemingly impossible large. But efficient testing algorithms have been and are being developed.[4]

In presenting your models and their results, you should always include the variations in the main variables of interest with changes in the critical parameters. This display reveals to the critical observer that you have a respect for the trouble that can be caused by what is still unknown about the process you study. The best thing that can happen to a modeler is to have one of their models used to aid important decision making. No important decision maker will use a model that has not been screened for its error potential.

---

[4]For a review of this approach and the general strategy of developing, sensitivity testing, and using these large spatial models, see: http://ice.gis.uiuc.edu, generally and http://ice.gis.uiuc.edu/TortModel/tortoise.html, specifically for the error-budgeting process.

3

Risky Population

> I know that history at all times draws
> strangest consequence from remotest cause.
>
> T.S. Eliot,
> *Murder in the Cathedral,* Part I, 1935.

In the previous chapter we saw how to model simple deterministic and random processes that influence population dynamics. We emphasized the need to thoroughly test your models before you move on and expand them. This chapter provides a novel expansion to the traditional model of population dynamics. Other expansions follow in the next chapters. Each of those expansions is kept to a minimum complexity, yet the resulting dynamics can be rather surprising.

For the following model assume that a population grows exponentially by virtue of a birth rate and dies according to a death rate. If the birth rate is 7% and the death rate is 4%, then the population grows exponentially at the rate of 3%. That is a simple matter and we have addressed this model in our most elementary examples of the previous chapters.

But we all know that as the population size (or density) grows and no migration can take place, the chance of a sudden unanticipated rise in the death rate increases. Perhaps the calamity is due to an infectious disease or to war. We leave out natural causes, such as earthquake and meteor impacts, because these disasters are not brought on by the population itself. Likewise, as some argue, a larger population causes an inventive focus to alleviate some of the human dilemma. We need to express the death rate in such a way that both the negative and the salutory effects of a rising population can be felt. The question is, When the birth rate is higher than the nominal death rate, what are the population dynamics when higher populations can have the effect of increasing and at other times decreasing the death rate? The answer is rather surprising.

Let us set the birth rate to a constant rate of 7%. To mimic unanticipated changes in the death rate, we introduce a nominal death rate, NOMINAL DR, which we set to

$$\text{NOMINAL DR} = .04 \tag{1}$$

We make use of this nominal death rate as the mean value in a normal distribution, DR DISTRIBUTION. To specify a normal distribution with a mean of nominal death rate, we make use of the built-in function NORMAL.

We must specify the mean and standard deviation as its arguments. We define

$$\text{DR DISTRIBUTION} = \text{NORMAL(NOMINAL DR,0.005*POPULATION)} \quad (2)$$

with an arbitrarily set standard deviation of 0.005*POPULATION. Thus, as the POPULATION increases, so does the standard deviation around the mean NORMAL DR.

Next we define a death rate signal, DR DIST CONTROL, that constrains the normally distributed DR DISTRIBUTION between 0.01 and 1:

$$\text{DR DIST CONTROL} = \text{IF (DR DISTRIBUTION} \geq 0.01\text{) AND (DR DISTRIBUTION} \leq 1\text{) THEN DR DISTRIBUTION ELSE } 0.01 \quad (3)$$

We assume that the absolute minimum of the death rate is 0.01. Finally, the DEATH RATE itself is set up to allow three test cases: the DR DIST CONTROL as noted earlier, a case where the standard deviation is allowed to only increase the death rate, and an exponentially declining nominal death rate whose standard deviation (0.005*POPULATION) is allowed to only increase the death rate. First, we define

$$\text{DEATH RATE} = \text{DR DIST CONTROL} \quad (4)$$

Consequently, the death rate varies with normal distribution from 0.01 to 1 in every time period (e.g., year, decade, generation), with the nominal death rate (here 4%) as the mean of this distribution. The STELLA diagram is shown in Figure 3.1. It contains a module that calculates the aver-

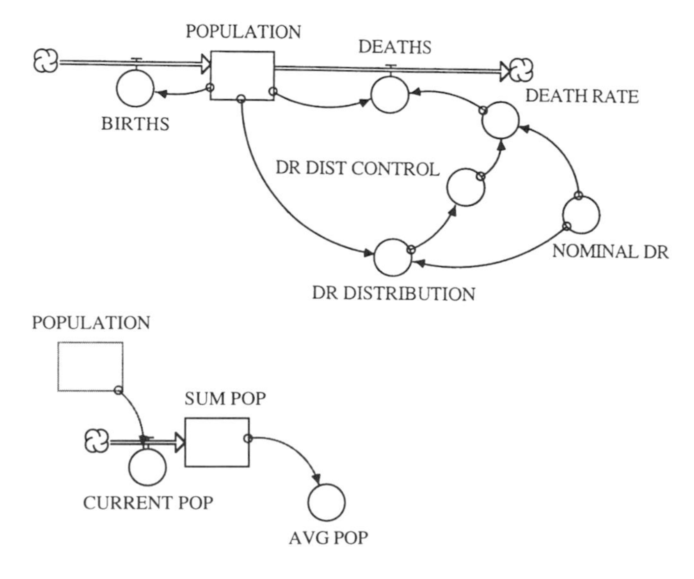

FIGURE 3.1

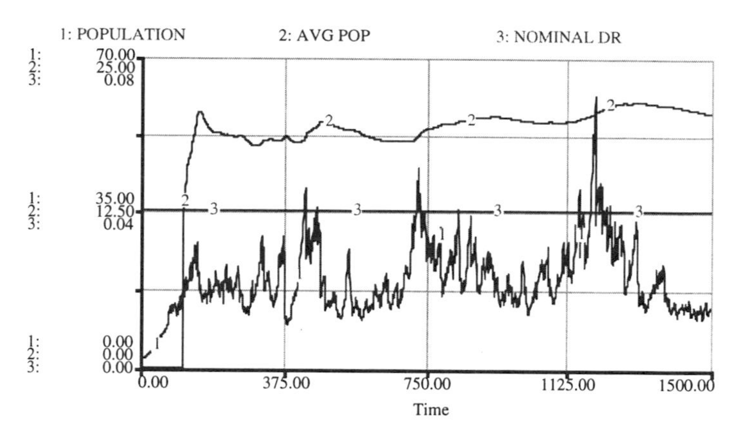

FIGURE 3.2

age population size AVG POP over time by summing the population over time and dividing by time. The results of this model are shown in Figure 3.2. Note that we have made use of the ghost here to connect POPULATION to the calculation of CURRENT POP.

The remarkable result is that the population reaches a somewhat steady condition; it is not expanding off to infinity at a net 3% growth rate as implied by the net birth rate. At first the population grows exponentially, but it then hits a kind of limit shown by the average population after about 200 time units. The population becomes large enough to be constrained by the sudden switches in its death brought on by the sheer size of the population, even allowing that many of these switches are death rate reducing.

How do the results of our model change if the DEATH RATE is allowed only to increase above the NOMINAL DR (up to 1.0) from period to period? To investigate this case, we change the definition of the DEATH RATE in equation (4) to

DEATH RATE = (IF DR DIST CONTROL > NOMINAL DR THEN DR DIST CONTROL ELSE NOMINAL DR) (5)

Here we find a lower mean population size of approximately 15, down from the previous model run of 18 (Figure 3.3). Can you explain why?

The results in Figure 3.3 were all derived for a fixed nominal death rate. Yet, the nominal death rate may slowly decline over time as the population grows. This is certainly the case for the human population, which, through advancements in medicine and political treaties, has significantly reduced the death rate in parts of the world. How do our results change if we have, on the one hand, a decline in the nominal death rate and, on the other hand, an incease in the standard deviation above the mean DR DISTRIBUTION? We model the decline in the nominal death rate as

NOMINAL DR = EXP(–.01*TIME)*.03 + .01 (6)

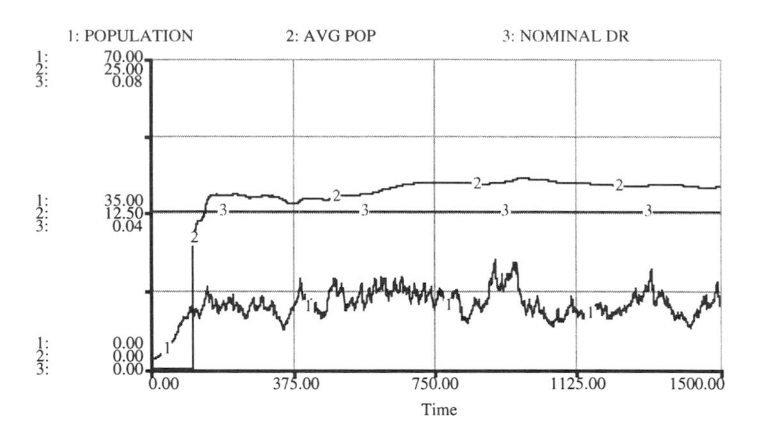

FIGURE 3.3

where EXP is Euler's number and TIME is a built-in function that takes on the same value as the current period of the model run. With this specification, the nominal death rate exponentially declines from 4% to 1%. The results are shown in Figure 3.4. Now, the mean population size is 26.

We note that at first the population rises in the expected exponential way and then the effect of the standard deviation takes over, causing an average population of about 26. None of these populations continues to grow. After about $t = 200$, a mean value is achieved that holds for the rest of the run, in all cases although the first and third cases clearly show a greater variation.

Can you develop any more scenarios? Should the death rate be allowed to vary so greatly from period to period? Perhaps the death rate should be

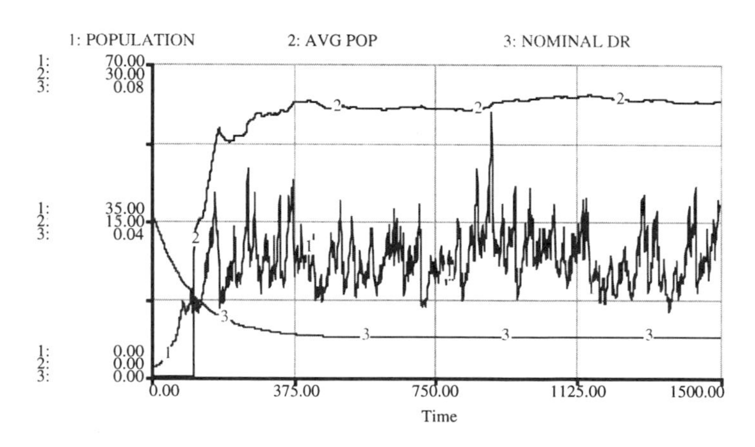

FIGURE 3.4

a state variable and thus change more slowly. Can you set up such a model? Should this distribution be normal? Is it a bit unreal to cut the normal distribution off at 0.01 on the low side and 1.0 on the high side? How about a normal distribution that has different standard deviations for the different sides of the mean? What about the idea of making the standard deviation a nonlinear function of the population size? What about the possibility of a delay of death rate decreases with no delay of death rate increases?

Now that we have seen some of the effects of randomness on population dynamics, we will return to deterministic models, models that yield the same results from run to run. The absence of randomness, however, does not necessarily mean that we will be able to make precise forecasts of a system's behavior once we know its history and current state. Rather, unforseen, chaotic events may occur. This is the topic of the following chapter.

RISKY POPULATION

```
POPULATION(t) = POPULATION(t - dt) + (BIRTHS - DEATHS)
* dt
INIT POPULATION = 2
INFLOWS:
BIRTHS = .07*POPULATION
OUTFLOWS:
DEATHS = DEATH_RATE*POPULATION

SUM_POP(t) = SUM_POP(t - dt) + (CURRENT_POP) * dt
INIT SUM_POP = 0
INFLOWS:
CURRENT_POP = IF TIME > 100 THEN POPULATION ELSE 0

AVG_POP = IF TIME ≠ 100 then SUM_POP/(time-100) else 0
DEATH_RATE = (IF DR_DIST_CONTROL > NOMINAL_DR THEN
DR_DIST_CONTROL ELSE NOMINAL_DR)*1 +0*DR_DIST_CONTROL
+0*NOMINAL_DR
DR_DISTRIBUTION = NORMAL(NOMINAL_DR,0.005*POPULATION)
DR_DIST_CONTROL = IF (DR_DISTRIBUTION ≥ 0.01) AND
(DR_DISTRIBUTION ≤ 1) THEN DR_DISTRIBUTION ELSE 0.01
NOMINAL_DR = (EXP(-.01*TIME)*.03 + .01)*1 + .04*0
```

4

Steady State, Oscillation, and Chaos in Population Dynamics

And then so slight, so delicate is death
That there's but the end of a leaf's fall
A moment of no consequence at all.

Mark Swann,
as quoted by Alfred Lotka, 1956,
The Elements of Physical Biology, Dover, NY, p. 376

4.1 The Emergence of Chaos in Population Models

Let us return to the simple population models of Chapter 2, in the absence of randomness, and explore the behavior of a simple deterministic model as a parameter value gets pushed outside the realm that is typically considered in these models. Denote the size of the population in time period t as N(t) and the net change in the population size during that period as ΔN. The exogenous parameter influencing the net flow is R. The net flow ΔN updates the stock N:

$$\Delta N = N(t + DT) - N(t). \tag{1}$$

The reproductive rule in this model is

$$N(t + DT) = R{*}N(t){*}(1 - N(t)), \tag{2}$$

and consequently,

$$\Delta N = R{*}N(t){*}(1 - N(t)) - N(t). \tag{3}$$

The STELLA model shown in Figure 4.1 has as its main component the ΔN equation updating the stock N at each period of time. We also calculate the stock N delayed by DT as LAG N. To calculate this lagged value, we make use of the built-in function DELAY, which requires as its input the variable of which a delay should be calculated, in our case N, the lag length DT, and the initial value. If you do not specify an initial value, STELLA will assume the initial value of the delayed variable is zero. Here, we specify

$$\text{LAG N} = \text{DELAY(N,DT)} \tag{4}$$

The STELLA model is shown in Figure 4.1.

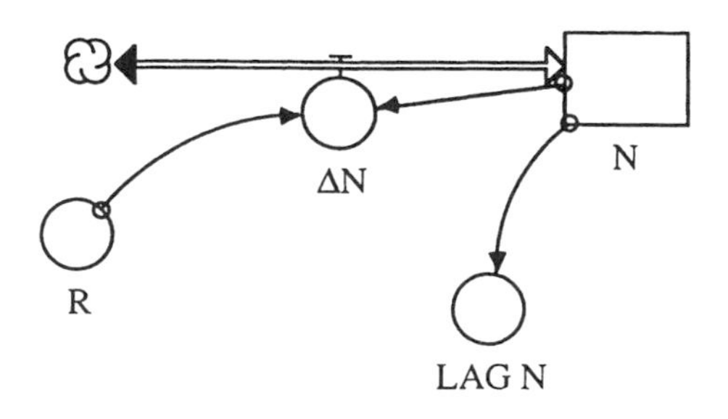

FIGURE 4.1

Set up the model with an initial value for N of 0.1, DT = 1, and R = 1. Make an educated guess before you run the model. What you should get is illustrated in Figure 4.2. The population size declines along the logistic curve.

Next increase R for subsequent runs to 2, then 3. You can do that with STELLA's sensitivity methods or in MADONNA. Again make a guess before you run the model. The results are plotted in Figure 4.3.

The first of these runs generates a steady-state population of N = 0.5, and the second run yields a damped oscillation. With R = 2 we have a situation in which the population size can be thought of overshooting some carrying capacity, then gets corrected and falls below that carrying capacity, only to overshoot again, albeit to a smaller extent, in the next period.

Increase R to 4. You should find that the population leaves its regular pattern and becomes chaotic—the curve never repeats itself.[1] Pause

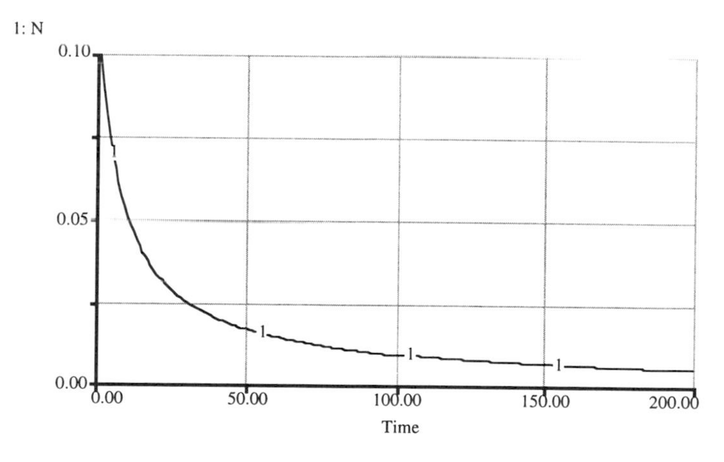

FIGURE 4.2

[1]Chaos can occur in continuous functions when they are described by nonlinear second order differential equations.

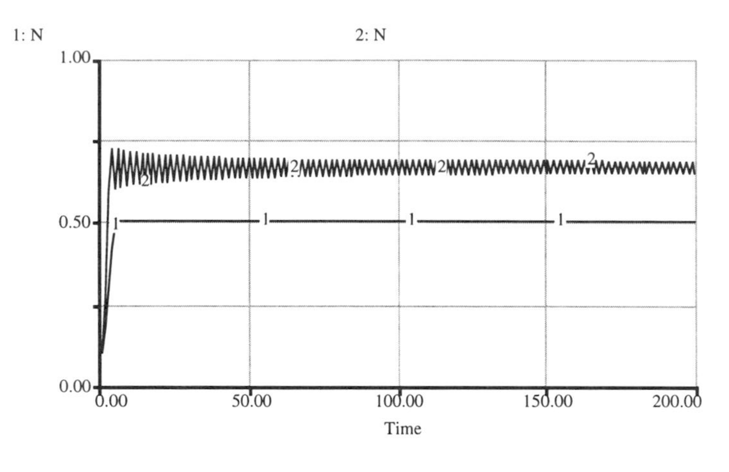

FIGURE 4.3

the model halfway through its run and make a guess on the future path. Can you predict where it is going, solely based on your observation of its past behavior and current position? Such prediction is impossible. Run the model over and over again, and you will find that always the same path is chosen. The system that we are dealing with here is not random, it is deterministic; yet, prediction from past and current states is impossible. Recognize also that if you change even very slightly the initial conditions, for example, N = 0.101 instead of N = 0.1, a very different path emerges. This should illustrate the sensitivity of nonlinear dynamic systems to initial conditions, and sensitize you to the limitations of real data, which always comes with measurement errors, in forecasting a system's behavior (Fig. 4.4).

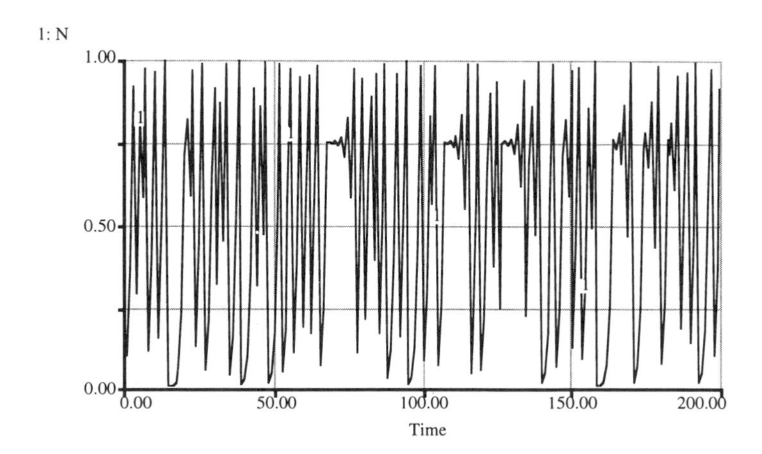

FIGURE 4.4

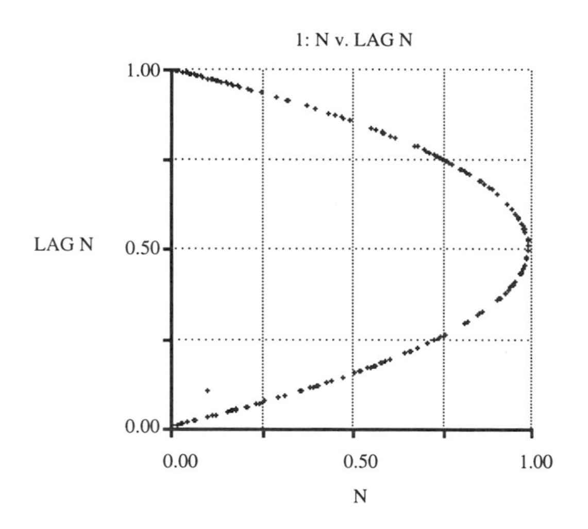

FIGURE 4.5

Even though there is seemingly no regularity in the system's behavior, all values lie within a well-defined range. Generate a scatter plot—a diagram of the system's phase space—by plotting LAG N against N, and observe the results. To set up the scatter plot, create a graph and choose Scatter as the Graph Type. The results are illustrated in Figure 4.5.

Can you find the value for R at which chaos seems to begin? Now lower the time step to DT = 0.5 and find the value for R at which chaos begins again. Keep shortening DT and you will find that there is a relation between the size of DT and the smallest A necessary to produce chaotic behavior (Table 4.1).

A pattern emerges. When DT is halved, R is doubled and lessened by one. That is, if DT(n + 1) = .5*DT(n), then R(n + 1) = 2*R(n) – 1. A function R(DT) to calculate the critical R is

$$R(DT) = (4.57/DT) - [1 + 2 + 4 + 8 + \ldots 1/DT] \tag{5}$$

or

$$R\ (DT) = (R(1) + 1)/DT - (1 + 2 + 4 + 8 + \ldots 1/DT), \tag{6}$$

TABLE 4.1

DT	R necessary for chaos
1	3.58
.5	6.12
.25	11.29
.125	21.56
.0625	42.13
.03125	83.24

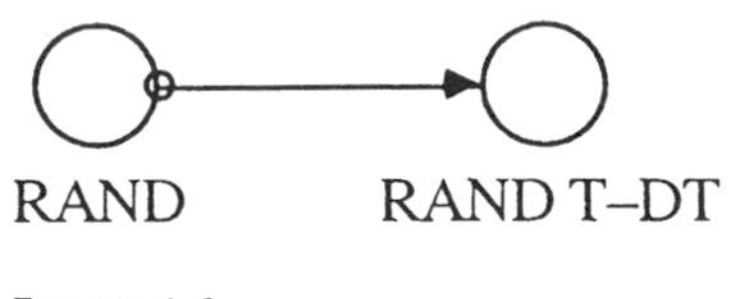

FIGURE 4.6

the boundary between chaos and finitely numbered solutions for the spectrum of discrete steps.

This model shows you that the R for DT = 0 is infinity. This result is correct because chaos is typically not noticed on the continuous level. Chaos in first order differential equations occurs on the continuous level only if you are stuck with a specific DT in your particular problem and the parameters lie within the critical range.[2]

Compare the chaotic time paths of your model to a truly random number. We defined such a number in the model as RAND and calculated its delayed value in Figure 4.6.

The random number, plotted against its delayed value, is shown in Figure 4.7. Rerun the model several times and see how the graph changes. The difference between chaotic, but deterministic, behavior and random behavior should become apparent.

At this point we should ask if chaos occurs in nature. We find that indeed it does. Water drips from a faucet chaotically, heart beat and brain wave variations show chaos. Both living and nonliving systems seem to show

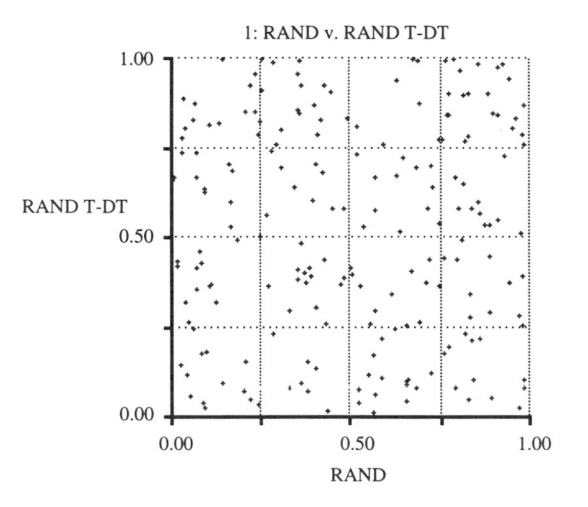

FIGURE 4.7

[2]For a full discussion of this and other versions of chaos, see R.V. Jenson, Classical Chaos, *American Scientist*, 75:168–181, 1987.

choas. Why? To what advantage is such a result to these systems? Stuart Kaufman,[3] among others, has proposed that all systems seem to evolve toward higher and higher efficiencies of operation. Many systems are so highly disturbed by variations in their environment that their efficiencies are not ever very high. However, if these disturbances can be held to minimum, then the evolution of the system becomes more complete and more efficient but closer to the border of chaotic behavior. Earthquakes and avalanches are examples of energy-storing systems that continuously redistribute the incoming stresses more and more efficiently until a breaking point is reached and the border to chaos is opened. Does this mean that the brain and the heart have somehow evolved close to some maximum efficiency for such organisms? We don't know the answer to this. We do know that the scale of measurement matters here. For example, if we were to watch the pattern on a patch of natural forest over many centuries, we would see the rise and sharp fall of the biomass levels, unpredictably. Forest fires and insects find ample host in such forest patches once they have developed a large amount of dry biomass bound up in relatively few species. The patch evolves or succeeds to greater and greater efficiency of light energy conversion by getting larger and fewer species. But the patch also becomes more vulnerable to fire and pests, and eventually collapses. Yet if we look at the total biomass on a large collection of such biomasses, whose collapses are not synchronized, this total biomass remains relatively constant. Thus chaotic-like behavior in the small is not seen in the large. Could this mean that natural systems have "found" chaos in their search for greater efficiencies and have "learned" to stagger the chaotic events, allowing faster rebound and large scale stablility? We don't know the answers here but we think the implications are fascinating. We will return to these questions with our models on self-organization and catastrophe, presented in Part VII of this book.

4.2. Simple Oscillator

The model of the previous section assumed that changes in population size occur instantaneously in response to the current population. Alternatively, we may assume that those changes are a function of the population size that is one time period delayed. Oscillations can occur in simple first-order differential equations, provided they are nonlinear and contain lags. Let us define the differential equation that guides the change in population size as

$$\Delta N = R^* \text{ LAG } N^* (1 - \text{LAG } N) \tag{7}$$

with

$$\text{LAG } N = \text{DELAY}(N,1) \tag{8}$$

[3]S. Kaufman, *The Origins of Order: Self Organization and Selection in Evolution*, New York: Oxford University Press, 1993.

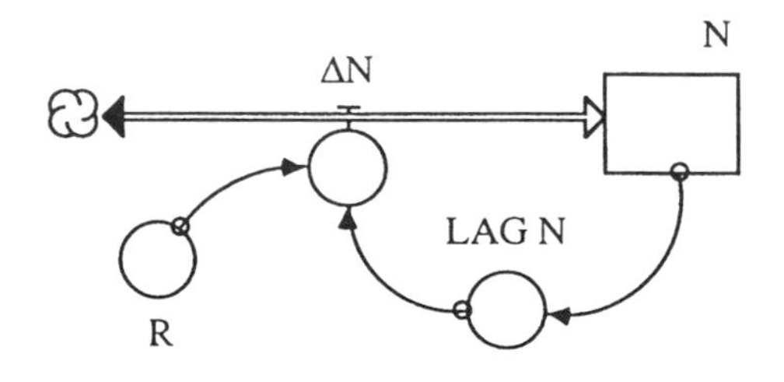

FIGURE 4.8

The STELLA model is seen in Figure 4.8.

Set R = 0.5 and DT = 1 and the model will generate population dynamics that oscillate but quickly settle down to a steady-state level. The results of the model are shown in Figures and 4.9 and 4.10. In steady state, LAG N = N, and thus the graph in phase-space collapses to a point.

Now increase R to 1, make an educated guess of the results of the model, and run it. Here is what you should get. Did you expect this result and can you explain it? Here a pattern emerges and yet is it chaos? Note how the arcs of the limit curve fill in with a definite pattern. This pattern of filling in arcs is due to the lagged N (Fig. 4.11).

Keep increasing R for subsequent runs and observe the results. The plot in Figure 4.12 is for R = 1.3 and it shows chaos.

If the DT is changed to 0.5 and R set to 1, you will find again an oscillation, shown in Figures 4.13 and 4.14. But if you increase R to 1.3, you will not find chaos. Rather, the number of limbs increases to 10 from 6 and the

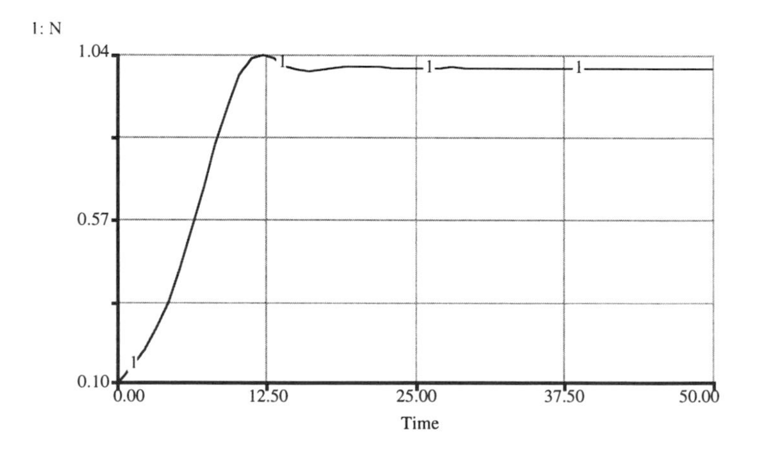

FIGURE 4.9

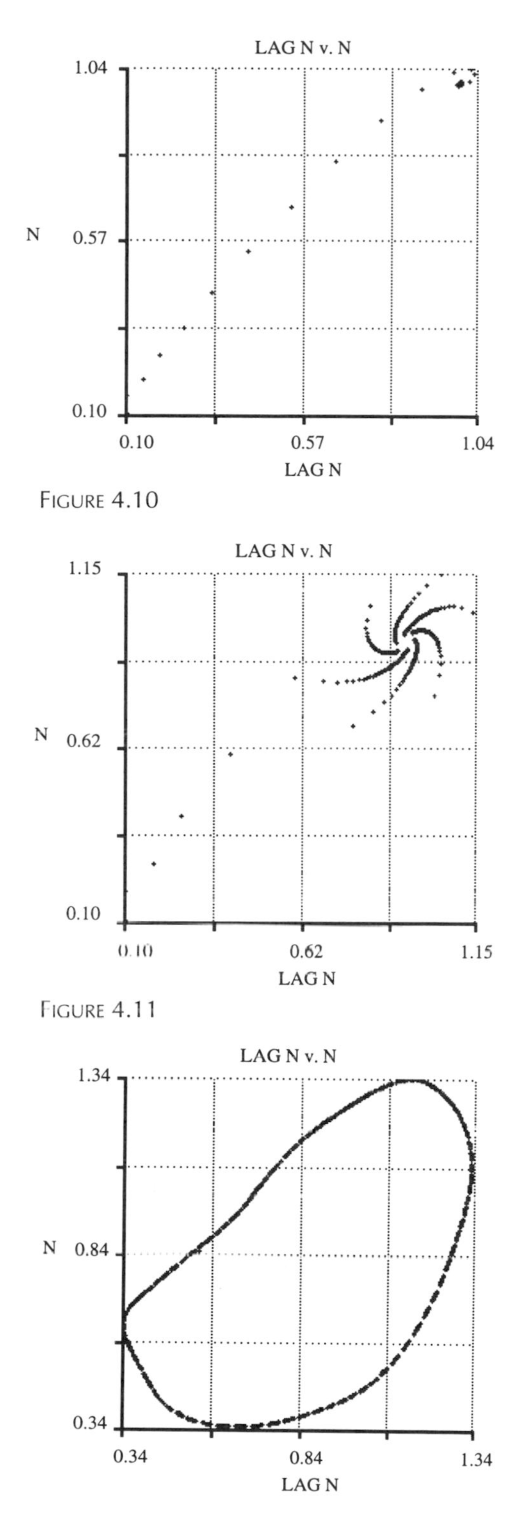

FIGURE 4.10

FIGURE 4.11

FIGURE 4.12

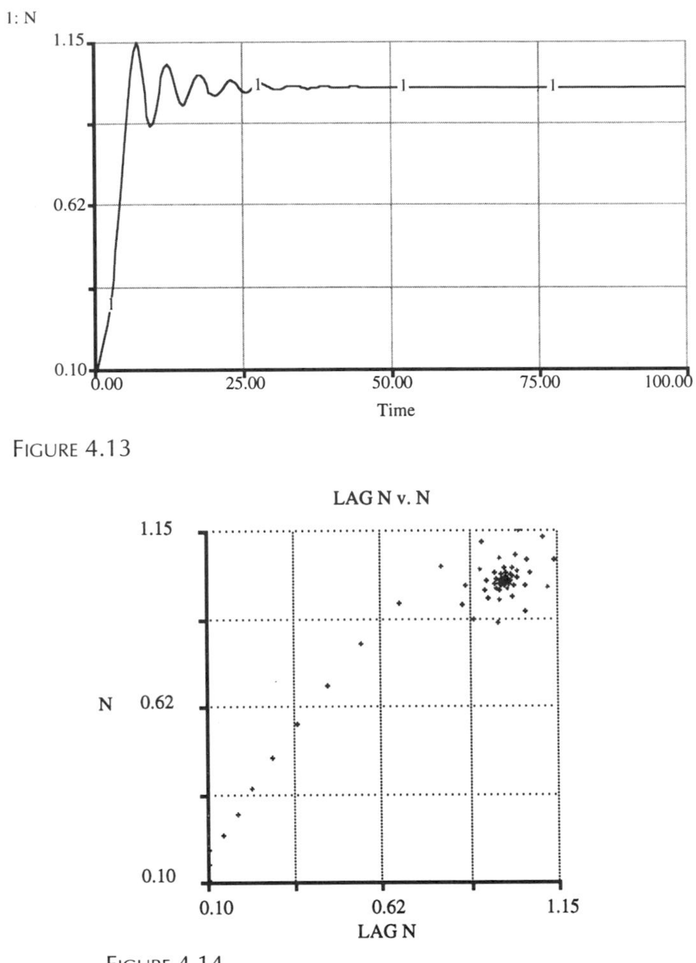

FIGURE 4.13

FIGURE 4.14

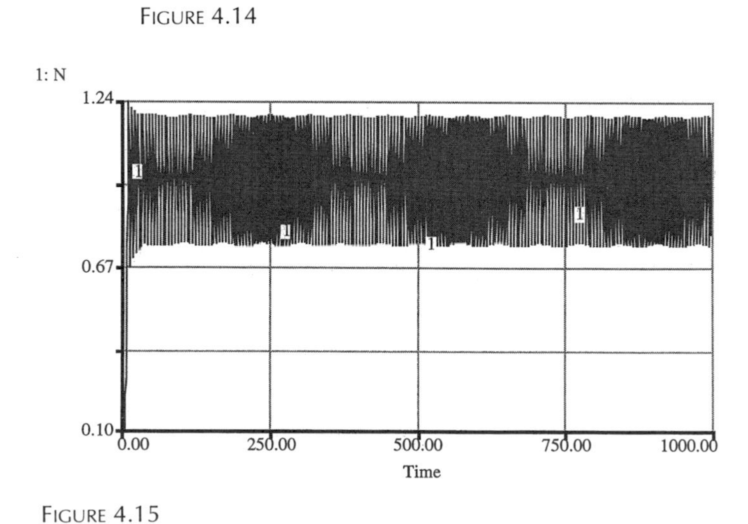

FIGURE 4.15

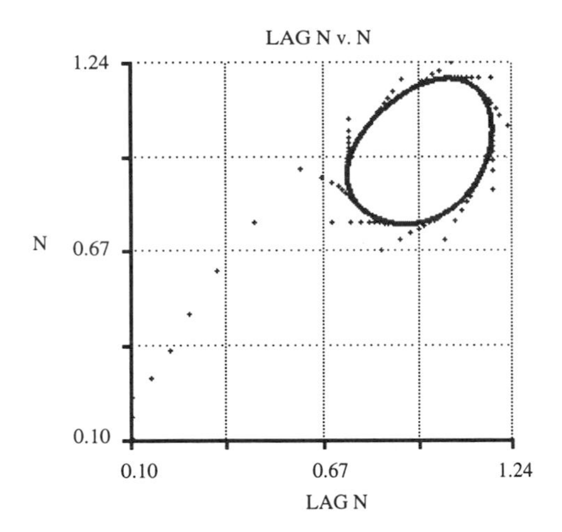

FIGURE 4.16

limit loop becomes smaller. See the last two graphs for this case (Figs. 4.15 and 4.16). What is the origin of these limbs?

Can you find the value of R at which chaos emerges? If the lag time is increased to 2 the model becomes unstable. Could you then stablize it somehow?

Oscillatory behavior has been observed in chemical systems. We will model one of the most prominent chemical oscillators, the Brusselator in Chapter 10. It has far-reaching implications for the understanding of real-world systems, and we will discuss these implications in that chapter.

SIMPLE OSCILLATOR

```
N(t) = N(t - dt) + (ΔN) * dt
INIT N = .1
INFLOWS:
ΔN = R*LAG_N*(1-LAG_N)

LAG_N = DELAY(N,1)
R = 1.3
```

5

Spatial Dynamics

> There is no coming into being of aught that perishes, nor any end for it. . . . but only mingling, and separation of what has been mingled.
>
> Empedocles

In the previous chapters we have modeled population dynamics in response to births and deaths, and we have assumed that the respective flows originate or disappear in "clouds." If we deal with migration in the context of population dynamics, we may want to explicitly model the source and recipients of the flows of migrants. This is the topic of this chapter. The model developed here is quite general and provides a basis for our discussion of spatial dynamics in Chapters 32, 36, 40, and 41.

Here is what might be called a *mobility framework*, in which migration from one state to another depends only on the current status of the donor state. Mobility might be the movement of animals between various resource points, chemical diffusion, the circulation of money, or the location of people in towns on a landscape. Mobility is defined by flows that are calculated as the difference between the value of the state variables at two points on the landscape. This difference is then multiplied by a fixed coefficient that reflects the strength of migration. The STELLA model is shown in Figure 5.1.

This is the egalitarian or the strict diffusion solution—no matter what the initial status of each of the states, all stocks find the same equilibrium. This is due to the fact that each of the exchanges is multiplied by the same coefficient. The final level is just the average of the four initial stock values, because there is no loss from the system (Fig. 5.2).

As long as all six of these coefficients are less than 1, the equilibrium solution is somewhere below the largest initial value. Multiply just one of the flows between any two states by a unique coefficient. Will the same equilibrium be reached? Change one of the coefficients to a value larger than 1 and observe the results.

To quickly compare the results of each model run without plotting a table, choose the numeric display—you find it among the STELLA icons to the right of the graph and table symbols—and place it in the STELLA diagram.

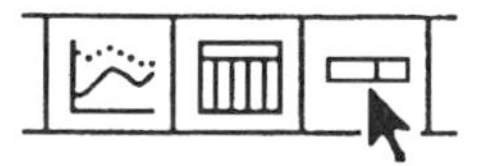

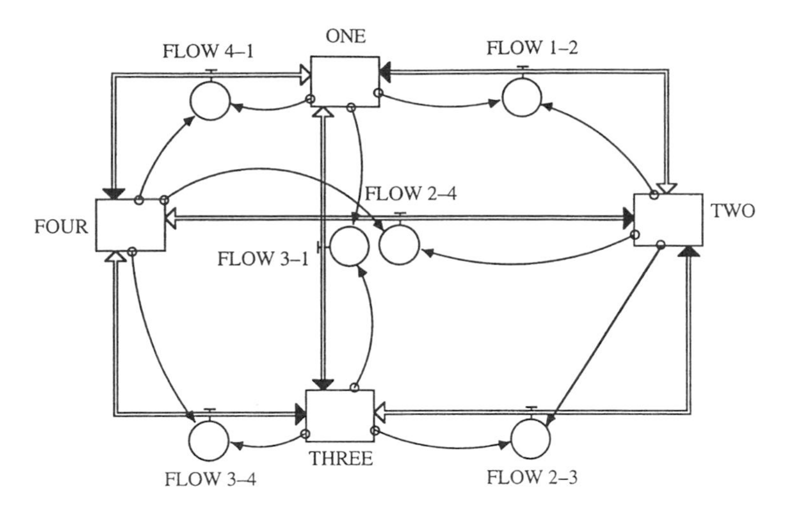

FIGURE 5.1

Double-click on the numeric display icon, select one of the system's state variables, and click on OK. Repeat this procedure for the other state variables. These numeric displays function like a counter and show you the value of a parameter as the model runs. If you specified the displays to maintain the ending balance, you can quickly compare the results from model run to model run. For example, the egalitarian result is shown in our model for coefficients equal to 0.03 as illustrated in Figure 5.3.

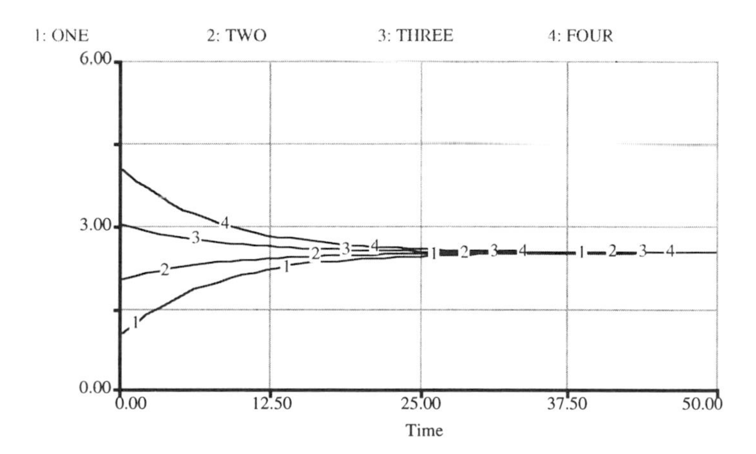

FIGURE 5.2

ONE	2.5
TWO	2.5
THREE	2.5
FOUR	2.5

FIGURE 5.3

Return to the case of coefficients smaller than 1. What are the effects of time lags on the time it takes for the system to equilibrate? Will the same equilibrium be reached as in the absence of those lags?

SPATIAL DYNAMICS

```
FOUR(t) = FOUR(t - dt) + (FLOW_3-4 + FLOW_2-4 -
FLOW_4-1) * dt
INIT FOUR = 4
INFLOWS:
FLOW_3-4 = .03*(THREE-FOUR)
FLOW_2-4 = .03*(TWO-FOUR)
OUTFLOWS:
FLOW_4-1 = .3*(FOUR-ONE)

ONE(t) = ONE(t - dt) + (FLOW_4-1 + FLOW_3-1 - FLOW_1-2)
* dt
INIT ONE = 1
INFLOWS:
FLOW_4-1 = .3*(FOUR-ONE)
FLOW_3-1 = .03*(THREE-ONE)
OUTFLOWS:
FLOW_1-2 = .03*(ONE-TWO)

THREE(t) = THREE(t - dt) + (FLOW_2-3 - FLOW_3-4 -
FLOW_3-1) * dt
INIT THREE = 3
INFLOWS:
FLOW_2-3 = .03*(TWO-THREE)
OUTFLOWS:
FLOW_3-4 = .03*(THREE-FOUR)
FLOW_3-1 = .03*(THREE-ONE)
```

```
TWO(t) = TWO(t - dt) + (FLOW_1-2 - FLOW_2-3 - FLOW_2-4)
* dt
INIT TWO = 2
INFLOWS:
FLOW_1-2 = .03*(ONE-TWO)
OUTFLOWS:
FLOW_2-3 = .03*(TWO-THREE)
FLOW_2-4 = .03*(TWO-FOUR)
```

Part 2

Physical and Biochemical Models

6

Law of Mass Action

> Let us now consider the character of the material Nature whose necessary results have been made available . . . for a final cause.
>
> Aristotle

The law of mass action is a powerful concept that describes the average behavior of a system that consists of many interacting parts, such as molecules, that react with each other, or viruses that are passed along from a population of infected individuals to nonimmune ones. The law of mass action has been derived first for chemical systems but subsequently found high use in epidemiology and ecology. In this chapter, we discuss the law of mass action in the context of a simple chemical system. In later chapters we apply it to issues as diverse as enzyme–substrate interactions, the spread of a disease, or the colonization of landscape patches.

Let us consider the case of oxygen O reacting with hydrogen molecules H to form water H_2O. The stocks of each substance is given as a concentration, measured in moles per cubic meter. For simplicity, we assume initial conditions of 200 moles of hydrogen per cubic meter and 100 moles of oxygen atoms per cubic meter. Thus, there is enough of each initial stocks to just form 100 molecules of water. The stochiometric equation for this reaction is

$$2\,O + H \rightarrow H_2O \qquad (1)$$

Obviously, not all of the oxygen molecules will instantaneously find hydrogen molecules to engage in the reaction. Rather, the reaction will take place over time. The reaction velocity will be high early on because the concentrations of each of the reactants is high. With declining concentrations, the reaction velocity declines. The law of mass action expresses the rate of formation of the product H_2O from the reactants O and H:

$$\Delta H_2O = K * H * O \qquad (2)$$

with K as the reaction rate constant, measured in 1/second. It is captured in our model in Figure 6.1.

The concentrations of the reactants decrease as the reaction proceeds. To calculate the respective outflows from the stocks O and H, we need to specify the number of moles that enter the reaction to form one mole of the product. For O and H these are, respectively, 2 and 1 (Fig. 6.2).

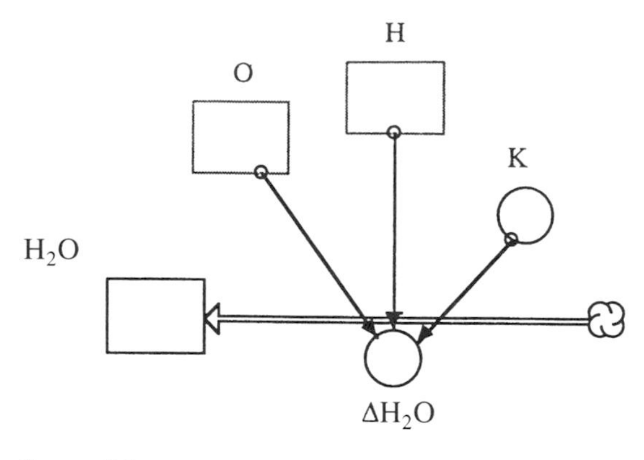

FIGURE 6.1

The results of our model are shown in Figure 6.3 for a hypothetical value of K = 0.005. As we would expect, the concentrations of both reactants decline as the product is formed. Also, the rate of reaction declines. Will eventually all the reactants be used up and exactly 100 moles of water per cubic meter be formed? Make an educated guess and then run the model for a longer time frame and observe the results.

The model in Figure 6.3 implicitly assumes that temperatures and pressures remain constant throughout the reaction. However, the formation of water from oxygen and hydrogen produces a significant amount of heat that, in turn, increases the reaction velocity. Can you introduce this effect in the model? Expand the model to the case of a simple photosynthetic process in which carbon dioxide and water react to form a glucose and oxygen:

$$6\ CO_2 + 6\ H_2O \rightarrow C_6H_{12}O_6 + 6\ O_2 \tag{3}$$

Can you change your model to capture the metabolic process in which the glucose reacts with oxygen to form water and carbon dioxide? Find in the

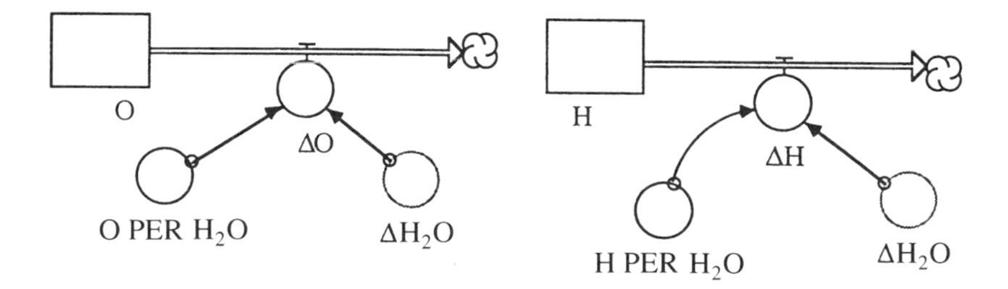

FIGURE 6.2

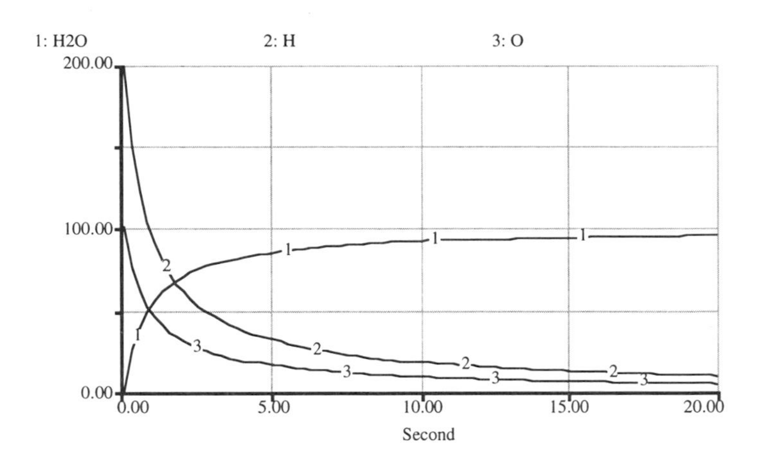

FIGURE 6.3

literature characteristic reaction rate constants for both the formation and metabolism of various types of glucose and model the respective chemical reactions.

LAW OF MASS ACTION

```
H(t) = H(t - dt) + (- ΔH) * dt
INIT H = 200 {Moles per Cubic Meter}
OUTFLOWS:
ΔH = ΔH20*H_PER_H20 {Decrease in H concentration as a
result of H2O formation; measured in Moles per Cubic
Meter per Second}

H2O(t) = H2O(t - dt) + (ΔH20) * dt
INIT H2O = 0 {Moles per Cubic Meter}
INFLOWS:
ΔH20 = K*H*O {Increase in H2O concentration; simple
first-order reaction; measured in Moles per Cubic Meter
per Second}

O(t) = O(t - dt) + (- ΔO) * dt
INIT O = 100 {Moles per Cubic Meter}
OUTFLOWS:
ΔO = O_PER_H20*ΔH20 {Decrease in O concentration as a
result of H2O formation; measured in Moles per Cubic
Meter per Second}
```

```
H_PER_H20 = 2 {Moles H required per Mole H2O formed -
from stoichiometry of reaction}
K = .005 {1/Second}
O_PER_H20 = 1 {Moles O required per Mole H2O formed -
from stoichiometry of reaction}
```

7

Catalyzed Product

> Many bodies . . . have the property of exerting on other bodies an action which is very different from chemical affinity. By means of this action they produce decomposition in bodies, and form new compounds into the composition of which they do not enter. This new power, hitherto unknown, is common both in organic and inorganic nature. I shall call it catalytic power. I shall also call Catalysis the decomposition of bodies by this force.
>
> Berzelius, Edin. New Phil. Jrnl. XXI., 1836

In the previous chapter we have used the law of mass action to describe chemical change of two substances interacting with each other and forming a product that is chemically distinct from the two reactants. In this chapter we expand on that model and deal with the case in which, after a series of reactions, one of the reactants reemerges to enter the reaction anew. For example, some enzyme E may enter a chemical reaction from which an intermediate product I results. This intermediate product, in turn, may enter a reaction from which E is released in unchanged form together with a new product F. An additional substance active in this process is the substrate D, which is converted into the product F by action of the enzyme E.

Such catalyzed reactions are common in biological processes. One example of these reactions is the production of fructose (F) from dextrose (D). In this process, the enzyme (E) mechanically locks onto the substrate molecule, breaks it into a new molecule, fructose, and is released again after the chemical reaction occurred[1].

Denote the reaction rate constants as K1, K2, and K3. With this notation, the basic reaction equation is

$$E + D \underset{K2}{\overset{K1}{\leftrightarrow}} I \overset{K3}{\rightarrow} E + F \tag{1}$$

The four basic differential equations that define the rate expressions, or flows, are

$$dD/dt = K1 * D * E - K2*I \tag{2}$$

$$dI/dt = K1 * S * E - (K2 + K3)*I \tag{3}$$

[1]See J.D. Spain, *Basic Microcomputer Models in Biology*, Reading, MA: Addison-Wesley Publishing, 1982.

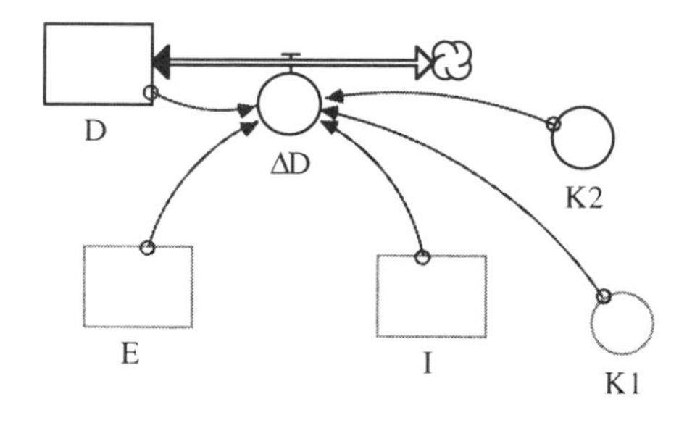

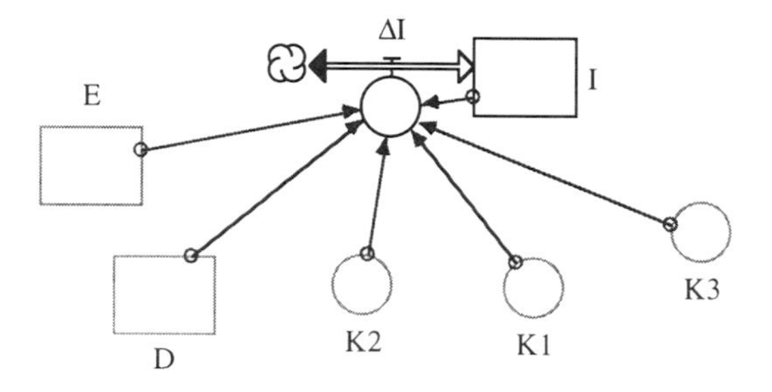

FIGURE 7.1

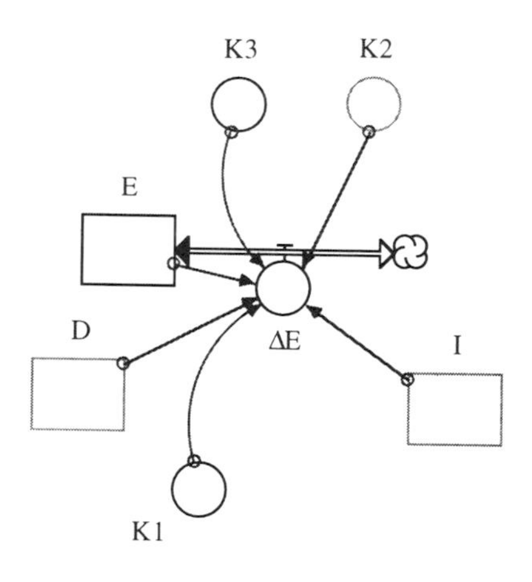

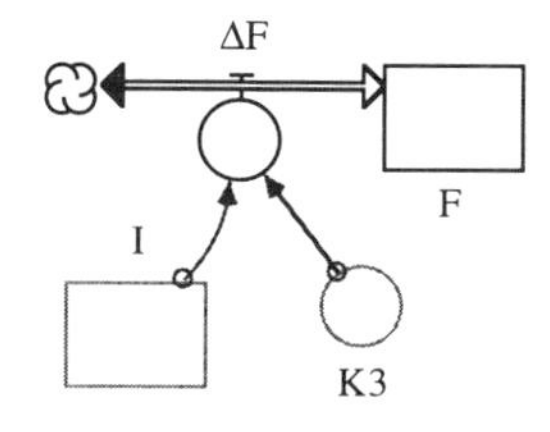

FIGURE 7.2

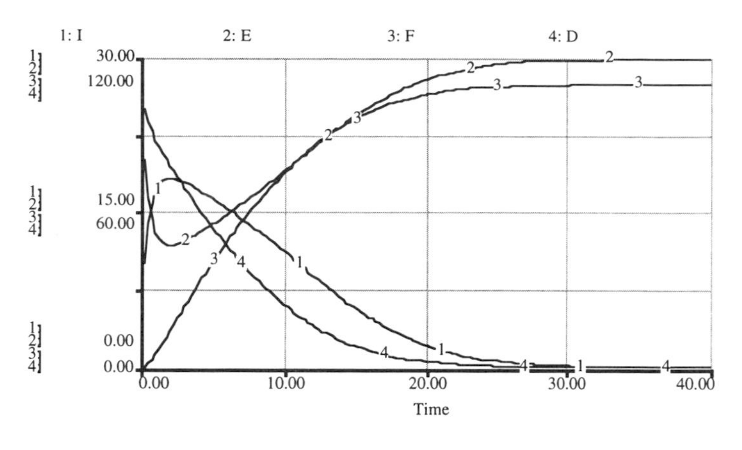

FIGURE 7.3

$$dE/dt = - dI/dt \tag{4}$$

$$dF/dt = K3*I \tag{5}$$

The differential equations are used in the following four modules of Figures 7.1 to 7.2. Each of the state variables is treated as a concentration.

The results show a temporary decline in the concentration of the enzyme as it gets locked up in the production of I (Fig. 7.3). As the concentration of the intermediate product declines, in response to the formation of the final product F, the concentration of E increases again. As the substrate D gets depleted, no more reactions take place and the system settles down to a set of equilibrium concentrations for E and F. What is the continuous addition rate of D such that a withdrawal rate of F is 0.1 units per time step?

For the model results above, we have arbitrarily set K1 = K2 = 0.1 and K3 = 0.5. Change the reaction rates and observe the results. Can you make this model with fewer stocks, say, just ones for I and F?

In this chapter we modeled in a general way the catalyzed reactions of enzymes and substrates. In the following chapter, we broaden our focus and deal with an entire cell that takes up nutrients from its environment and excretes waste products into its environment.

CATALYZED REACTION

```
D(t) = D(t - dt) + (- ΔD) * dt
INIT D = 100 {Moles per Cubic Meter}
OUTFLOWS:
ΔD = K1*D*E-K2*I {Moles per Cubic Meter per Time
Period}
```

```
E(t) = E(t - dt) + (- ΔE) * dt
INIT E = 20 {Moles per Cubic Meter}
OUTFLOWS:
ΔE = K1*D*E-(K2+K3)*I {Moles per Cubic Meter per Time
Period}

F(t) = F(t - dt) + (ΔF) * dt
INIT F = 0 {Moles per Cubic Meter}
INFLOWS:
ΔF = K3*I {Moles per Cubic Meter per Time Period}

I(t) = I(t - dt) + (ΔI) * dt
INIT I = 10 {Moles per Cubic Meter}
INFLOWS:
ΔI = K1*D*E-(K2+K3)*I {Moles per Cubic Meter per Time
Period}

K1 = .01 {1/Time Period}
K2 = .01 {1/Time Period}
K3 = .5 {1/Time Period}
```

8

Two-Stage Nutrient Uptake Model

In the nutrient-rich waters of the Thames type, a burst of algal growth may sometimes cease before any serious depletion of the mineral nutrient in the water has apparently taken place.

Nature 21 Sept. 1946

In the previous model we investigated the chemical reaction that may take place within a cell. In this chapter, we model in more detail the activities of an entire cell that receives nutrients from its envronment, uses these nutrients for growth and maintenance, and then excretes waste products back into its environment.

Assume a cell with an internal nutrient concentration Q is immersed in a media with a nutrient concentration N[1]. The growth of the cell biomass X is directly dependent on the internal rather than the external nutrient concentration. Nutrient uptake is proportional to the cell biomass. The proportionality factor is MU. An outline of the structure of this system is given in Figure 8.1.

Through respiration and mortality, the nutrient is passed back into solution outside the cell. The rate of return from the cell is proportional to the biomass of the cell. The proportionality factor is R. There is a minimum level of the internal nutrient concentration, Q_0, which is needed before the cell will grow at all. Thus, the change in cell biomass, ΔX, is

$$\Delta X = \text{IF } Q \geq Q\_0 \text{ THEN } (MU - R)*X \text{ ELSE } 0. \quad (1)$$

The change in the nutrient concentration outside the cell, ΔN, depends on the nutrient passing through the cell wall,

$$\Delta N = \text{IF } N > 0 \text{ THEN } R*Q*X - V*X \text{ ELSE } 0. \quad (2)$$

R*Q*X is the return of the nutrients to the external environment, depending on the internal nutrient concentration Q. The nutrient balance equation for the internal concentration of the nutrient is

$$\Delta Q = \text{IF } Q \geq Q\ 0 \text{ THEN } V - MU*Q \text{ ELSE } 0. \quad (3)$$

Equation (3) is a form of the Monod equation used to predict the change in the internal concentration.

---

[1]For a reference on the origins of this model see J.D. Spain, *Basic Microcomputer Models in Biology*, Reading, MA: Addison-Wesley Publishing, 1982, p. 242. The model is called the Caperon–Droop model.

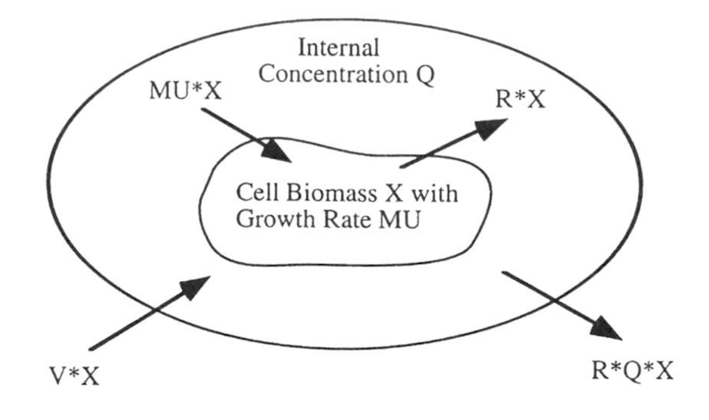

FIGURE 8.1

The rates V and MU are calculated from a now-standard Michaelis–Menten formula. The general form for this equation is derived from the enzyme–substrate equation discussed in the previous chapter[2]:

$$V = VM*N/(KN + N) \tag{4}$$

$$MU = MU\ BAR*(Q - Q0)/(KQ + (Q - Q0)), \tag{5}$$

with VM the maximum rate of nutrient uptake per unit biomass, KN the half-saturation constant for nutrient uptake, KQ the half-saturation constant for growth, and MU BAR the maximum biomass growth rate.

Recognize that this model (Fig. 8.2) differs from the specification of the reaction rate in the previous chapter in that the reaction rate here is not based on the product of the concentrations of the compounds. Also, note the way in which MU is used in the ΔX and ΔQ equations. It is an interesting use of a common variable. The ΔQ equation makes more sense if you multiply through by X. Curiously, R*X and R*Q*X are not included in the ΔQ equation. The R factor is apparently designed to represent the communication of the cell biomass to the main nutrient source only. This fact, as with the definition of Q, is the likely result of experimental measurement problems of the time.

In this model N and Q are concentrations, but they are different kinds of concentrations. N is measured in mg N/l, a volumetric concentration, and Q is measured in mg N/mg X, a mass-based concentration. It must be too hard to measure the volume of a cell. The units are tricky. The units of the uptake rate for N, V, are mg N/mg X/time, while the rate of formation and

[2]L. Edelstein-Keshet, *Mathematical Models in Biology*, New York: Random House, 1988, p. 278.

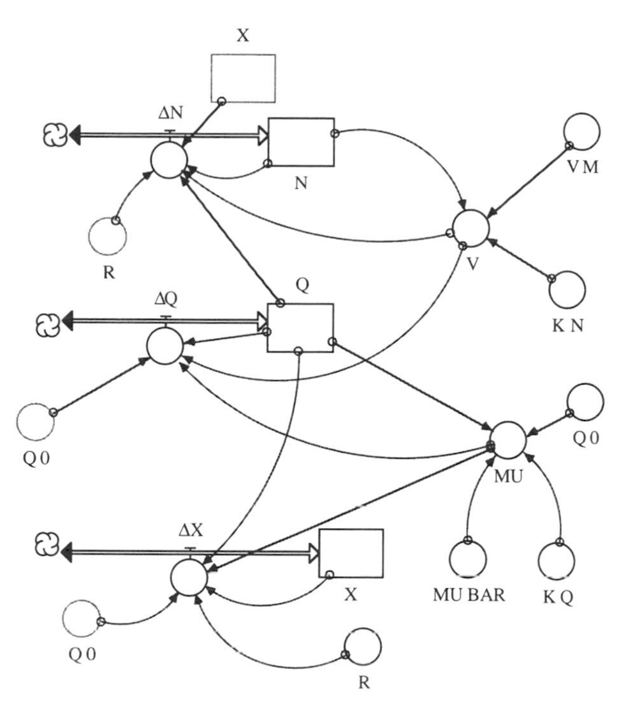

FIGURE 8.2

mortality of the biomass X are 1/time. The units of MU are 1/time. Work out the units of ΔN, ΔQ, and ΔX to make sure that they are consistent.

Figure 8.2 shows the three differential equations and the supporting parameters. Figure 8.3 shows how the concentrations and the biomass levels change with time and how the cell growth rate depends on the internal nutrient concentration. To obtain such results, we had to run the time step at DT = 0.125 and use the Runga–Kutta 4 solution technique. Try running the model with Euler and DT = 1.0, and note the difference in results. Try shifting the parameters and note their effect on the concentration trajectories.

How is it possible in our model that the biomass can maintain itself in the steady state forever? Does this cell know something about perpetual motion that it is not telling us or have we missed something? Actually, the cell is using energy all the time. It has to be taking up high quality energy (ATP) and giving off low quality energy (heat). We are simply not modeling that part of the cell activity. We model only the use of a single nutrient. Neither are we modeling at the smaller level where things are inevitably falling apart and being replaced, and that replacement process is not faultless.

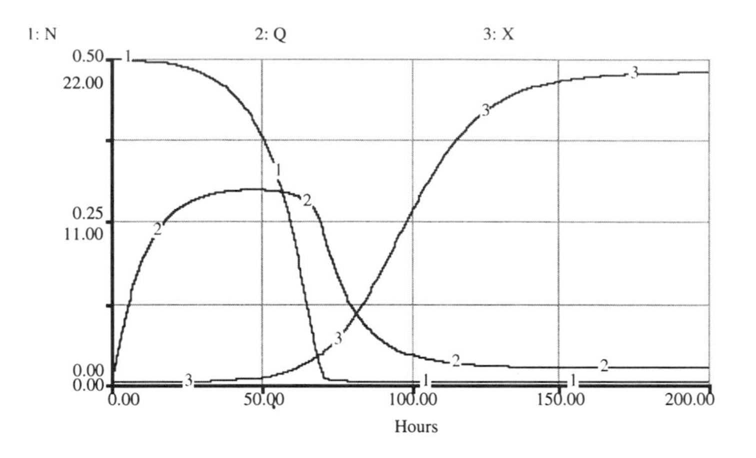

FIGURE 8.3

The following chapter expands the boundaries of systems processes beyond an individual cell to the level of an organism. There, we distinguish different compartments among which a substance is being distributed. Models of organs and entire organisms are presented in Part IV of the book.

TWO STAGE NUTRIENT UPTAKE MODEL

```
N(t) = N(t - dt) + (ΔN) * dt
INIT N = 0.5 {mg/liter}
INFLOWS:
ΔN = IF N > 0 THEN R*Q*X -V*X ELSE 0

Q(t) = Q(t - dt) + (ΔQ) * dt
INIT Q = 0.02 {mg/mg of X}
INFLOWS:
ΔQ = IF Q ≥ Q_0 THEN V - MU*Q ELSE 0

X(t) = X(t - dt) + (ΔX) * dt
INIT X = 0.01 {mg/liter. Total cell biomass per liter.}
INFLOWS:
ΔX = IF Q ≥ Q_0 THEN (MU -R)*X ELSE 0

K_N = 0.05 {mg/liter}
K_Q = 0.03 {mgN/mgX}
```

```
MU = MU_BAR*(Q - Q_0)/(K_Q + (Q - Q_0)) {1/hour}
MU_BAR = 0.1 {1/hour}
Q_0 = 0.02 {mgN/mgx}
R = 0.01 {1/hour}
V = V_M*N/(K_N + N) {mgN/mgX/hour}
V_M = 0.03 {mgN/mgX*hour}
```

9

Iodine Compartment Model

> Iodine was discovered accidentally, about the beginning of the year 1812, by M. Courtois, a manufacturer of saltpetre at Paris.
>
> Henry, *Elementary Chemistry I*, 1826

The previous two models focused, respectively, on chemical reactions within a cell and on the activities of an entire cell. In this chapter we model the flow of a substance, iodine, among different parts of an organism.

Iodine is found in three places in the body: the thyroid gland, tissue connected to and surrounding the thyroid gland, and inorganic iodine in the circulatory system. The following figure (Fig. 9.1) shows how the flow balance is established for each of the stocks representing each of the places that iodine can be found. For our model (Fig. 9.2), inorganic iodine is constantly injected into the system at 150 micrograms per day, for example, through certain foods, or iodized salt. This exogenous iodine input is denoted DI.

The thyroid gland is consuming inorganic iodine and some leaves the system with urine. Ultimately, the remainder leaves the digestive system as feces. Conversion rates required to specify the flows from compartment to compartment are assumed to be dependent on the amount of iodine in

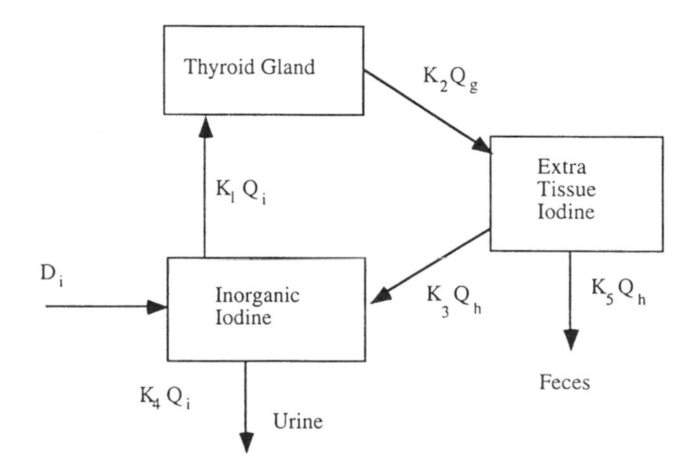

FIGURE 9.1

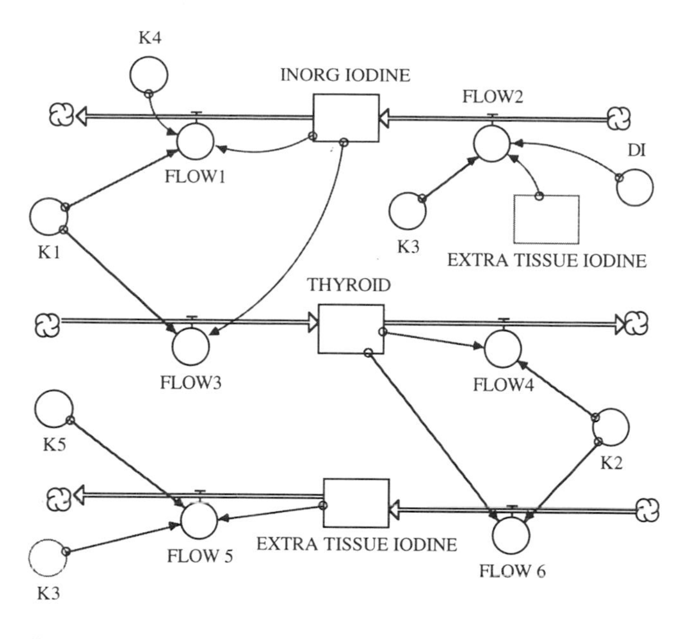

FIGURE 9.2

each place. This is a questionable assumption that nevertheless is supported by experiment.

The thyroid level declines to a stable level with the input rate (Fig. 9.3). Change the external input rate to one tenth of the stable level and plot the result. This level represents an extremely deficient iodine level.

This very simple model of a complex and important process demonstrates how one can model simply yet accurately. Note that we did not use the idea

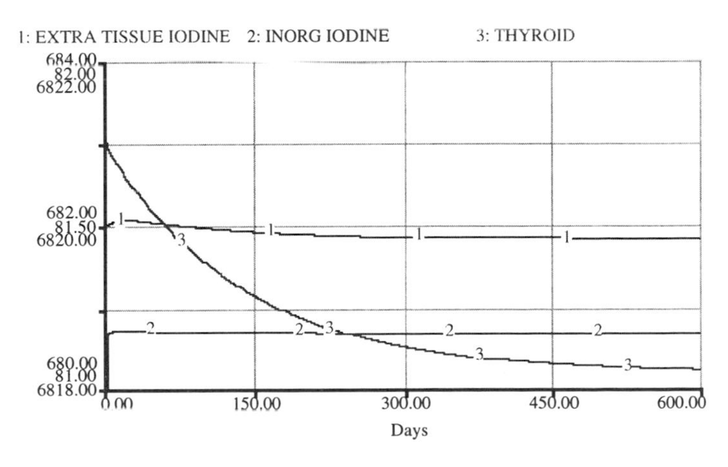

FIGURE 9.3

of the mass action law here. Why not? Can you see here where some flows are recipient controlled and some are donor controlled? This identification of the state variable controls early in the modeling process can be very helpful.

We began this part of the book on physical and biochemical systems with simple chemical reactions and proceeded to catalyzed processes, the activities of individual cells, and the distribution of chemical substances among different compartments. We return in the following chapter to chemical reactions and combine our insight into oscillatory system behavior of Chapter 4 with our knowledge of chemical processes.

IODINE COMPARTMENT MODEL

```
EXTRA_TISSUE_IODINE(t) = EXTRA_TISSUE_IODINE(t - dt) +
(FLOW_6 - FLOW_5) * dt
INIT EXTRA_TISSUE_IODINE = 682 {micrograms}
INFLOWS:
FLOW_6 = K2*THYROID
OUTFLOWS:
FLOW_5 = (K3+K5)*EXTRA_TISSUE_IODINE

INORG_IODINE(t) = INORG_IODINE(t - dt) + (FLOW2 -
FLOW1) * dt
INIT INORG_IODINE = 81 {micrograms}
INFLOWS:
FLOW2 = DI+K3*EXTRA_TISSUE_IODINE
OUTFLOWS:
FLOW1 = (K1+K4)*INORG_IODINE

THYROID(t) = THYROID(t - dt) + (FLOW3 - FLOW4) * dt
INIT THYROID = 6821 {micrograms}
INFLOWS:
FLOW3 = K1*INORG_IODINE
OUTFLOWS:
FLOW4 = K2*THYROID

DI = 150 {micrograms/day}
K1 = 0.84 {1/day}
K2 = 0.01 {1/day}
K3 = 0.08{1/day}
K4 = 1.68{1/day}
K5 = 0.02 {1/day}
```

10

The Brusselator

In the Beginning the Heav'ns and Earth Rose out of Chaos.

P.L. Milton, 1667

One important charactersitic of real-world processes is that interactions among system components are nonlinear. As we have seen in the previous chapters, nonlinearities can give rise to rich temporal patterns. An example of nonlinearities in chemical reactions is the autocatalytic process,

$$A + X \rightarrow 2X, \tag{1}$$

whereby X stimulates its own production from A. Such autocatalytic reactions may involve a series of intermediate sates, for example, if X produces a substance Y, which in turn accelerates the production of X. One such cross-catalytic reaction has been studied extensively by a group of scientists around the Nobel Laureate Ilya Prigogine in Brussels and is known as the Brusselator[1]. It involves the following series of reaction steps:

$$A \rightarrow X \tag{2}$$

$$2X + Y \rightarrow 3X \tag{3}$$

$$B + X \rightarrow Y + D \tag{4}$$

$$X \rightarrow E \tag{5}$$

These reactions can be maintained far from equilibrium by continually supplying the substances A and B, and extracting D and E. These additions and subtractions eliminate the back reactions by holding the concentrations A, B, C, D, and E constant. This assumption allows us to capture in the STELLA model the concentrations of those products as transforming variables rather than reservoirs. In contrast, the two intermediate components (X and Y) may have concentrations that change in time.

For simplicity, and without loss of generality, we set the kinetic constants equal to one. The following system of nonlinear equations results, after eliminating D, which does not *enter* any of the reactions, and is continuously removed from the system:

---

[1] I. Prigogine, *From Being to Becoming: Time and Complexity in the Physical Sciences*, New York: W.H. Freeman and Company, 1980.

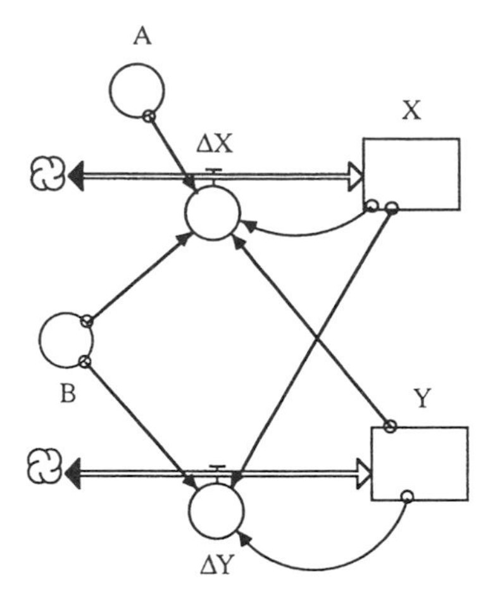

FIGURE 10.1

$$dX/dt = A + X^2*Y - B*X - X \tag{6}$$

$$dY/dt = B*X - X \wedge 2*Y \tag{7}$$

The corresponding STELLA diagram is seen in Figure 10.1. The assumption that the products A, B, C, D, and E are held constant through either removal or addition allows us to model those products as constants.

Set A = 0.7 and B = 2, and run the model. Choose the initial conditions X(t=0) = A and Y(t=0) = B/A, run the model at a DT = 0.0625, and you will find the steady state of this system. However, this steady state is unstable. To confirm this observation, run the model for two initial conditions close to the ones that yield steady state, for example, X(t=0) = 1, and X(t=0) = 0.8, with Y(t=0) = 2.5 in each of the two cases. Here the results are plotted in the same graph (Fig. 10.2).

Small changes in the initial conditions lead away from the (unstable) steady state but ultimately lead the system to a steady limit cycle. Will this cycle be chosen by the system for initial conditions that are significantly different from those that we already chose? The assumptions of the following five runs are listed in Table 10.1.

TABLE 10.1 Initial conditions for 5 sensitivity runs

Run	X(t=0)	Y(t=0)
1	0	0
2	0	1
3	0	2
4	3	0.5
5	1.5	0

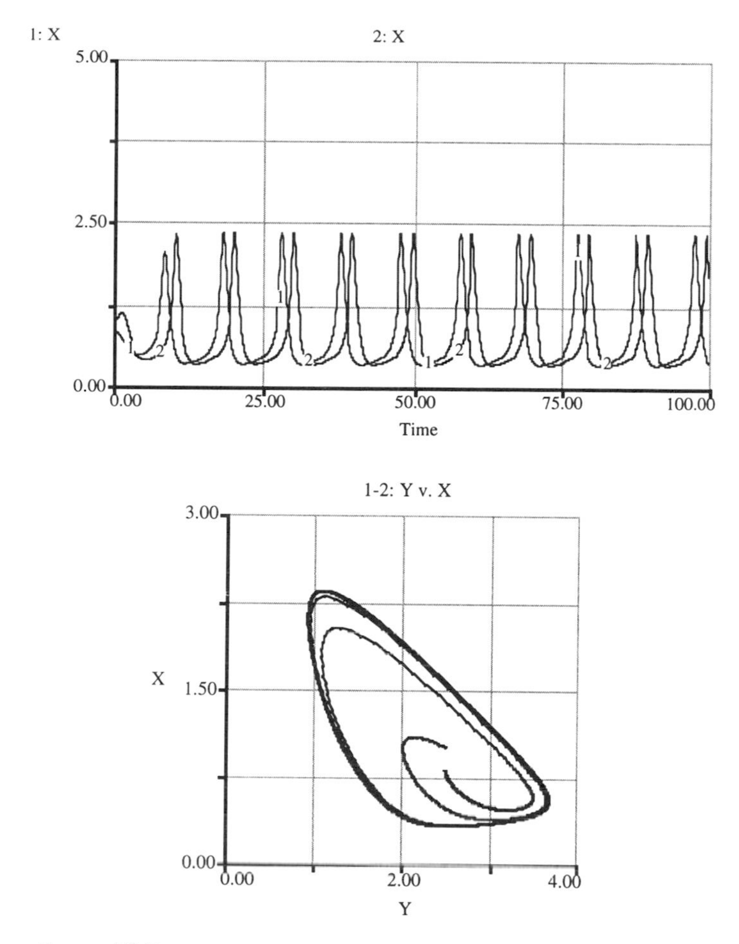

FIGURE 10.2

These runs were performed with STELLA's sensitivity analysis, making use of the *ad hoc values* option and plotting X against Y in a *comparative* scatter plot. The results in Figure 10.3 show that irrespective of the initial conditions, the system converges to the same limit cycle.

Two conditions are necessary for the limit cycle to occur—the system must be open and interactions among system components must be nonlinear. The first of these conditions is fullfilled by withdrawing and adding the products A, B, C, D, and E, effectively leaving their concentrations constant. As a result, the system is maintained away from an equilibrium at which reactants get used up and the chemical reactions come to a halt. The second condition is met by equations (6) and (7). Prigogine and his coworkers argue that virtually any real-world system is open, characterized by nonlinearities, and

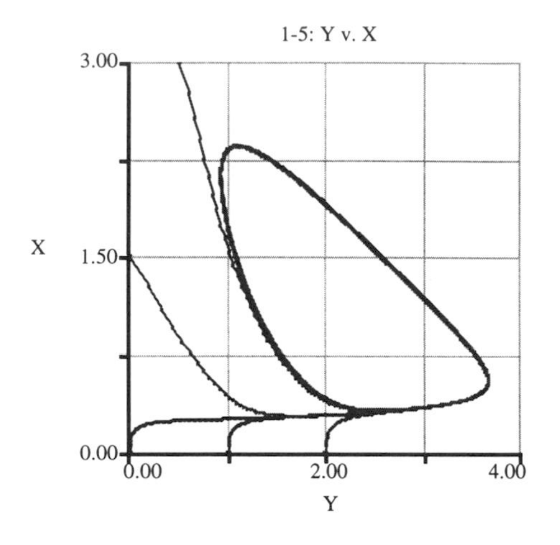

FIGURE 10.3

maintained out of equilibrium with their surroundings. Individual organisms receive material and energy inputs from their surroundings and excrete waste products and waste heat. Similarly, entire ecosystems channel materials and energy through their systems. The constant influx of "reactants" and energy into these systems, and the constant removal of waste materials and heat, makes it possible for those systems to function. They are clearly open and not in equilibrium with their surroundings.

Change the values for A and B, and rerun the model for alternative initial conditions. Can you find the steady-state conditions? How does the limit cycle change? How are the results affected by the choice of DT and integration methods? Save the equations as text, open them with MADONNA, and perform sensitivity tests on the choice of DT.

Once you explored the dynamics of this system and familiarized yourself sufficiently with the models of chemical processes discussed in this part of the book, move on to learn more about the application of physical principles and tools to the understanding of biological processes. This is the topic of the following chapter.

BRUSSELATOR

```
X(t) = X(t - dt) + (ΔX) * dt
INIT X = 1
INFLOWS:
ΔX = A+X^2*Y-B*X-X
```

```
Y(t) = Y(t - dt) + (ΔY) * dt
INIT Y = 2.5
INFLOWS:
ΔY = B*X-X^2*Y

A = .7
B = 2
```

11

Fitzhugh–Nagumo Neuron Model

Yet ha we A Braine that nourishes our Nerues.
Shakespeare, *Antony & Cleopatra* iv. viii. 21–1606

The previous chapter showed rich dynamics for the case of *chemical* reactions in an open system. In that chapter, we stress the sensitivity of system responses to initial conditions and the role of nonlinearities in determining system behavior. The model of this chapter captures nonlinearities in the changes of the *physical* state of a cell in response to electrical impulses from its surroundings.

Figure 11.1 depicts a neurone or nerve cell. Neurons consist of a cell body, picking up possible signals from one or all of its several dendritic branches (signals from other neurones) and transmitting these received (electrical) signals down a relatively long axon pathway to its terminal. At the terminal the signal is amplified and transmitted across a synapse to the next nerve cell. The cell remains inactive until the collective input from the dendrites reaches a critical level, whereupon the cell fires—it reacts in such a way as to amplify the collective input signals into a signal potential at its terminal end. If the collective input signal is not great enough, the signal dies out in the cell due to the action of a recovery mechanism.

The Fitzhugh–Nagumo model is of a nerve cell under special laboratory conditions, in which all dendritic receivers are kept at the same potential. The space-change in potential along the axon and throughout the cell is thus ignored. The only way to cause a cell reaction, then, is to boost the external voltage to a sufficiently high level or to establish an input signal through the dendritic connections.

The Fitzhugh–Nagumo equations are[1]

$$dV/dt = -V^*(V - V1)^*(V - V2) - W + E \qquad (1)$$

$$dW/dt = EPSILON^*(V - C^*W), \qquad (2)$$

where V is the departure of the membrane potential from its equilibrium and W is the recovery variable reflecting conductance of ions depending on

[1]Adapted from D. Brown, P. Rothery, *Models in Biology: Mathematics, Statistics and Computing*, New York: Wiley, 1993, pp. 320–326.

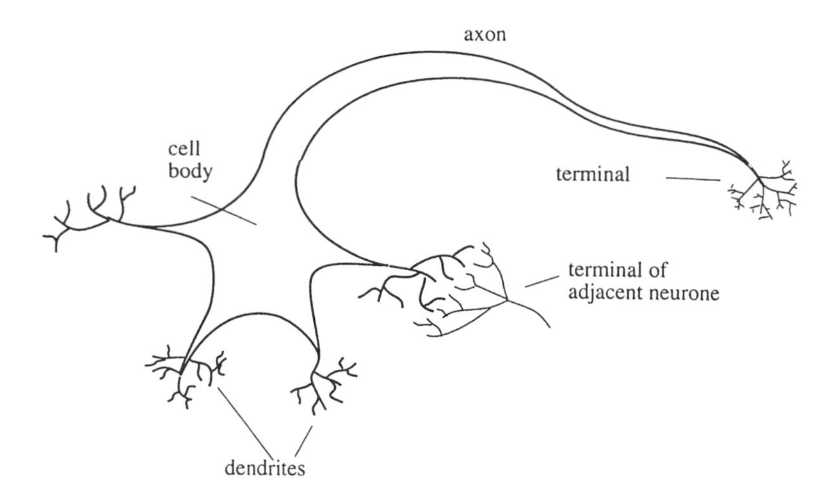

FIGURE 11.1

voltage. These two variables are the state variables of the system. The parameters V1 and V2 capture the influence of V on the rate of change of V and are held constant at 0.2 and 1.0, respectively. The parameter E reflects the electrical current to which the neuron is exposed.

The rate of change of the recovery variable W, defined in equation (2), is dependent on the difference between the departure of the membrane potential from its equilibrium, V, and the recovery variable W that decays at a constant rate C. In our model, we arbitrarily set C = 0.5. The change in W is assumed to be proportional to (V-C*W), with a proportionality factor of EPSILON. We set EPSILON = 0.02.

The STELLA model of the Fitzhugh–Nagumo equations is shown in Figure 11.2. Note that in this STELLA model, the control flows are set up as biflows and the stocks are set to allow negative values. We run this model at a DT = 0.1, using the Euler integration method.

If you set V(t=0) = 0 and W(t=0) = 0, both the membrane potential and recovery mechanism are at equilibrium. If you set further E = 0, the cell does not respond because all potentials are zero.

When the neurone is exposed to either a dendritic voltage or an external potential, the cell membrane potential builds and fires an amplification of the combined received voltages. The dendritic signals are simulated by initial values of the membrane potential. The membrane potential response shows a critical threshold voltage potential is required. A quick examination of the differential equations shows that this amplification threshold is 0.2 (V1) and amplification ceases when the voltage exceeds 1.0 (V2) — our choice of parameters for the model.

When the dendritic potential is zero [V(t = 0) = 0, W(t = 0) = 0], the effect of applied voltages (E) can be seen. For applied voltages E ≥ 0.23, the

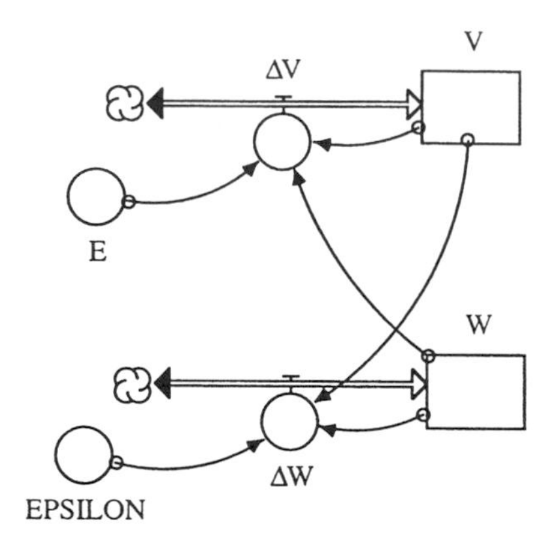

FIGURE 11.2

membrane potential cycles represent an unstable but experimentally observed condition. The results for E = 0.23 are shown in Figure 11.3. This is the lowest applied voltage that will produce cycling. Can you find the value for E that leads to the largest amplification of the input signal?

Figure 11.4 illustrates the case of amplification of the dendritic signal: E = 0, V(t = 0) = 0.4, with all other variables the same as earlier. The cell responds when the initial membrane potential is set to some positive potential. The membrane potential asymptotically approaches zero from its initial value, after rising, then overshooting zero, and dropping below the equilibrium potential (zero on our scale). Try EPSILON = 0.002 and note the different level of response. Demonstrate that the amplification range of the cell is actually $0.2 \leq V(t = 0) \leq 1.0$. Why do you suppose the nerve cell amplifies the combination of the input signals?

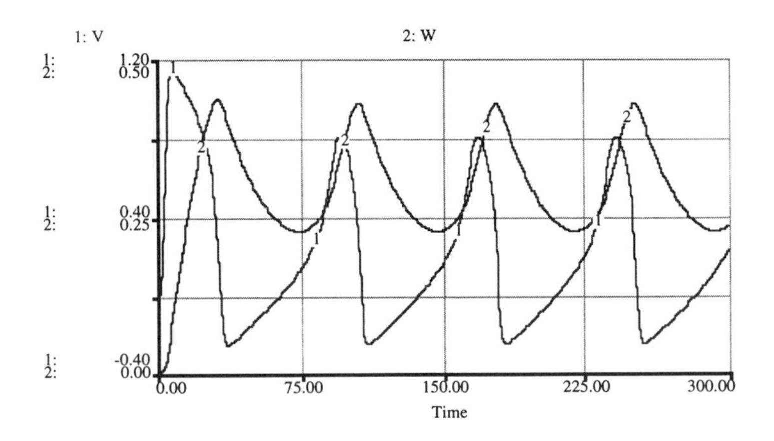

FIGURE 11.3

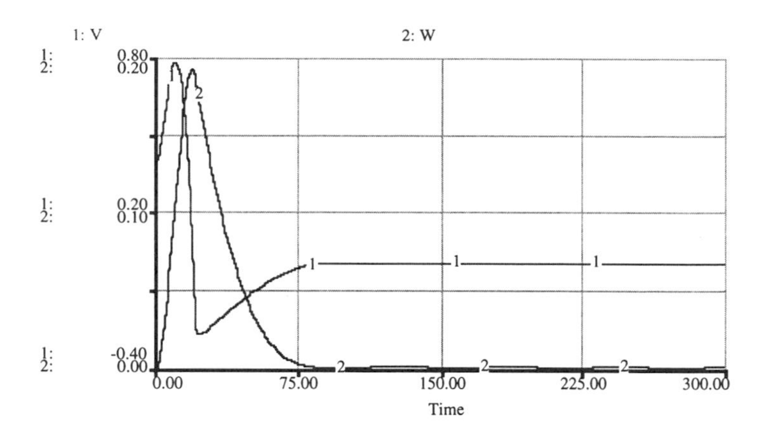

FIGURE 11.4

The first model run in Figure 11.3 showed cycling of V and W for E = 0.23. Keep increasing E for subsequent runs and observe the results. You will find that when the applied voltage in the model is greater than 1.3, cycling ceases (Fig. 11.4), only to return again when this potential reaches the range of 25 to 32; first cycling returns and then chaos ensues. This chaos phenomenon is probably only of interest to modelers. Could the nerve cell actually become chaotic if it could withstand these apparently high voltages? It depends on whether or not the actual function of the cell is discrete in time. If the cell has some operating period required below which no action takes place, then chaos of the membrane potential should be experimentally demonstrable.

The results for E = 32 are illustrated in Figures 11.5 and 11.6, with all other parameters as in the first run. The neuron experiences chaos. The

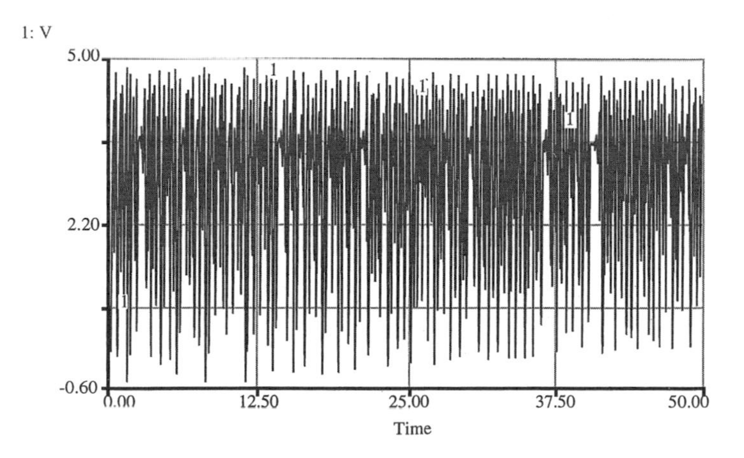

FIGURE 11.5

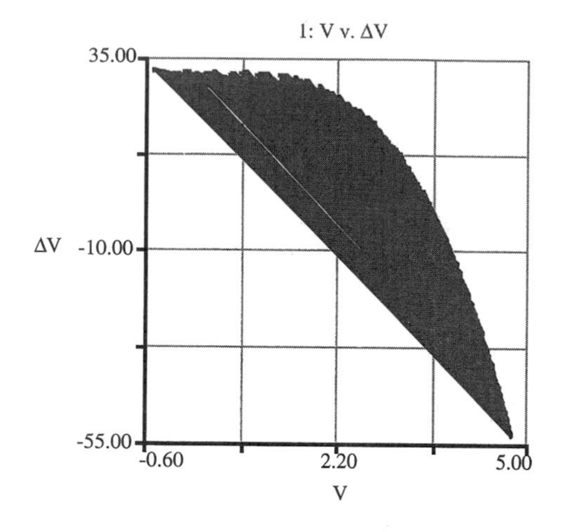

FIGURE 11.6

applied voltage is very high, probably too high to be applied to a living cell.

Following this model of the behavior of an individual cell in response to the impulses from its environment, we model the behavior of an entire organsim in response to changes in its physical surroundings. This is the topic of Part 4.

FITZHUGH-NAGUMO MODEL

```
V(t) = V(t - dt) + (ΔV) * dt
INIT V = 0
INFLOWS:
ΔV = -V*(V-.2)*(V-1)-W+E

W(t) = W(t - dt) + (ΔW) * dt
INIT W = 0
INFLOWS:
ΔW = EPSILON*(V-.5*W)

E = 32
EPSILON = .02
```

12

Solar Radiation to Nonobstructed Inclined Surfaces

We must, as Ficinus adviseth us, get us solar eyes, spectacles as they that looke on the Sunne.

Robert Burton, *Anatomy of Melancholy*, iii. iv. i. i. 1621

Researchers and students in physical climatology are interested in radiative, sensible heat, and latent heat exchanges between the atmosphere and the Earth and how to use models to compute these energy exchanges in natural and agricultural systems. The models can be used, for example, to assess different responses of plants to diurnal or seasonal radiation changes. In this chapter we present a typical application of one such model.

Most growth and reproductive processes of plants are temperature dependent. For this reason, plants that inhabit cold environments tend to grow slowly, reproduce infrequently, and live long lives. Because temperature is such an important limiting factor in the cold alpine tundra, many plants have developed interesting strategies for maximizing the temperature of their important body parts. The most common physiological response is the short growth form. By remaining small and growing along the tundra surface in high mountain environments, plants can stay out of the strong winds and maintain warm body temperatures.

Hymenoxys gradiflora (the Old-Man-of-the-Mountain) appears to have developed a very different strategy for maximizing temperature based on radiative energy exchange. This sunflower rotates its inflorescence during the day, following the Sun across the sky. Assuming that this behavior is driven exclusively by the plant's need to maximize incoming solar radiation, we can calculate the benefit *Hymenoxys* gains by following this strategy. In other words, what is the difference in the daily insolation to an inflorescence that continuously rotates to maintain maximum insolation throughout the day and an inflorescence that stays in one position (e.g., the optimal fixed position)?

The following model can be used to calculate direct (solar beam), diffuse sky (scattered), and total insolation to inclined surfaces on the Earth's surface for any day of the year from measurements of insolation to a nonobstructed horizontal surface. The time step is an hour and the flux density units are W/m^2 or $J/m^2/s^2$. The model is based on at least three important assumptions: (1) the surface is "nonobstructed," (2) an isotropic approximation for diffuse radiation, and (3) the standard correlation between diffuse

TABLE 12.1 Typical insolation values for alpine tundra

Hour	Insolation	Hour	Insolation
1	0	13	830
2	0	14	700
3	0	15	550
4	0	16	355
5	0	17	160
6	50	18	50
7	160	19	0
8	355	20	0
9	550	21	0
10	700	22	0
11	830	23	0
12	900	24	0

and total insolation[1]. The model can thus be used to give an answer to the above-mentioned question, assuming that *Hymenoxys* is found on horizontal surfaces with no mountain peaks close enough to obstruct the solar beam. Indeed, *Hymenoxys* inhabits large, horizontal, meadows at the crest of Niwot Ridge in the Colorado Front Range (latitude = 40.1N), and we use measurements from Niowot Ridge to initialize the model. Typical insolation values for this alpine tundra site during the growing season (e.g., on the 170th day of the year) are listed in Table 12.1.

The model calculates the angles of incidence of the solar beam for horizontal and inclined surfaces from earth-sun geometrical relationships. These relationships are shown in Figures 12.1 and 12.2 and Table 12.2.

Inputs to this model are the solar constant (set at 1353.8 W/m^{-2}), latitude (degrees), slope from horizontal (degrees), and direction of maximum slope (aspect) for the site, Julian date, and the hourly measurements of insolation to a nonobstructed horizontal surface (the average of the flux density measurements for the hour in W/m^{-2}). Given these inputs, the solar declination, SOLAR DEC, the angle between the zenith and the solar beam, INC ANGLE HOR, and the slope, INC ANGLE SLP, can be calculated as (Fig. 12.3)

SOLAR DEC = 23.5*COS(2*pi*(JULIAN DATE–172)/365) (1)

INC ANGLE HOR = SIN(LATITUDE*pi/180) * SIN(SOLAR DEC*pi/180)
+ COS(LATITUDE*pi/180)
* COS(SOLAR DEC* pi/180)
* COS(HOUR ANGLE*pi/180) (2)

[1]David M. Gates provides a good discussion of ways to compute insolation to plants and animals in his text, *Biophysical Ecology* (1980). *An Introduction to Solar Radiation* by Muhammad Iqbal (1983) is an excellent treatment of the physics of solar radiation. It contains many theoretical and empirical relationships that could be used to modify this model.

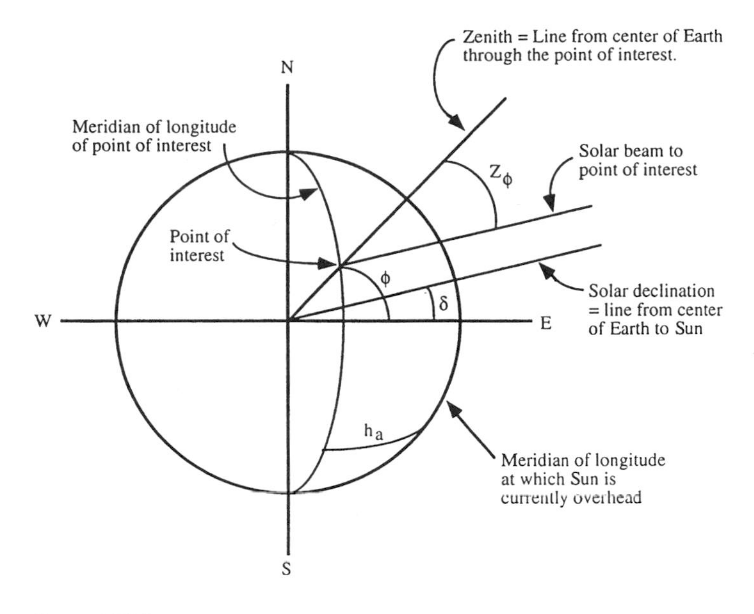

FIGURE 12.1

INC ANGLE SLP = (COS(LATITUDE*pi/180)*COS(HOUR ANGLE*pi/180)
*COS(SLOPE*pi/180) – SIN(HOUR ANGLE*pi/180)
*SIN(ASPECT*pi/180) *SIN(SLOPE*pi/180)
– SIN(LATITUDE*pi/180) *COS(HOUR ANGLE*pi/180)
*COS(ASPECT*pi/180)*SIN(SLOPE*pi/180))
*COS(SOLAR DEC
*pi/180)+(COS(LATITUDE*pi/180)
*COS(ASPECT*pi/180))
*SIN(SLOPE*pi/180)+SIN(LATITUDE*pi/180)
*COS(SLOPE*pi/180))*SIN(SOLAR DEC*pi/180) (3)

TABLE 12.2 Definition of model variables

STELLA II Variable	Definition
Solar Dec	δ = solar declination angle (+ in N. hemis.); between ± 23.5°
Latitud	ϕ = latitude of site (+ in N. hemis.); between ± 90°
Hour Angle	h_a = hour angle (Earth rotates 15° per hour; – in morning, + in afternoon
Inc Angle Hor	Z_ϕ = angle between zenith (line perpindicular to horizontal surface) and solar beam

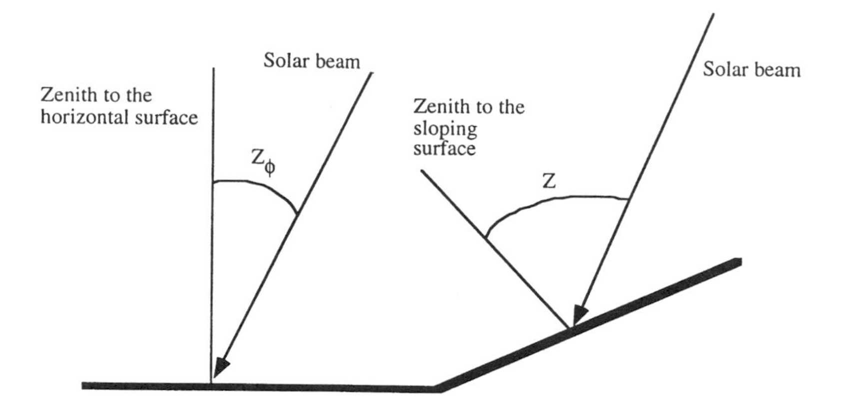

β is the ground slope relative to the horizontal.
θ is the angle that the normal to the plane of interest makes with the compass points. North is zero. Measure counterclockwise.

FIGURE 12.2

The insulation values from Table 12.1 are captured in the INSOL HOR translation variable as a graph and are used to calculate the transmission coefficient for solar radiation to the horizontal surface, INSOL TRANS (Fig. 12.4):

$$\text{INSOL TRANS} = \text{INSOL HOR}/(\text{SOL CONSTANT}/\text{RAD VECT}^2 * \text{INC ANGLE HOR}) \quad (4)$$

To capture the relationship between the transmission coefficient INSOL TRANS and diffusion, we must develop and use an empirical correlation between the ratio of diffuse radiation to a horizontal surface and total solar radiation to a horizontal surface, The INSOL DIFUSE RATIO. This relationship is shown in Figure 12.5 and captured in Figure 12.6,

$$\text{DIRECT TRANS} = (1-\text{INSOL DIFUSE RATIO})*\text{INSOL TRANS} \quad (5)$$

$$\text{DIFUSE TRANS} = \text{INSOL DIFUSE RATIO}*\text{INSOL TRANS} \quad (6)$$

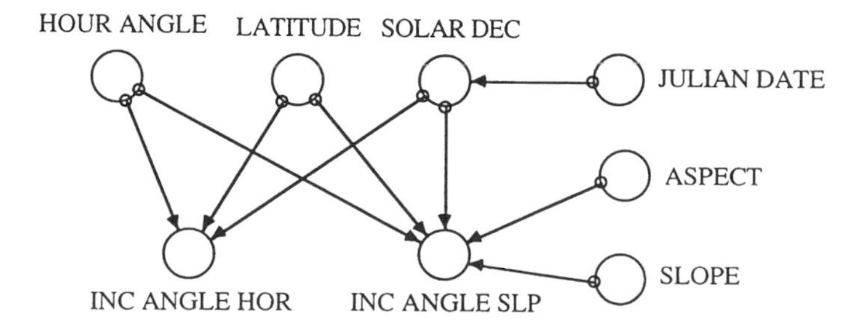

FIGURE 12.3

INSOL DIFUSE RATION = IF INSOL TRANS >.8 THEN 0.165
ELSE IF INSOL TRANS ≤ 0.22 THEN 1.0
– 0.09*INSOL TRANS ELSE 0.9511 – 0.1604
*INSOL TRANS +4.388*INSOL TRANS^2
–16.638*INSOL TRANS^3+12.336
*INSOL TRANS^4 (7)

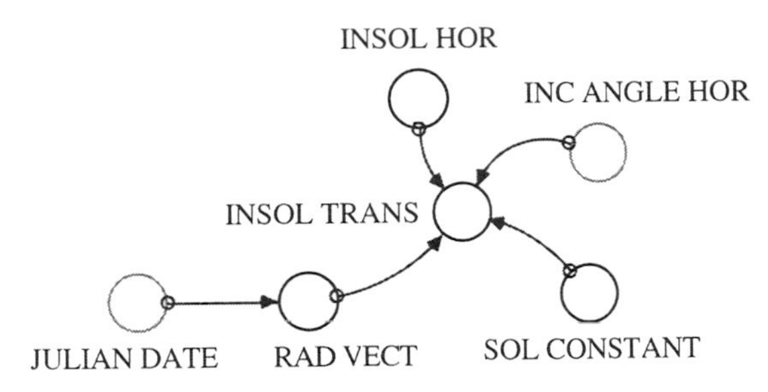

FIGURE 12.4

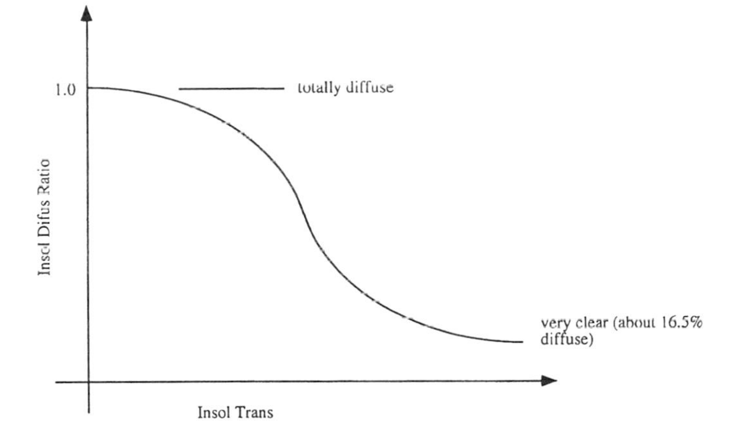

FIGURE 12.5

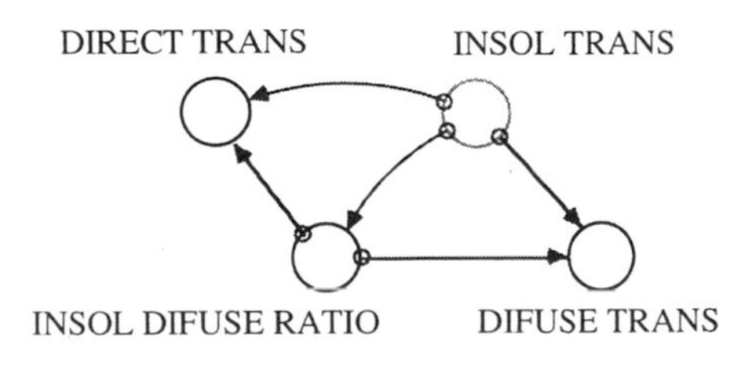

FIGURE 12.6

The atmospheric transmission coefficient is computed as the ratio of the insolation calculated for a horizontal surface at the top of the atmosphere to insolation measured at the Earth's surface. The standard correlation between diffuse radiation and total radiation is used to calculate transmission coefficients for diffuse and direct solar radiation. The flux densities of direct and diffuse radiation are calculated for horizontal and inclined surfaces at the Earth's surface and then summed to yield total insolation to the inclined surface (Fig. 12.7):

DIFUSE HOR = SOL CONSTANT/RAD VECT^2*DIFUSE TRANS
*INC ANGLE HOR (8)

DIRECT HOR = SOL CONSTANT/RAD VECT^2*DIRECT TRANS
*INC ANGLE HOR (9)

DIRECT SLP = SOL CONSTANT/RAD VECT^2*DIRECT TRANS
*INC ANGLE SLP (10)

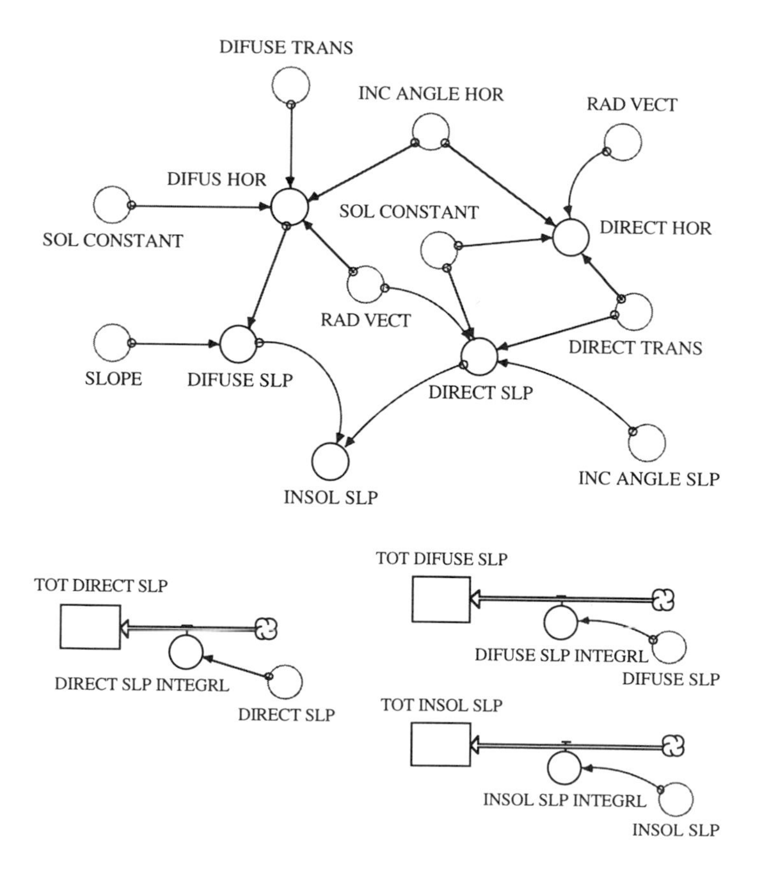

FIGURE 12.7

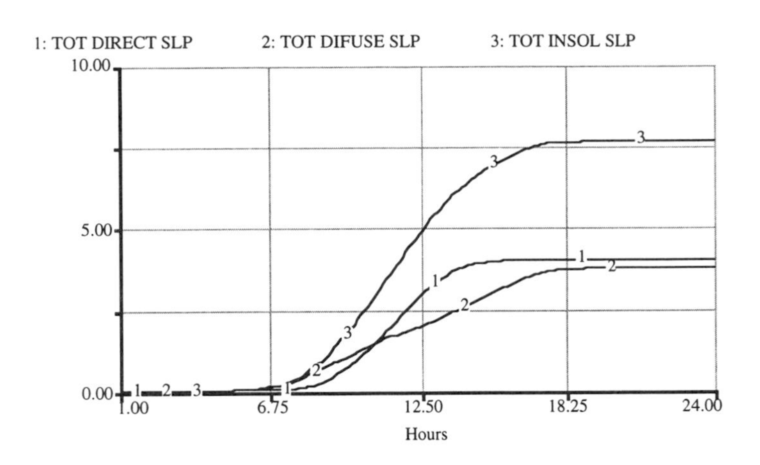

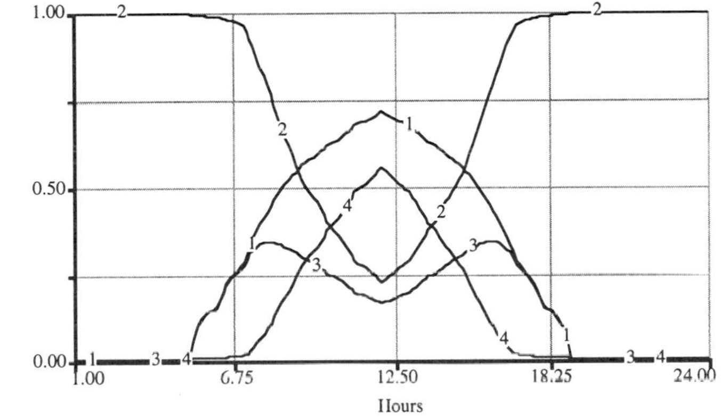

FIGURE 12.8

$$\text{INSOL SLP} = \text{DIFUSE SLP} + \text{DIRECT SLP} \tag{11}$$

The radiation components could be integrated over meaningful time periods (e.g., hours, days, seasons, years), but this step is not included in the model.

Total solar radiation, diffusion, and insolation to the declined surface are shown in Figures 12.8 and 12.9. Each are zero early in the day and monotonically increase to level off towards the end of the day.

With this model, find the optimal position assuming the plant does not adjust to changes in the solar declination. Then determine the optimal rotation of *Hymenoxis* and provide an answer for the question above. How does this answer change if the plant cannot adjust instantaneously to changes in the solar declination but has a 1 hour time lag?

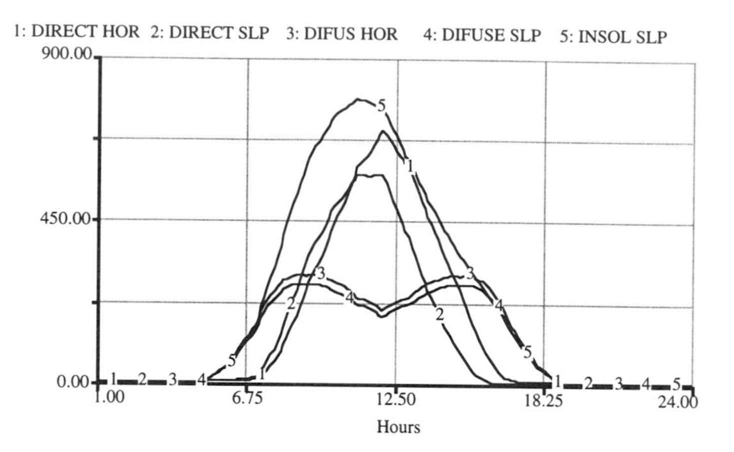

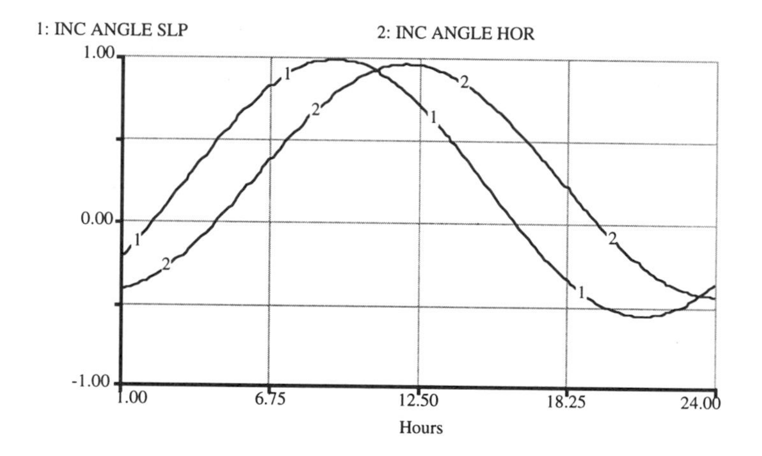

FIGURE 12.9

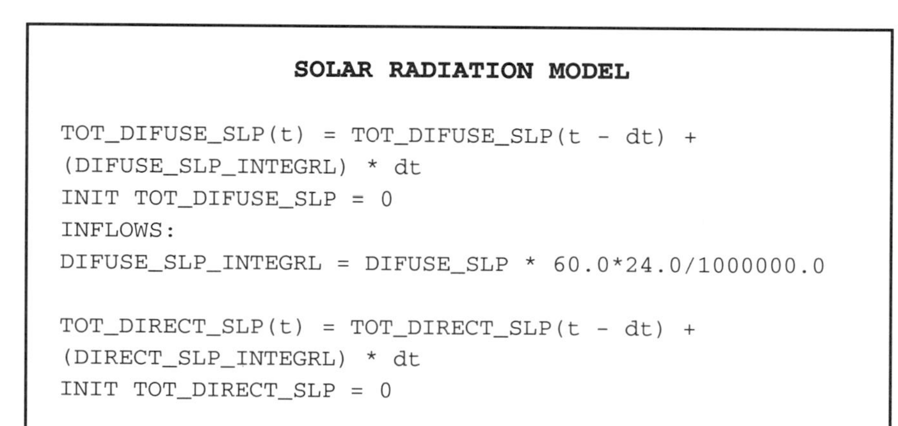

SOLAR RADIATION MODEL

```
TOT_DIFUSE_SLP(t) = TOT_DIFUSE_SLP(t - dt) +
(DIFUSE_SLP_INTEGRL) * dt
INIT TOT_DIFUSE_SLP = 0
INFLOWS:
DIFUSE_SLP_INTEGRL = DIFUSE_SLP * 60.0*24.0/1000000.0

TOT_DIRECT_SLP(t) = TOT_DIRECT_SLP(t - dt) +
(DIRECT_SLP_INTEGRL) * dt
INIT TOT_DIRECT_SLP = 0
```

```
INFLOWS:
DIRECT_SLP_INTEGRL = DIRECT_SLP * 60.0*24.0/1000000.0

TOT_INSOL_SLP(t) = TOT_INSOL_SLP(t - dt) +
(INSOL_SLP_INTEGRL) * dt
INIT TOT_INSOL_SLP = 0
INFLOWS:
INSOL_SLP_INTEGRL = INSOL_SLP * 60.0*24.0/1000000.0

ASPECT = 90
DIFUSE_SLP = 0.5*DIFUS_HOR*(1+Cos(SLOPE*pi/180))
DIFUSE_TRANS = INSOL_DIFUSE_RATIO*INSOL_TRANS
DIFUS_HOR =
SOL_CONSTANT/RAD_VECT^2*DIFUSE_TRANS*INC_ANGLE_HOR
DIRECT_HOR =
SOL_CONSTANT/RAD_VECT^2*DIRECT_TRANS*INC_ANGLE_HOR
DIRECT_SLP =
SOL_CONSTANT/RAD_VECT^2*DIRECT_TRANS*INC_ANGLE_SLP
DIRECT_TRANS = (1-INSOL_DIFUSE_RATIO)*INSOL_TRANS
HOUR_ANGLE = (Time-12)*15.0
INC_ANGLE_HOR =
SIN(LATITUDE*pi/180)*SIN(SOLAR_DEC*pi/180)+COS(LATITUDE
*pi/180)*COS(SOLAR_DEC*pi/180)*COS(HOUR_ANGLE*pi/180)
INC_ANGLE_SLP =
(Cos(LATITUDE*pi/180)*Cos(HOUR_ANGLE*pi/180)*Cos(SLOPE*
pi/180)-
Sin(HOUR_ANGLE*pi/180)*Sin(ASPECT*pi/180)*Sin(SLOPE*pi/
180)-
Sin(LATITUDE*pi/180)*Cos(HOUR_ANGLE*pi/180)*Cos(ASPECT*
pi/180)*Sin(SLOPE*pi/180))*Cos(SOLAR_DEC*pi/180)+(Cos(L
ATITUDE*pi/180)*Cos(ASPECT*pi/180)*Sin(SLOPE*pi/180)+Si
n(LATITUDE*pi/180)*Cos(SLOPE*pi/180))*Sin(SOLAR_DEC*pi/
180)
INSOL_DIFUSE_RATIO = If INSOL_TRANS >.8 then 0.165 Else
If INSOL_TRANS ≤ 0.22 then 1.0 - 0.09*INSOL_TRANS else
0.9511-0.1604*INSOL_TRANS+4.388*INSOL_TRANS^2-
16.638*INSOL_TRANS^3+12.336*INSOL_TRANS^4
INSOL_SLP = DIFUSE_SLP+DIRECT_SLP
INSOL_TRANS =
INSOL_HOR/(SOL_CONSTANT/RAD_VECT^2*INC_ANGLE_HOR)
JULIAN_DATE = 170

LATITUDE = 40
```

```
RAD_VECT = 0.01676*Cos(pi-0.0172615*(JULIAN_DATE-3))+1
SLOPE = 35

SOLAR_DEC = 23.5*Cos(2*pi*(JULIAN_DATE-172)/365)
SOL_CONSTANT = 1353.8
INSOL_HOR = GRAPH(time)
(1.00, 0.00), (2.00, 0.00), (3.00, 0.00), (4.00, 0.00),
(5.00, 0.00), (6.00, 50.0), (7.00, 160), (8.00, 355),
(9.00, 550), (10.0, 700), (11.0, 830), (12.0, 900),
(13.0, 830), (14.0, 700), (15.0, 550), (16.0, 355),
(17.0, 160), (18.0, 50.0), (19.0, 0.00), (20.0, 0.00),
(21.0, 0.00), (22.0, 0.00), (23.0, 0.00), (24.0, 0.00)
```

Part 3

Genetics Models

13

Mating and Mutation of Alleles

O world! But that thy strange mutations make us hate thee
Life would not yield to age.

Shakespeare, *King Lear*, 1605

Genetic theory provides us with insight into the changes of the genetic makeup of organisms. These insights are of particular interest in the context of conservation biology, where genetic diversity is seen as one advantage of a species to survive in a changing environment. The more different genetic information is present, the better prepared the species to deal with a range of environmental factors.

Several processes influence the genetic makeup of organisms in a species. Among those influences is the combination of various alleles into a genotype from "parents" that carry those alleles. Other factors determining genetic makeup are the random mutations of one type of allele into another. These two cases are dealt with in the model presented in this chapter. The following chapter explores the impacts of natural selection and fitness in conjunction with mutation on genotype distribution.

To model the process of genotype mixing and mutation we restrict ourselves, without loss of generality, to the case of two alleles, A and B, which are drawn randomly from an initial pool of 200 A alleles and 300 B alleles. The results of the simple mating of two alleles are explained by the Hardy–Weinberg law, which states that the genotype frequencies are determined in a random mating process. These genotype frequencies are for AA, p^2; for AB, 2*p*q; and for BB, q^2, where p and q are the A and B allele frequencies, respectively. In our sample problem, p = 200/500 or 0.4 and q = 300/500 or 0.6 for our initial pool of 200 A alleles and 300 B alleles. From 500 alleles we can have 250 genotypes, so the Hardy–Weinberg law tells us that we should end up with 0.4*0.4*250 or 40 AA genotypes, 2*0.4*0.6*250 or 120 AB genotypes, and, finally, 0.6*0.6*250 or 90 BB genotypes.

Of course, the allele frequencies change from generation to generation as mutation occurs. Here, we assume that mutation is a random process and can occur in both directions. For the model we assume that a random fraction of A alleles turns into B alleles and vice versa, and we only deal with the net of the mutation in both directions. We define

$$\text{NET MUTATION} = \text{RANDOM}(-0.05, 0.05) * (\text{A ALLELES} + \text{B ALLELES}) \tag{1}$$

where A ALLELES and B ALLELES are the stocks of alleles of each type that are available for mating to form the next generation of genotypes. Each of these stocks is emptied when a new generation of genotypes is formed. We call the respective outflows A MATE and B MATE. The inflows into the stocks are based on the alleles that have been temporarily tied up in genotypes. For example, one AA GENOTYPE releases two A alleles and one AB GENOTYPE releases only one A allele. As alleles are dumped into the respective stocks, mutation occurs as specified in equation (1). The STELLA model that deals with this part of the process of allele mixing and mutation is shown in Figure 13.1.

The alleles are combined according to the Hardy–Weinberg law. The Hardy–Weinberg law is used here to calculate the inflows into the stocks of the three genotypes. The outflows are set equal to the stocks, reflecting the fact that after a generation was formed they will be available again for re-mating (Fig. 13.2).

The results of the model are different from run to run because of the random mutation process that takes place. What will happen when the total number of alleles decreases? Run the model several times and observe the results. How do the results change if mutation is more likely to occur in one direction than another? What are the implications of a decreasing number of alleles and asymmetric mutation for conservation biol-

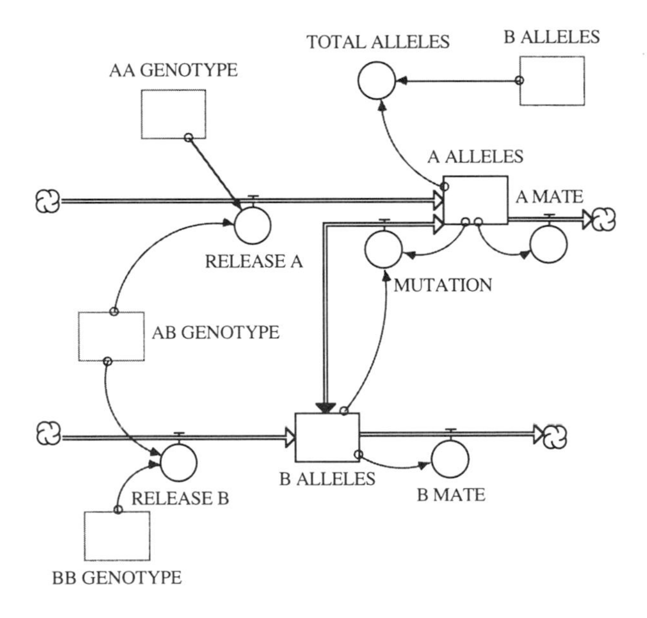

FIGURE 13.1

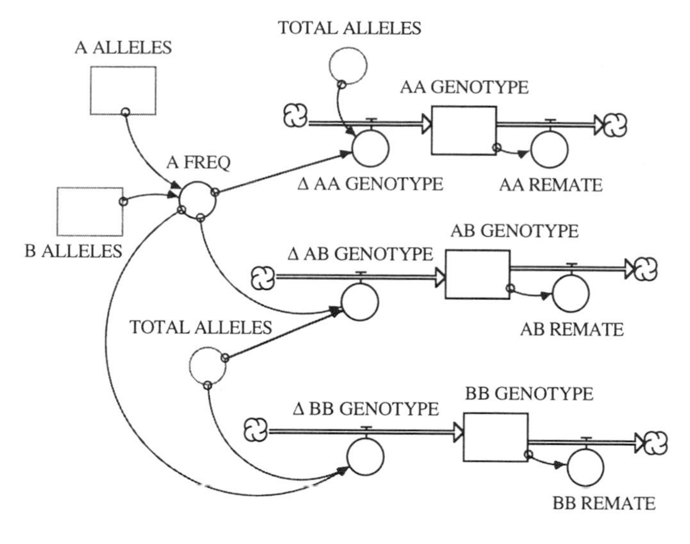

FIGURE 13.2

ogy? Can you set up the problem without making explicit use of the Hardy–Weinberg law?

The model expands on the one developed in this chapter by including the impacts of natural selection and fitness on the abundance of the three genotypes in a population. The results are shown in Figure 13.3.

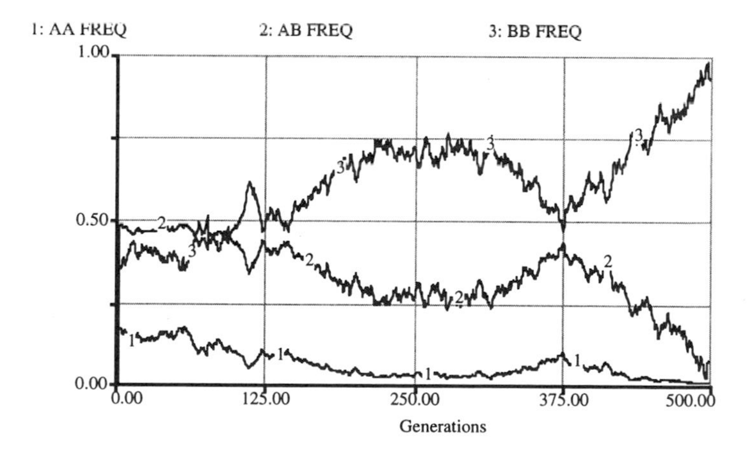

FIGURE 13.3

MATING AND MUTATION OF ALLELES

```
AA_GENOTYPE(t) = AA_GENOTYPE(t - dt) + (Δ_AA_GENOTYPE -
AA_REMATE) * dt
INIT AA_GENOTYPE = 0
INFLOWS:
Δ_AA_GENOTYPE = TOTAL_ALLELES/2*A_FREQ^2
OUTFLOWS:
AA_REMATE = AA_GENOTYPE

AB_GENOTYPE(t) = AB_GENOTYPE(t - dt) + (Δ_AB_GENOTYPE -
AB_REMATE) * dt
INIT AB_GENOTYPE = 0
INFLOWS:
Δ_AB_GENOTYPE = 2*TOTAL_ALLELES/2*A_FREQ*(1 - A_FREQ)
OUTFLOWS:
AB_REMATE = AB_GENOTYPE

A_ALLELES(t) = A_ALLELES(t - dt) + (RELEASE_A +
MUTATION - A_MATE) * dt
INIT A_ALLELES = 200
INFLOWS:
RELEASE_A = 2*AA_GENOTYPE +AB_GENOTYPE
MUTATION = RANDOM(-0.05,0.05)*(A_ALLELES+B_ALLELES)
OUTFLOWS:
A_MATE = A_ALLELES

BB_GENOTYPE(t) = BB_GENOTYPE(t - dt) + (Δ_BB_GENOTYPE -
BB_REMATE) * dt
INIT BB_GENOTYPE = 0
INFLOWS:
Δ_BB_GENOTYPE = TOTAL_ALLELES/2*(1 - A_FREQ)^2
OUTFLOWS:
BB_REMATE = BB_GENOTYPE

B_ALLELES(t) = B_ALLELES(t - dt) + (RELEASE_B - B_MATE
- MUTATION) * dt
INIT B_ALLELES = 300
INFLOWS:
RELEASE_B = 2*BB_GENOTYPE + AB_GENOTYPE
OUTFLOWS:
B_MATE = B_ALLELES
```

```
MUTATION = RANDOM(-0.05,0.05)*(A_ALLELES+B_ALLELES)
AA_FREQ = IF TIME > 0 THEN AA_GENOTYPE/TOTAL_GENOTYPE
ELSE 0
AB_FREQ = IF TIME > 0 THEN AB_GENOTYPE/TOTAL_GENOTYPE
ELSE 0
A_FREQ = A_ALLELES/(A_ALLELES+ B_ALLELES)
BB_FREQ = IF TIME > 0 THEN BB_GENOTYPE/TOTAL_GENOTYPE
ELSE 0
TOTAL_ALLELES = A_ALLELES + B_ALLELES
TOTAL_GENOTYPE = AA_GENOTYPE+AB_GENOTYPE+BB_GENOTYPE
```

14

Natural Selection, Mutation, and Fitness

> There is such an unerring power at work, or Natural Selection, which selects exclusively for the good of each organic being.
>
> Darwin, *Letters*, 1857

This problem is an extension of the genetics models discussed in the previous chapter. Here we introduce survival and fertility rates for the genotypes and combine these rates to form a fitness measure. Together with mutation, these effects consort to give the new genotype frequencies. Geneticists and conservation biologist think the existence of the heterozygote, the AB genotype, is a measure of health of the system of geneotypes. The strength of its presence is sometimes referred to as *hybrid vigor* and is obviously desired because it carries both alleles.

In this problem, we examine the requirements for equilibrium. We define equilibrium as that condition when the heterotype reaches some steady or approximately steady state over many generations. Ideally, we want that steady state to be nonzero, and in fact we want it to be high enough to overcome the spectrum of environmental challenges that confront the geneotypes. To facilitate an experiment to determine the necessary conditions for equilibrium, we create a convertor for the concept of relative fitness. Here we define the raw fitness of the AB genotype as the AB fertility rate/(1 - AB survival rate). *Ceteris paribus*, a higher fertility rate and a higher survival rate both produce greater fitness. We normalize the raw fitness of the AA and BB genotypes by the raw fitness of the AB genotype and define the relative fitness of the AB genotype as one. Otherwise our model is the same as that of the previous chapter. We shut off both mutation rates and explore for the conditions that produce equilibrium (Figs. 14.1 and 14.2).

After a little experimenting we find that as long as the fitness of AB is greater than the least of the fitnesses of AA or BB, equilibrium can be reached. If AB fitness is greater than both AA and BB fitness, then the frequency of AB will be greater than either of the other two. If the fitness of AB lies between the other two, then the homozygotes with the larger fitness will be dominant. We find that if the AB fitness is lower than the lowest of all, the elimination of one of the alleles can be stopped and

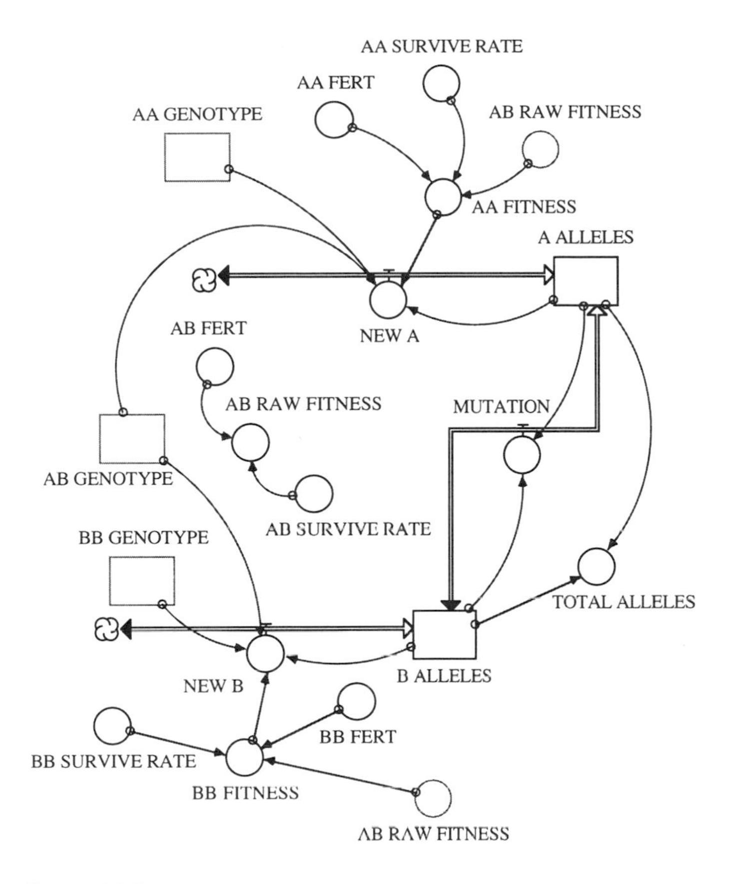

FIGURE 14.1

even reversed by a high enough relative rate of mutation to that allele (see Fig. 14.3). We also learn that to stamp out a recessive allele takes many generations, giving those who impose observational eugenics great problems.

Let us suppose that blond hair (dominant allele) was revealed by people with BB and AB genotypes, and that black hair (recessive allele) was revealed by those bearing the AA genotype. Eliminating black-haired members of the population will not eliminate the tendency for black hair as offspring of the AB blonds will sometimes have black hair. Only pertinent DNA tests on all blonds would allow the demise of the black haired.

This model can be extended to cover the cases of sex-linked development of the genotypes. If males carry one type of fitness ratings and females

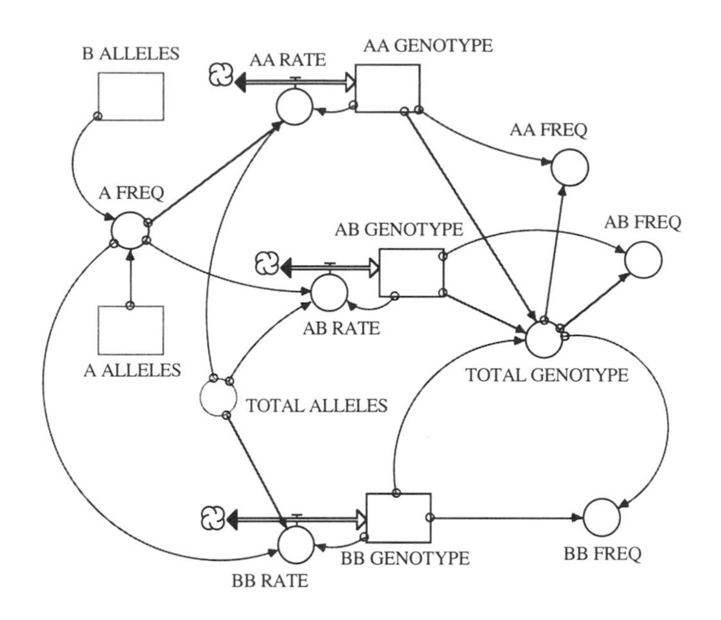

FIGURE 14.2

another, separate male and female models can be developed and the genotype frequencies can be developed from the combined result. Inbreeding is the result of sex-linked development in which the mating is not random. Can you modify this model to meet the specified inbreeding criteria?

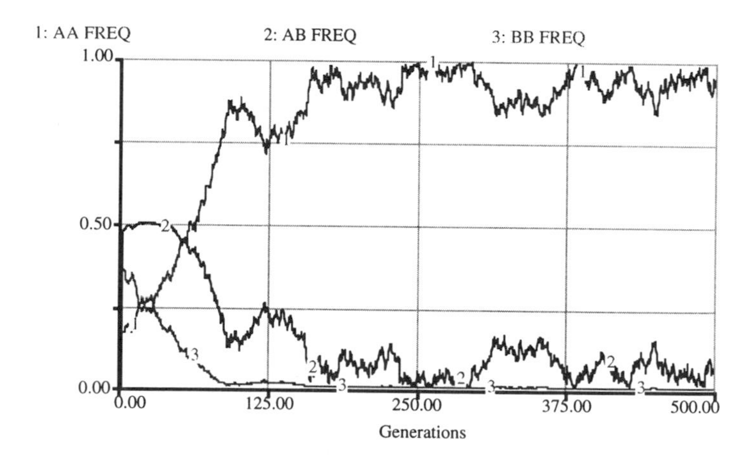

FIGURE 14.3

```
NATURAL SELECTION, MUTATION, AND FITNESS

AA_GENOTYPE(t) = AA_GENOTYPE(t - dt) + (AA_RATE) * dt
INIT AA_GENOTYPE = 1
INFLOWS:
AA_RATE = TOTAL_ALLELES/2*A_FREQ^2 - AA_GENOTYPE

AB_GENOTYPE(t) = AB_GENOTYPE(t - dt) + (AB_RATE) * dt
INIT AB_GENOTYPE = 1
INFLOWS:
AB_RATE = 2*TOTAL_ALLELES/2*A_FREQ*(1 - A_FREQ) -
AB_GENOTYPE

A_ALLELES(t) = A_ALLELES(t - dt) + (NEW_A + MUTATION) *
dt
INIT A_ALLELES = 200
INFLOWS:
NEW_A = 2*AA_GENOTYPE*AA_FITNESS + AB_GENOTYPE -
A_ALLELES

MUTATION = RANDOM(-0.05,0.05)*(A_ALLELES+B_ALLELES)
BB_GENOTYPE(t) = BB_GENOTYPE(t - dt) + (BB_RATE) * dt
INIT BB_GENOTYPE = 1
INFLOWS:
BB_RATE = TOTAL_ALLELES/2*(1 - A_FREQ)^2 - BB_GENOTYPE

B_ALLELES(t) = B_ALLELES(t - dt) + (NEW_B - MUTATION) *
dt
INIT B_ALLELES = 300
INFLOWS:
NEW_B = 2*BB_GENOTYPE*BB_FITNESS + AB_GENOTYPE -
B_ALLELES
OUTFLOWS:
MUTATION = RANDOM(-0.05,0.05)*(A_ALLELES+B_ALLELES)

AA_FERT = 2 {Fraction per timestep of one.}
AA_FITNESS = AA_FERT/(1-AA_SURVIVE_RATE)/AB_RAW_FITNESS
AA_FREQ = IF TIME > 0 THEN AA_GENOTYPE/TOTAL_GENOTYPE
ELSE 0
AA_SURVIVE_RATE = .83 {Fraction per timestep of one.}
AB_FERT = 2 {Fraction per timestep of one.}
AB_FREQ = IF TIME > 0 THEN AB_GENOTYPE/TOTAL_GENOTYPE
ELSE 0
```

```
AB_RAW_FITNESS = AB_FERT/(1-AB_SURVIVE_RATE) {AB
fitness is one by defintion.}
AB_SURVIVE_RATE = .81 {Fraction per timestep of one.}
A_FREQ = A_ALLELES/(A_ALLELES+ B_ALLELES)
BB_FERT = 2 {Fraction per timestep of one.}
BB_FITNESS = BB_FERT/(1-BB_SURVIVE_RATE)/AB_RAW_FITNESS
BB_FREQ = IF TIME > 0 THEN BB_GENOTYPE/TOTAL_GENOTYPE
ELSE 0
BB_SURVIVE_RATE = .81
TOTAL_ALLELES = A_ALLELES + B_ALLELES
TOTAL_GENOTYPE = AA_GENOTYPE+AB_GENOTYPE+BB_GENOTYPE
```

Part 4

Models of Organisms

15

Odor-Sensing Model

> The only defect of our senses is, that they give us disproportion'd images of things.
>
> Hume, 1739.

In this chapter we develop a model of a "nose" with three receptors. Each of the three receptors is tuned to a particular type of odor, either 1, 2, or 3. The goal of the "nose" is for the dominant receptor to suppress the other two odors in order to send a clear signal to the brain regarding the type of odor being sensed. The odor changes over time according to the following rule:

$$\text{ODOR} = \text{IF TIME} \leq 15 \text{ THEN NORMAL(1,.4) ELSE NORMAL(3,.4)} \quad (1)$$

Our "nose" can select the mean value out of a normally distributed odor. The mean value of the odor is thought to be an integer, and it is assumed that integers above 3 would merely require additional receptors. In order for our nose to work, each receptor has to have a way of forgetting its present signal, should that signal decrease. We assume that the receptor is constantly forgetting or "zeroing out" the effect of the signal. Thus, when the strength of a signal to a particular receptor declines, the stength of the signal transmitted by that receptor declines. The three receptors are set up accordingly as

$$\begin{gathered}\text{RECEPTOR 1} = \text{IF (ODOR} \geq .5) \\ \text{AND (ODOR} \leq 1.5) \text{ THEN 1/DT ELSE} - \text{A}\end{gathered} \quad (2)$$

$$\begin{gathered}\text{RECEPTOR 2} = \text{IF (ODOR} > 1.5) \\ \text{AND (ODOR} \leq 2.5) \text{ THEN 1/DT ELSE} - \text{B}\end{gathered} \quad (3)$$

$$\begin{gathered}\text{RECEPTOR 3} = \text{IF (ODOR} > 2.5) \\ \text{AND (ODOR} \leq 3.5) \text{ THEN 1/DT ELSE} - \text{C}\end{gathered} \quad (4)$$

The model is shown in Figure 15.1. Our resulting signal to the brain from the three receptors is unambiguous in spite of the fact that the incoming odor signal was mixed and overlapped the boundaries of the receptors (Fig. 15.2).

The model does not depict a perfect nose. There are too few receptors. Add more receptors and increase the range of the possible odors. Change

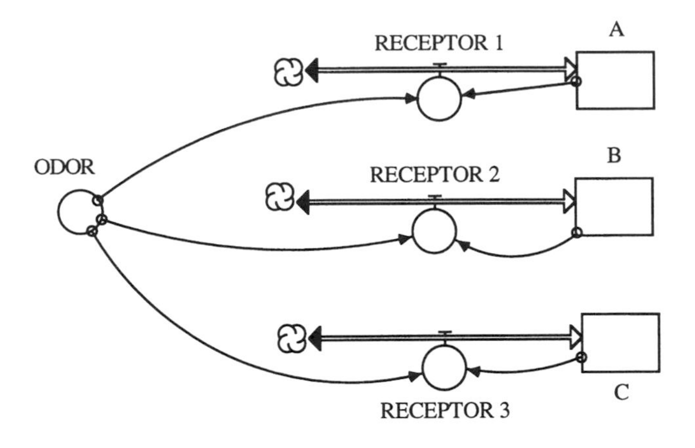

FIGURE 15.1

the sensitivity of some receptor such that if there is too much odor of a particular strength or for a particular duration, that receptor gets "numb" and provides the wrong signals to the brain. How can the brain minimize those errors?

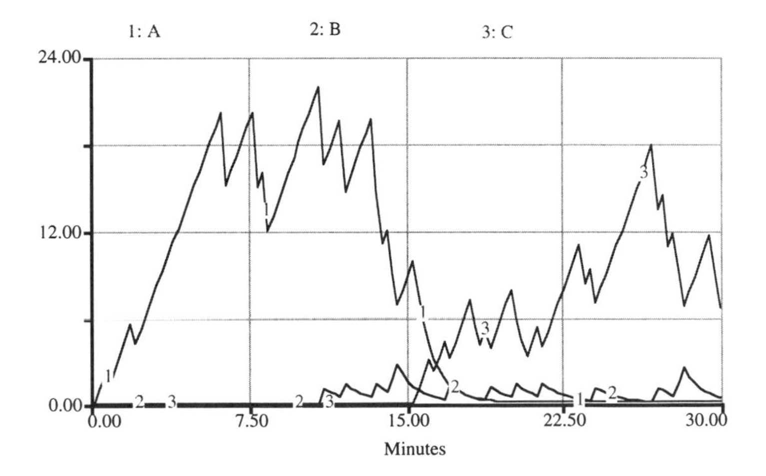

FIGURE 15.2

ODOR-SENSING MODEL

```
A(t) = A(t - dt) + (RECEPTOR_1) * dt
INIT A = RECEPTOR_2 {Signals}
INFLOWS:
RECEPTOR_1 = IF (ODOR ≥ .5) AND (ODOR ≤ 1.5) THEN 1/DT
ELSE - A {Signals per Minute}

B(t) = B(t - dt) + (RECEPTOR_2) * dt
INIT B = 0 {Signals}
INFLOWS:
RECEPTOR_2 = IF (ODOR > 1.5) AND (ODOR ≤ 2.5) THEN 1/DT
ELSE -B {Signals per Minute}

C(t) = C(t - dt) + (RECEPTOR_3) * dt
INIT C = 0 {Signals}
INFLOWS:
RECEPTOR_3 = IF (ODOR > 2.5) AND (ODOR ≤ 3.5) THEN 1/DT
ELSE -C {Signals per Minute}
ODOR = IF TIME≤15 THEN NORMAL(1,.4) ELSE NORMAL(3,.4)
{Odor Units}
```

16

Stochastic Resonance

The life of the senses has its deep poetry.

M. Arnold, 1865

In the previous chapter we modeled a "nose" that detects some odor. Let us model in this chapter an "ear" with a limited range of amplitudes of acoustic signals that it can detect. The threshold of audible amplitudes is arbitrarily set at 0.03. Let us specify a harmonic signal whose amplitude (0.01) is too low to be heard.

$$\text{HARMONIC} = \text{SINWAVE}(.01,.1) + .01 \tag{1}$$

Additionally there is noise in the system.

$$\text{NOISE} = \text{RANDOM}(0,.02) \tag{2}$$

With the addition of noise, the peak apparent amplitude to the ear is doubled, goes above the audio threshold, and thus can be "heard":

$$\text{COMBINED SIGNAL} = \text{HARMONIC} + \text{NOISE} \tag{3}$$

The "frequency" of the noise is the DT. Every DT the random level of the noise is changed. If the noise is not generated often enough, the signals become impossibly masked by the noise. Try a DT of 0.001 and of 0.0625. The model and results are shown, respectively, in Figures 16.1 and 16.2. Recognize how little of the harmonic signal can be detected if the frequency is too low.

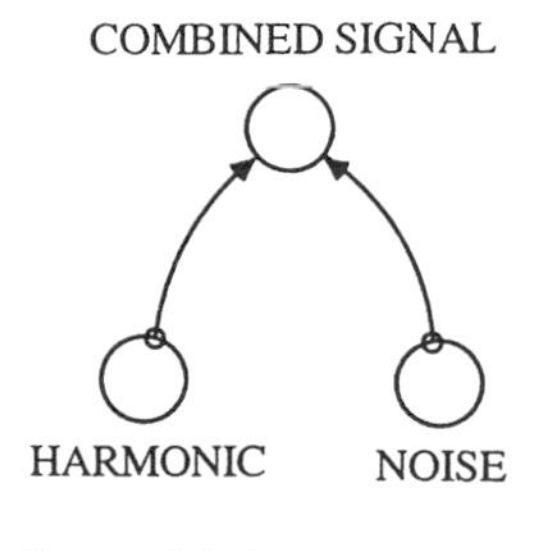

FIGURE 16.1

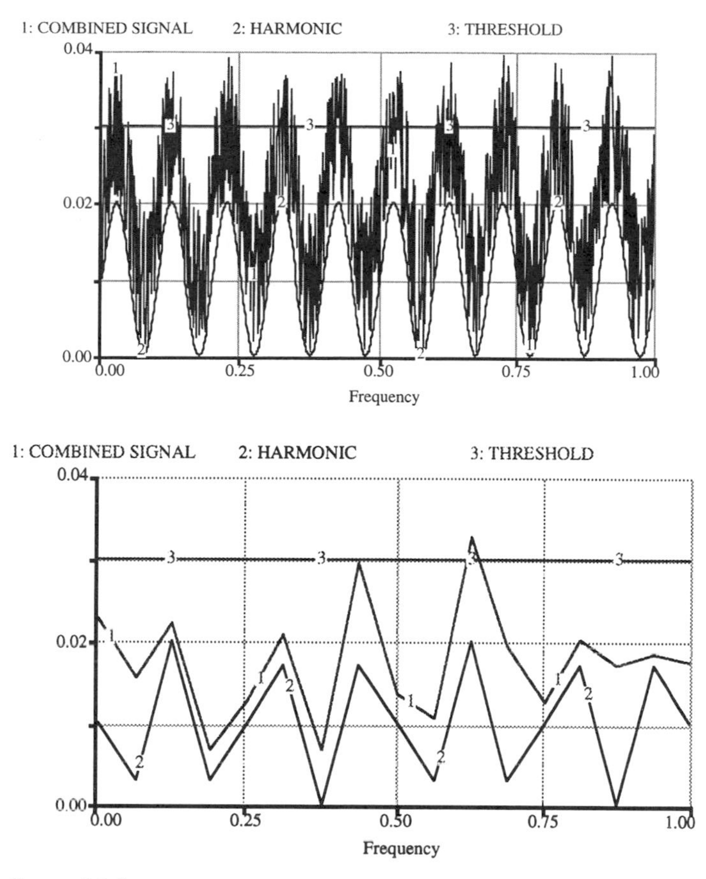

FIGURE 16.2

The human ear–brain smooths over the fact that the peaks above the threshold are composed of several tightly spaced signals of noise plus signal. The same phenomenon is present in the eye–brain system. Imagine a set of leafless branches fidgeting in the wind, seen against a bright sky. Assume we are looking for a mammal moving through the branches. The movement of the branches is the "white" visual noise in this system. Our eye–brain system is extremely well equipped to filter out the dark shapes formed at random by the intersection of several branches because this sort of shape does not persist. We easily see the mammal flitting though the branches. Adding leaves to these branches just makes the problem harder, but perhaps not impossible. So the eye–brain and ear–brain systems must learn to allow persistence of the input signal in order to appropriately average it and make decisions based on that average value. In the case of the combined audio signal, this quality of persistence allows one to compose the parts of this combine that exceed the audio threshold into a particular frequency. In the visual example, the persistence quality allows us to ignore those signals that are random and focus our attention on the more apparently purposeful motion.

Now that we have modeled a simple "nose" and an "ear," let us turn to a more elaborate model of a four-chambered heart. This is the topic of the following chapter.

STOCHASTIC RESONANCE

```
COMBINED_SIGNAL = HARMONIC+NOISE
HARMONIC = SINWAVE(.01,.1)+.01
NOISE = RANDOM(0,.02)
```

17

Heartbeat Model

Tell me where is fancie bred, Or in the heart, or in the head.

Shakespeare, 1596

In Chapter 4 we mentioned the fact that organs, or entire organisms, may be viewed as having evolved to a critical state in which they are seemingly close to chaotic behavior. Yet, at that stage of their evolution they may have optimized their behavior with regard to a specific task. The heart is such an organ, and we model it in a simplified form in this chapter.

The behavior of the four-chambered human heart follows a series of closely interrelated flow processes. Deoxygenated blood from the venous system is collected into the vena cava and then delivered into the right atrium. Blood then is pumped into the right ventricle past the bicuspid valve. Blood is pumped to the lungs via the pulmonary arteries, where it becomes oxygenated. Blood then is delivered into the left atrium via the pulmonary vein. From there the blood is pumped into the left ventricle. The oxygenated blood is then pumped to the body's arterial system through the aorta. This pumping rate is controlled by the heart's pacemaker. Special cells in both atrial chambers have the ability to send electrical impulses that cause the atria to contract. The same impulse is also carried to the A-V node, which causes ventricular contraction. The result is first an atrial contraction, and then, after a few millisecond delay, the ventricular contraction. The basic processes are laid out in Figure 17.1.

Our model of the four-chambered heart is constructed to respond to changes in blood demand and to disease. A volume of blood is pumped through each of the four chambers of the heart. At the core of the model are two pacemakers, each sending a Pulse that pumps the blood, causing a flow from one chamber to the next:

$$\text{PACEMAKER} = \text{PULSE(RIGHT ATRIUM,1,MEDULLA)} \tag{1}$$

$$\text{PACEMAKER 2} = \text{PULSE(LEFT ATRIUM,1,MEDULLA)} \tag{2}$$

The pulse's firing frequency is controlled by the medulla, part of the brain that stimulates some body functions, such as breathing and heartbeat. The activity of the medulla is a function of the activity of the heart, here arbitrarily set as

$$\text{MEDULLA} = 8*\text{ACTIVITY} \tag{3}$$

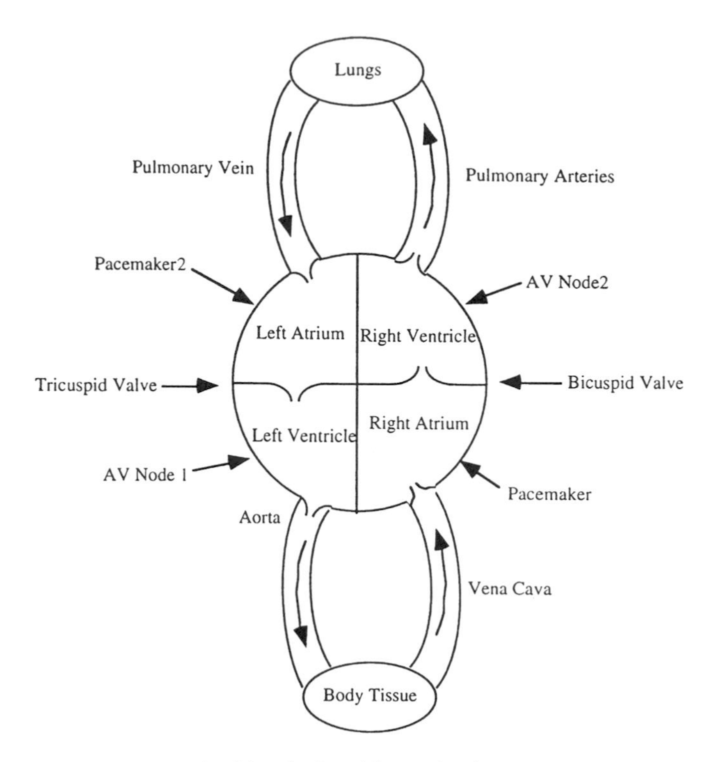

FIGURE 17.1 A highly idealized heart–body arrangement.

with ACTIVITY = 1 for a person at rest and ACTIVITY <1 for an active person. Increased activity increases the rate of pulses being sent by the pacemaker. FITNESS is also included in the model. The more fit an individual, the more efficient the heart is at pumping blood.

A Delay is used in both the Bicuspid and Tricuspid flows to create a short pause between atrial and ventricle firing.

$$\text{PULMONARY ART} = \text{DELAY(AV NODE 2,1)} \tag{4}$$

$$\text{AORTA} = \text{DELAY(AV NODE,1)} \tag{5}$$

To make the graph of the blood flow more realistic, a smooth function was used. The built-in SMTH1(A,X) calculates the first-order exponential smooth of a variable A, using an exponential averaging time of X. The smooth function gives the appearance of blood gradually flowing into its chamber. To capture the gradual flow of blood in our model, we define the flows connecting the atrium and ventricle of the left and right chambers, BISCUSPID VALVE and TRICUSPID VALVE, with the smooth function:

$$\text{BISCUSPID VALVE} = \text{SMTH1(PACEMAKER*FITNESS,MEDULLA*.1)} \tag{6}$$

$$\text{TRICUSPID VALVE} = \text{SMTH1(PACEMAKER 2*FITNESS,MEDULLA*.1)} \tag{7}$$

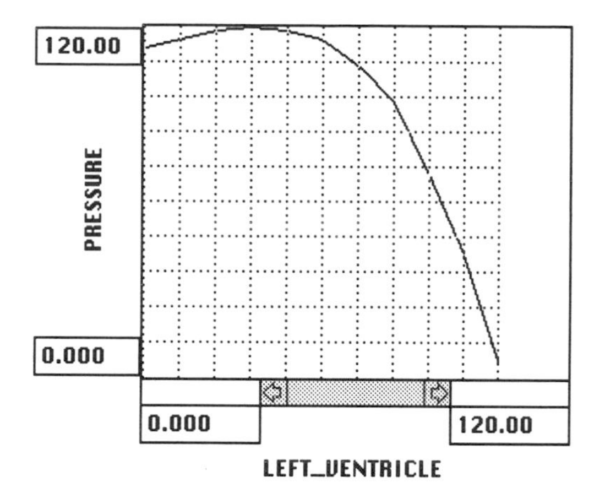

FIGURE 17.2

Blood pressure can be read by graphing the translation variable PRESSURE. In our model, PRESSURE is specified as a function of the amount of blood in the left ventricle (Fig. 17.2).

Heart disease was added into the model with the infarction factor, or I FACTOR. This value represents the quantity of heart tissue damage as a parameter. Increased damage influences the transmittance of the electical impulse. Run the model for alternative I FACTOR values and observe the result.

We annotated the STELLA building blocks in the diagram (Fig. 17.3) to visualize the magnitude of stocks and flows. Go to the Diagram menu and

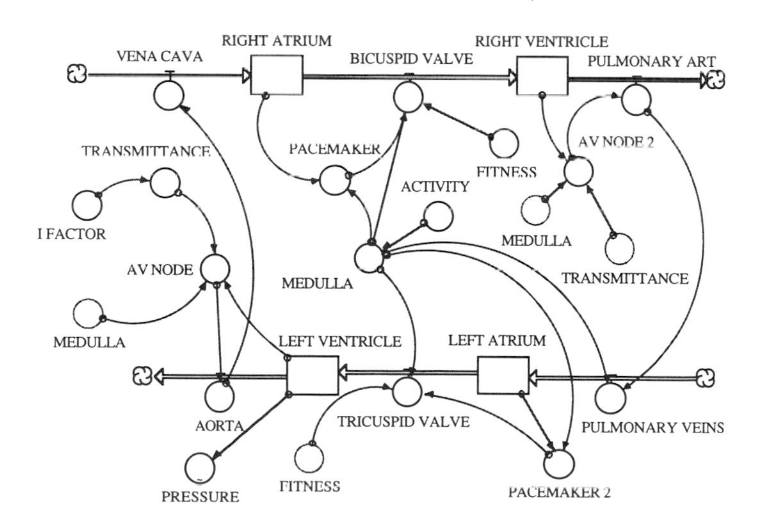

FIGURE 17.3

choose Diagram Prefs. . . . Here you can specify which model building blocks to annotate. Click once on the respective symbol to annotate it. Click again on it to deactivate the annotation.

The two graphs in Figure 17.4 show the pressure of the left atrium and ventricle over time for two different fitness factors. The first graph is for FITNESS = 1, the second for FITNESS = 1.7. Increasing fitness increases the heart's effectiveness in moving the blood from compartment to compartment. As a comparison of the graphs shows, the maxima and minima for the atrium and ventricle become more similar with an incease in FITNESS.

The pressure in the left ventricle is plotted in Figure 17.5 for the two fitness rates. At all times, the pressure is lower under higher fitness rates than

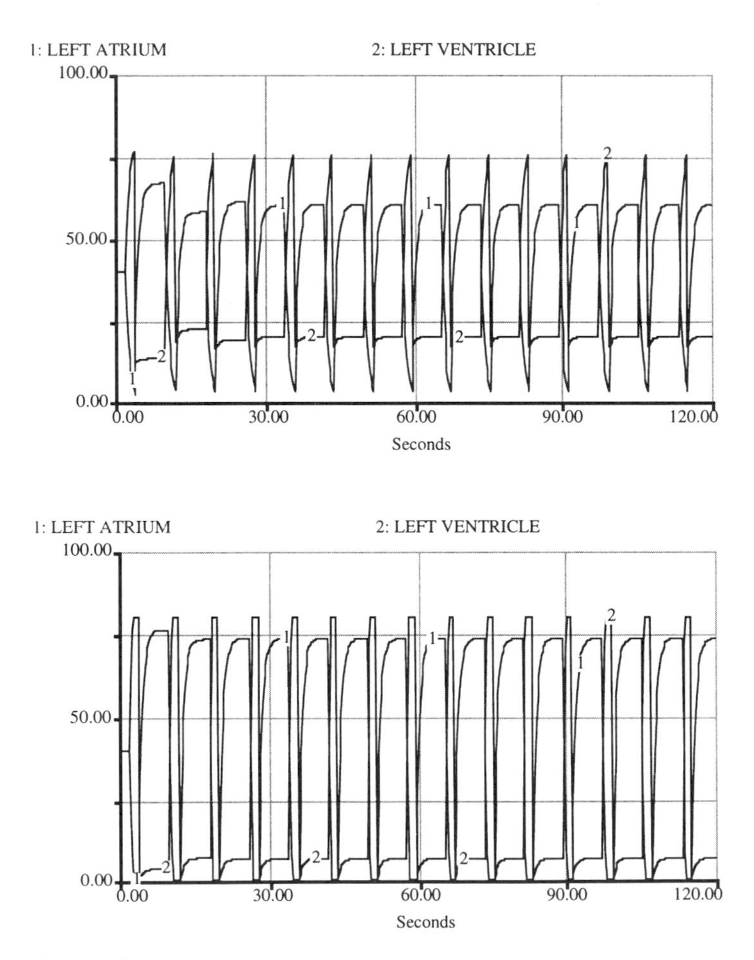

FIGURE 17.4

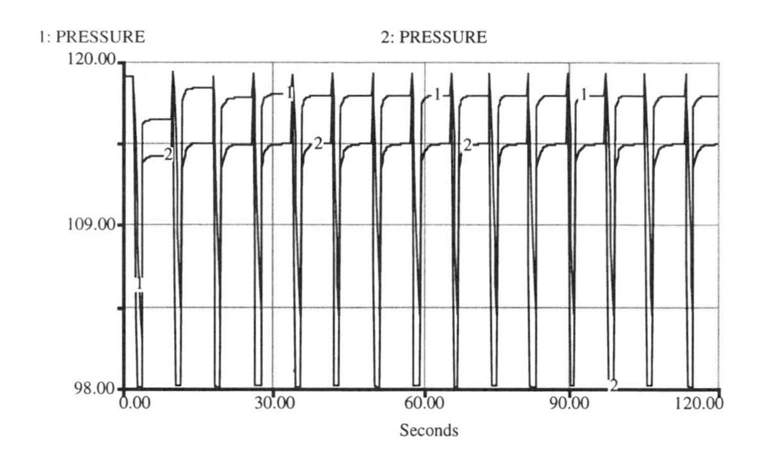

FIGURE 17.5

lower ones. Modify the model to allow for the gradual awakening of a sleeping person and a gradual increase in that person's stress level. How does a less-fit person's heart performance compare with that of a fitter person? What about a display of the effects of heart disease?

HEARTBEAT MODEL

```
LEFT_ATRIUM(t) = LEFT_ATRIUM(t - dt) + (PULMONARY_VEINS
- TRICUSPID_VALVE) * dt
INIT LEFT_ATRIUM = 40 {ml}
INFLOWS:
PULMONARY_VEINS = SMTH1(PULMONARY_ART,MEDULLA*.1) {ml
per second}
OUTFLOWS:
TRICUSPID_VALVE = SMTH1(PACEMAKER_2*FITNESS,MEDULLA*.1)
{ml per second}

LEFT_VENTRICLE(t) = LEFT_VENTRICLE(t - dt) +
(TRICUSPID_VALVE - AORTA) * dt
INIT LEFT_VENTRICLE = 40 {ml}
INFLOWS:
TRICUSPID_VALVE = SMTH1(PACEMAKER_2*FITNESS,MEDULLA*.1)
{ml per second}
OUTFLOWS:
AORTA = DELAY(AV_NODE,1) {ml per second}
```

```
RIGHT_ATRIUM(t) = RIGHT_ATRIUM(t - dt) + (VENA_CAVA -
BICUSPID_VALVE) * dt
INIT RIGHT_ATRIUM = 40 {ml}
INFLOWS:
VENA_CAVA = AORTA {ml per second}
OUTFLOWS:
BICUSPID_VALVE = SMTH1(PACEMAKER*FITNESS,MEDULLA*.1)
{ml per second}

RIGHT_VENTRICLE(t) = RIGHT_VENTRICLE(t - dt) +
(BICUSPID_VALVE - PULMONARY_ART) * dt
INIT RIGHT_VENTRICLE = 40 {ml}
INFLOWS:
BICUSPID_VALVE = SMTH1(PACEMAKER*FITNESS,MEDULLA*.1)
{ml per second}
OUTFLOWS:
PULMONARY_ART = DELAY(AV_NODE_2,1)

ACTIVITY = 1 {1 represents a person at rest, values
less than 1 represent activity}
AV_NODE = PULSE(LEFT_VENTRICLE,TRANSMITTANCE,MEDULLA)
{ml}
AV_NODE_2 =
PULSE(RIGHT_VENTRICLE,TRANSMITTANCE,MEDULLA) {ml}
FITNESS = 1.7
I_FACTOR = 1
MEDULA = 8*ACTIVITY
PACEMAKER = PULSE(RIGHT_ATRIUM,1,MEDULLA) {ml}
PACEMAKER_2 = PULSE(LEFT_ATRIUM,1,MEDULLA) {ml}
TRANSMITTANCE = 2*I_FACTOR
PRESSURE = GRAPH(LEFT_VENTRICLE)
(0.00, 113), (12.0, 116), (24.0, 119), (36.0, 119),
(48.0, 118), (60.0, 115), (72.0, 107), (84.0, 93.6),
(96.0, 69.0), (108, 42.0), (120, 5.00)
```

18

Bat Thermoregulation

Ere the Bat hath flowne His Cloyster'd flight.

Shakespeare, *Macbeth* iii. ii. 40

Objects of a particular temperature, surrounded by a cooler environment, tend to lose heat to their surrounding. Under the assumption that the environment is large in comparison with the object, the rate of change in the temperature of the object is determined by the difference between the object's temperature and the ambient temperature. The ambient temperature becomes the target final temperature of the object.

In this chapter we model the thermoregulatory process of a bat. The bat loses heat based on the Newton's law of cooling. This law states that the rate of change of a body temperature is linearly proportional to the temperature difference between the object and the environment. In our case, heat loss by a bat is

$$\text{HEAT LOSS} = K * (\text{BODY TEMP} - \text{AMBIENT TEMPERATURE}) \tag{1}$$

Unlike the standard setting for Newtonian cooling of inanimate objects, bats are able to influence the cooling coefficient K by adjusting their fur, rolling into a more nearly spherical shape (minimum surface per unit volume), and crowding. Thus, the cooling coefficient is a function of time. The relationship between the cooling coefficient and ambient temperature is shown in Figure 18.1.[1]

The heat gain is a function of the difference between the body temperature and a set temperature, TN. The chemical reaction rate is halved for every 10°C drop in body temperature. The metabolic rate maximum (10) sets the upper limit on basal metabolic rate, QR:

$$\begin{aligned} \text{QR} = {} & \text{IF } 2 + 2 * (\text{TN} - \text{BODY TEMP}) \le 10 \\ & \text{THEN } 2 + 2 * (\text{TN} - \text{BODY TEMP}) \text{ ELSE } 10 \end{aligned} \tag{2}$$

[1]The data on the bat are from J.D. Spain, *BASIC Microcomputer Models in Biology*, Reading, MA: Addison-Wesley, 1992.

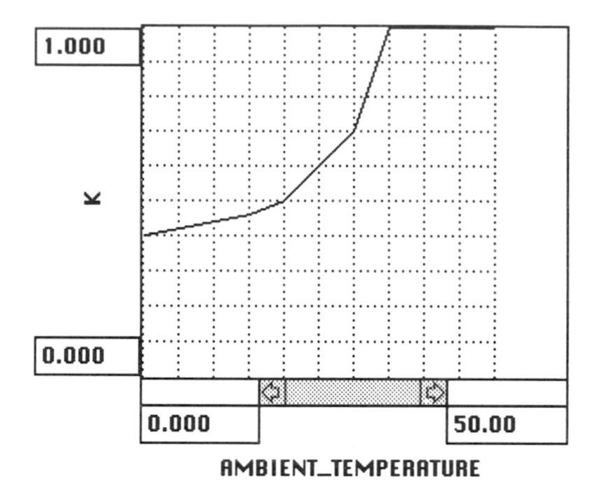

FIGURE 18.1

The complete system is shown in Figure 18.2. The parameters K and A translate heat flows into temperature changes for the bat.

Set the ambient and initial body temperature at different levels for consecutive model runs and watch the heat loss and heat gain converge while the body temperature approaches its final value. When this process is carried out for a variety of ambient temperatures, the graphs in Figures 18.3 and 18.4 result. We plotted the temperatures for eight model runs with AMBIENT TEMPERATURE ranging from 10° to 24°C. Note how the thermoregulation process is proceeding normally (nearly horizontal portion of the curve) and then the curve drops suddenly. Why is this break point occurring? What process is producing temperature equilibrium in the bat at ambient temperatures below 15°C?

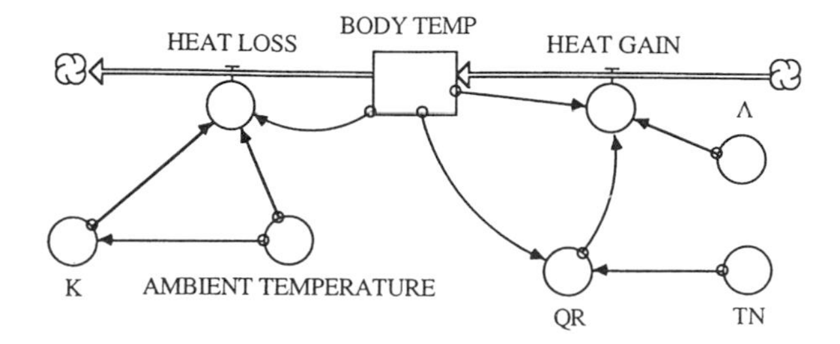

FIGURE 18.2

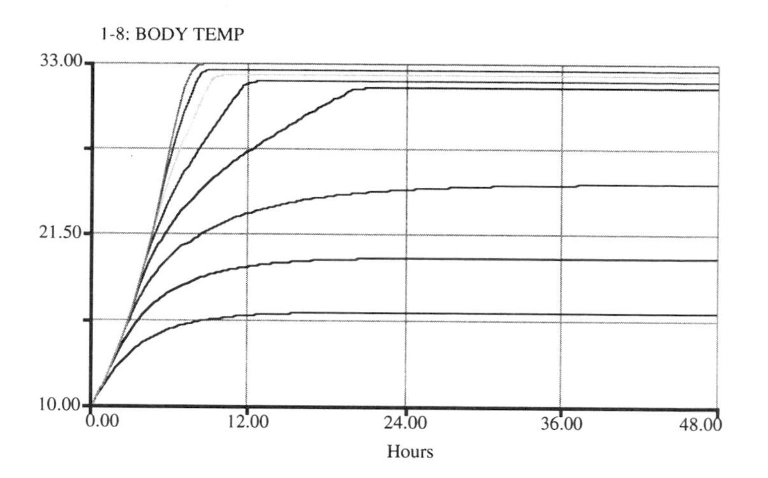

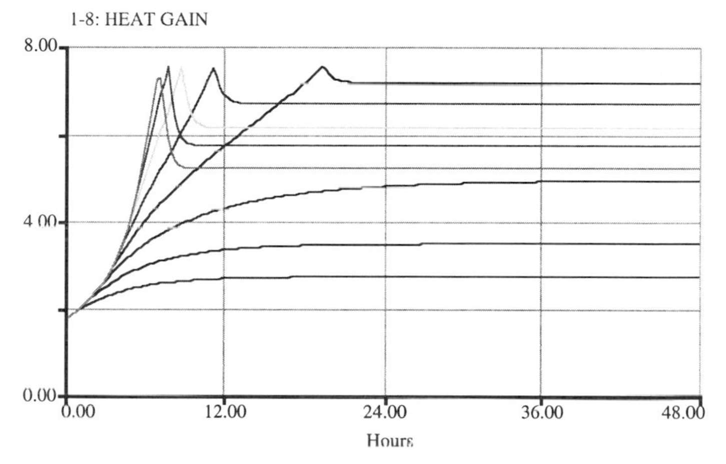

FIGURE 18.3

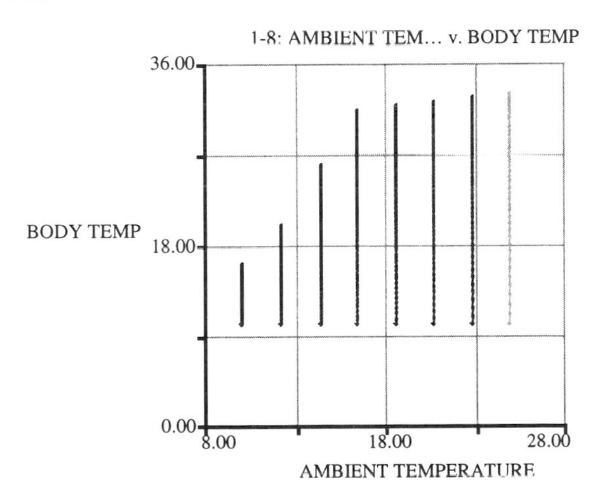

FIGURE 18.4

BAT THERMOREGULATION

```
BODY_TEMP(t) = BODY_TEMP(t - dt) + (HEAT_GAIN -
HEAT_LOSS) * dt
INIT BODY_TEMP = 10 {Deg C}
INFLOWS:
HEAT_GAIN = QR*EXP(-A*(35 - BODY_TEMP)) {Deg C per
Hour}
OUTFLOWS:
HEAT_LOSS = K*(BODY_TEMP-AMBIENT_TEMPERATURE) {Deg C
per Hour}

A = .0693 {Deg C}
AMBIENT_TEMPERATURE = 14 {Deg C}
QR = IF 2 + 2*(TN - BODY_TEMP) ≤ 10 THEN 2 +2*(TN -
BODY_TEMP) ELSE 10
TN = 35 {Deg C}
K = GRAPH(AMBIENT_TEMPERATURE)
(0.00, 0.4), (5.00, 0.42), (10.0, 0.44), (15.0, 0.46),
(20.0, 0.5), (25.0, 0.6), (30.0, 0.7), (35.0, 1.00),
(40.0, 1.00), (45.0, 1.00), (50.0, 1.00)
```

19

The Optimum Plant

Plant, a Natural Body that has a vegetable Soul.

1696 Phillips (ed. 5)

Just describing or simulating the change in living organisms is not good enough. We all want ultimately to predict what these organisms would do under prescribed circumstances. Scientists interested in predictions first need a good description of the behavior of the living organism. Towards that end, they frequently find it advantageous to set up optimality hypotheses of the organism's behavior and then compare the optimization results with results of experiments on the actual dynamics of the organism.

The work begun by Cohen in 1971 on the optimization of plants makes a good example of this kind of approach[1]. Cohen's model is the simplest possible of optimal control in plants. The basic hypothesis is that this plant strives to produce the maximum reproductive biomass by the end of the growing season, a period that is T units long. We assume that the plant is genetically "wired" for this growing season, that is, its genetics have been so shaped by the local environment that the plant acts as though it "knows" what the length of the growing season is. We further assume that the growing season lasts for five time units, and still further that the growth of the vegetative part ΔX in the time DT is given by

$$\Delta X = \frac{dX}{dt} = U * X, \tag{1}$$

and the growth of the reproductive part ΔY in the time DT is

$$\Delta Y = \frac{dY}{dt} = (1 - U) * X. \tag{2}$$

U is the control variable that we use to simulate a plant's shift of its resources from vegetative growth to reproductive growth. The control, U, is necessarily $0 \leq U \leq 1$. The optimality problem is one of maximizing Y(T).

1. D. Cohen, Maximizing Final Yield When Growth Is Limited by Time or by Limited Resources, *Journal of Theoretical Biology* 33:299–307, 1971. For a good summary of such processes, see J. Roughgarden, Models of Population Processes, *Lectures on Mathematics in the Life Sciences*, American Mathematical Society, 18:235–267, 1986.

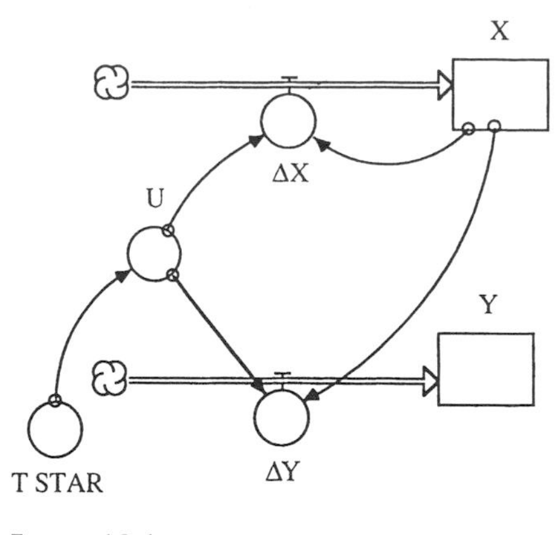

FIGURE 19.1

For this problem, assume that either the vegetative *or* reproductive portion is growing but not both at the same time. In such a case we say that the control is *bang-bang*, it is either 1 or 0. Under this assumption, the control becomes

$$\begin{aligned} &\text{if } t < \text{T STAR then } U = 1 \\ &\text{if } t > \text{T STAR then } U = 0, \end{aligned} \tag{3}$$

where T STAR is the shift time, the time when the plant's production shifts from vegetative to reproductive.

We can build a model of this process (Fig. 19.1) and change T STAR for successive runs until we find the maximum Y at t = T = 5. This value is 5.35, if we use a small enough DT, and we experimentally find that the optimal switch time, T STAR, is equal to 4.00 (Fig. 19.2).

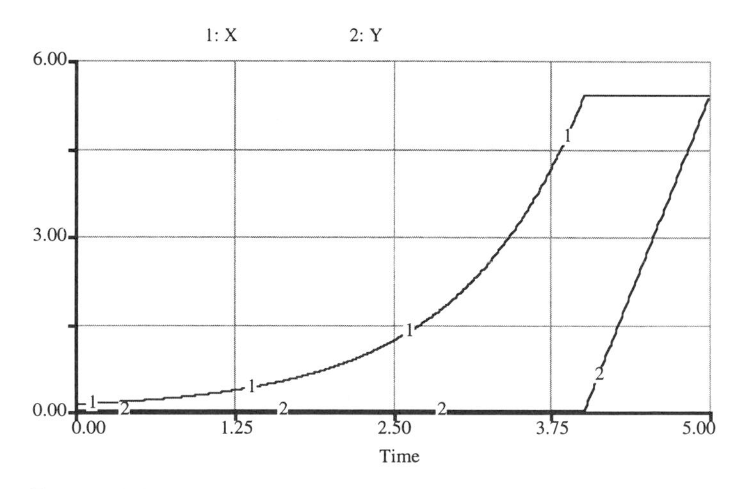

FIGURE 19.2

This model is a good example of how the correct DT must be found. DT = 1 is too large. Choices of DT = .01 and smaller are appropriate because they give the same answer of Y = 5 when T STAR = 4.

Our result can be verified by theory using the Pontryagin optimization theory[2]. For the situation outlined here, the Hamiltonian, H, is

$$H = A * U * X + B * (1 - U) * X \tag{4}$$

The variables A and B are called the costate or adjoint variables. These costate variables must conform to the following conditions according to the Pontryagin theory:

$$dA/dt = -\delta H/\delta X \tag{5}$$

$$dB/dt = -\delta H/\delta Y. \tag{6}$$

Thus,

$$dA/dt = -A * U - B * (1 - U) \tag{7}$$

and

$$dB/dt = 0. \tag{8}$$

Maximizing Y(T) means that the terminal conditions on the costate variables must be

$$A(T) = 0 \tag{9}$$

and

$$B(T) = 1. \tag{10}$$

From (10) and (8), we have

$$B(t) = 1. \tag{11}$$

In the region $t^* \leq t \leq T$, u = 0, so with (7) we have

$$A(t) = T - t. \tag{12}$$

A third and final Pontryagin condition is

$$\delta H/\delta u - 0 \tag{13}$$

at the optimal switch time only. So at t = t*, a = b; t* = T − 1; or in our case, t* = 4.0, which is what we found experimentally on the computer.

Does the hypothesis of optimal plant behavior yield the right answer? There is only one way to find out. Compare it with experimental results[3].

---

[2]M. Kamien and N. Schwartz. *Dynamic Optimization*, New York: North Holland, 1983, pp. 186–192.

[3]B. Hannon, The Optmal Growth of *Helianthus annuus*, *Journal of Theoretical Biology*, 165:523–531, 1993

Even if you successfully compare, the hypothesis may not be sufficiently general to cover the behavior of many different types of plants under different environmental conditions. Even if your model did predict correctly for several different kinds of plants, it is only a good suspect in the search for whether or not living organisms seem to follow any kind of optimal plan.

These optimal control problems in plants can be very difficult. Imagine that the growth equations are logistic rather than the simple ones given earlier. Further, imagine that the growth periods overlap, and finally think of the perennial plant that regrows from root extensions and seeds. Then the determination of the actual optimal path of the control and of X and Y may be accomplished only by numerical analysis. The control may not be bang-bang but graded, allowing both types of biomass to grow simultaneously for some part of the growing season. The best procedure to follow in most cases is to first do as much analytical work as possible to simplify the ensuing numerical analysis. Usually, one of the costate variables can be found in terms of the two biomasses, and perhaps the control. However, the actual solution frequently must be obtained numerically, even with a significant quantity of numerical analysis.

OPTIMUM PLANT

```
X(t) = X(t - dt) + (ΔX) * dt
INIT X = .1 {Kg}
INFLOWS:
ΔX = U*X {Kg per Time Period}

Y(t) = Y(t - dt) + (ΔY) * dt
INIT Y = 0 {Kg}
INFLOWS:
ΔY = (1 - U)*X {Kg per Time Period}

T_STAR = 4
U = IF TIME ≤ T_STAR THEN 1 ELSE 0 {1/Time Period}
```

20

Infectious Diseases

> The endemic and epidemic diseases in Scotland fall chiefly, as is usual, on the poor.
>
> Thomas Malthus, 1798

20.1 Basic Model

In this model we consider the spread of an infectious disease within a population. We assume that there is some initial number of individuals already infected with the disease. These individuals can pass on the disease to a group of susceptibles S. We do not model explicitly the agents that cause the disease, such as viruses or bacteria. Doing that would be rather impractical if we would want to apply our model to real-world diseases. Tracing the billions of agents that can cause the outbreak with a particular disease is virtually impossible. Therefore, we do not explicitly model the dynamics of individuals in a population of disease-causing agents but deal with their effects in an aggregate way.

The law of mass action discussed in Part 2 of this book has proven to be a powerful analogous way of capturing the spread of a disease in a population. The two reactants in our case are the susceptible individuals S and the infective ones I. We define a contact rate BETA at which these two groups of individuals make contact and propagate the disease. This contact rate BETA is analogous to the reaction rates in chemical reactions.

We also model (Fig. 20.1) an influx of susceptibles into the stock S. Additionally, we assume that once an individual has the disease that individual ultimately becomes immune to the disease. We therefore remove those infectives from the stock I and let them disappear in a "cloud" — they will not further affect the spread of the disease, and, therefore, we need not keep track of them.

This is a very simple epidemic model, but can you anticipate the resulting dynamics? Make an educated guess before you run the model. How do the dynamics change with a change in the rate of NONIMMUNE IMMIGRANTS? This rate is here set to 7 per time period, and the initial stocks for S and I are 1000 and 20, respectively. The results are shown in Figure 20.2.

What are the effects of a vaccine on the spread of the disease? Assume that only 20% of the population receives the vaccine and that it is only 90% effective among the susceptibles. Infectives are not immunized. Change the

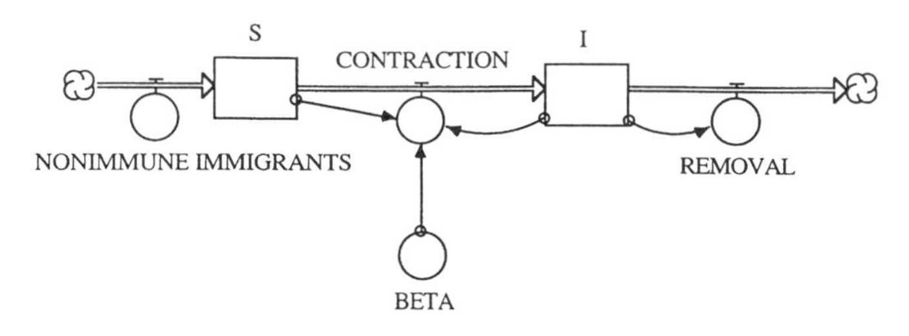

FIGURE 20.1

model to include an incubation period and reassess the ability of your vaccine to limit the spread of the disease.

INFECTIOUS DISEASE

```
I(t) = I(t - dt) + (CONTRACTION - REMOVAL) * dt
INIT I = 20
INFLOWS:
CONTRACTION = BETA*S*I
OUTFLOWS:
REMOVAL = I

S(t) = S(t - dt) + (NONIMMUNE_IMMIGRANTS - CONTRACTION)
* dt
INIT S = 1000
INFLOWS:
NONIMMUNE_IMMIGRANTS = 7
OUTFLOWS:
CONTRACTION = BETA*S*I

BETA = .002
```

20.2 Two Infective Populations

Let us expand on the model of the previous section and assume that an individual breaks out with the disease upon contact with a virus that can either be carried by members of the same population or by organisms of another species. Prominent examples are the Marburg and Ebola viruses, which can spread from monkeys to humans.[1]

[1]For a powerful description of the dynamics of these viruses see R. Preston, *The Hot Zone*, New York: Anchor Books, 1995.

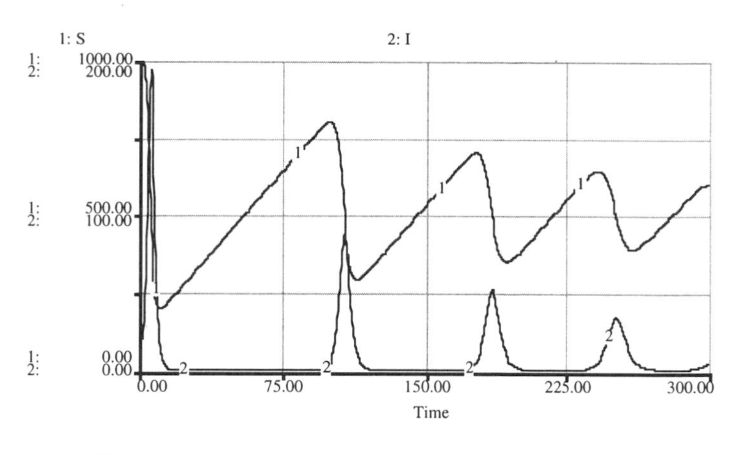

FIGURE 20.2

We begin our model with the set up of the previous section and duplicate it to capture the spread of the disease in the second population and from that population to the other one. To duplicate the STELLA model of the previous section, select the entire model by choosing Select All from the Edit menu and then copy it. Then make the necessary changes in the names of the variables and the connections among the two model parts. Notice that we only captured here the one-way movement of the virus from the infective stock I 2 to S 1. You can easily explore the case of the virus spreading from any of the two populations to the other.

For the first model run we set the parameters as indicated in Table 20.1.

We set the number of nonimmune immigrants for the two populations to 7 and 10, respectively. All of these numbers are hypothetical.

In Figure 20.3, we used the following specifications to calculate the number of individuals that contract the virus:

$$\text{CONTRACTION 1} = \text{BETA 1*S 1*I 1} + \text{BETA 1 2*S 1*I 2} \quad (1)$$

$$\text{CONTRACTION 2} = \text{RATE OF CONTACT 2*I 2*S 2} \quad (2)$$

TABLE 20.1 Model parameters

Variable	Value	Explanation
S 1(t = 0)	1000	Initial stock of susceptibles in population 1
I 1(t = 0)	20	Initial number of infective individuals in population 1
BETA 1	0.008	Contact rate of S 1 with I 1
SURVIVAL RATE 1	0.065	Rate of survival upon contact of S 1 with disease
S 2(t = 0)	1000	Initial stock of susceptibles in population 1
I 2(t = 0)	20	Initial number of infective individuals in population 1
BETA 2	0.003	Contact rate of S 2 with I 2
BETA 2 1	0.00015	Contact rate of S 1 with I 2
SURVIVAL RATE 2	0.2	Rate of survival upon contact of S 1 with disease

Can you anticipate the dynamics of the spread of the disease? The initial outbreak will be larger than that of the previous model because the number of susceptibles and infective individuals is larger, although the contact rates are quite a bit smaller. What matters, though, is the product of contact rates and sizes of the individual stocks.

Here are the results for the parameter settings listed in Table 20.1 and a choice of DT = 1. Following an initially severe outbreak, new episodes of the disease occur at relatively constant intervals and slightly increase in their amplitude. After only five episodes of outbreaks of the disease among S1, the disease "burns out" and disappears from the population (Fig. 20.4). Can you explain why in the very long run the disease disappears? Plot S 2 and I 2 in a separate graph to help you find an answer.

Let us assume that the parameters listed earlier are representative of one of two strains of the virus. The first strain, modeled earlier, does not move easily from one population to the other, but when it does, it results in only very small rates of survival. You may think of this strain as one that can only be passed on through direct contact with bodily fluids, such as saliva or blood. In constrast, the other strain of the virus can travel through air. It is therefore more easy to contract the disease, but this strain is also less lethal. Assume BETA 1 2 = 0.00025, SURVIVAL RATE 1 = 0.155, and all the other parameters are as listed earlier. The first outbreak

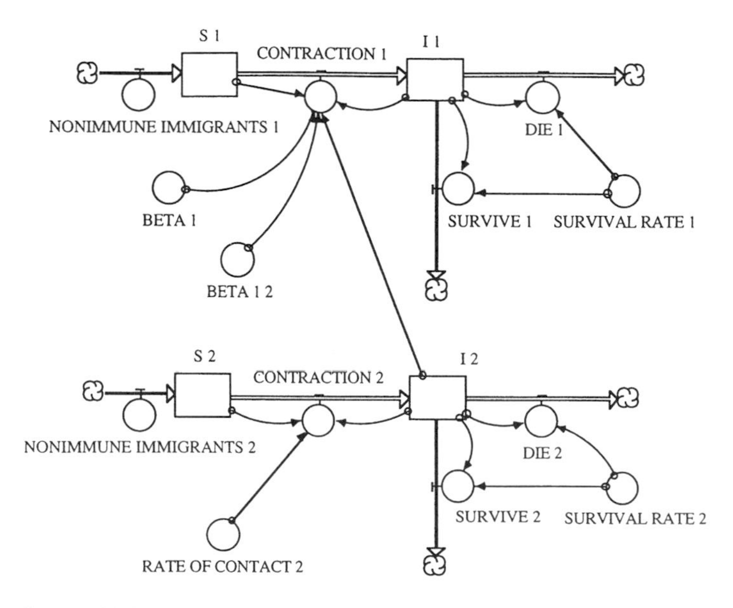

FIGURE 20.3

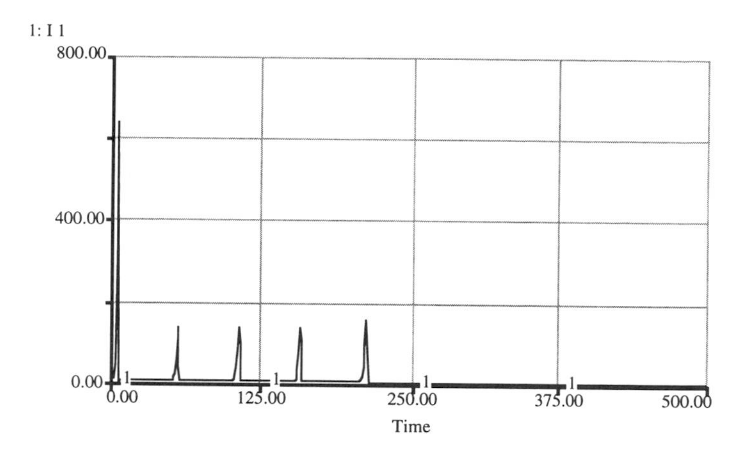

FIGURE 20.4

with the disease is virtually the same for both strains but the subsequent dynamics are very different (Fig. 20.5). The virus of the second strain can stay in the population for long times. Can you explain why? Perform a sensitivity analysis of the contact rates BETA 1, BETA 2 1 and BETA 2. How are the results affected by the choice of survival rates? How does the outbreak pattern of the disease change for smaller DT and for different integration methods?

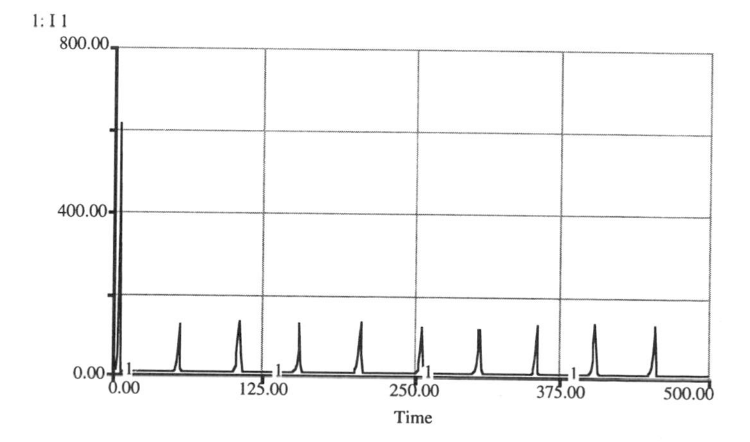

FIGURE 20.5

TWO INFECTIVE POPULATIONS

```
I_1(t) = I_1(t - dt) + (CONTRACTION_1 - SURVIVE_1 -
DIE_1) * dt
INIT I_1 = 5
INFLOWS:
CONTRACTION_1 = BETA_1*S_1*I_1 + BETA_1_2*S_1*I_2
OUTFLOWS:
SURVIVE_1 = SURVIVAL_RATE_1*I_1
DIE_1 = (1-SURVIVAL_RATE_1)*I_1

I_2(t) = I_2(t - dt) + (CONTRACTION_2 - DIE_2 -
SURVIVE_2) * dt
INIT I_2 = 20
INFLOWS:
CONTRACTION_2 = RATE_OF_CONTACT_2*I_2*S_2
OUTFLOWS:
DIE_2 = (1-SURVIVAL_RATE_2)*I_2
SURVIVE_2 = SURVIVAL_RATE_2*I_2

S_1(t) = S_1(t - dt) + (NONIMMUNE_IMMIGRANTS_1 -
CONTRACTION_1) * dt
INIT S_1 = 1000
INFLOWS:
NONIMMUNE_IMMIGRANTS_1 = 7
OUTFLOWS:
CONTRACTION_1 = BETA_1*S_1*I_1 + BETA_1_2*S_1*I_2

S_2(t) = S_2(t - dt) + (NONIMMUNE_IMMIGRANTS_2 -
CONTRACTION_2) * dt
INIT S_2 = 1000
INFLOWS:
NONIMMUNE_IMMIGRANTS_2 = 10
OUTFLOWS:
CONTRACTION_2 = BETA_2*I_2*S_2

BETA_1 = .008
BETA_1_2 = 0.00015
BETA_2 = .003
SURVIVAL_RATE_1 = .065
SURVIVAL_RATE_2 = .2
```

20.3 Temporary Immunity

In the previous sections of this chapter we modeled a disease that spreads in a population by turning susceptible, nonimmune individuals into infective ones. Following infection, the individuals were assumed to either die or remain immune for the rest of their lives. In both cases, the outflows from the stock of infective individuals disappeared into clouds—we did not keep track of them. We need to change this setup if we wish to model the case of a disease that only leads to temporary immunity. In that case, the flow of individuals who survive must end in a reservoir, rather than a cloud (Fig. 20.6). What are the effects of temporary immunity on the population modeled in Section 20.1?

Assume that once an individual becomes infected with the virus, that individual will either die from the disease or survive. The survival rate is 90% every t = 1, and there are no other causes of death. Those individuals that survive become temporarily immune. A fraction of the stock of temporarily immune individuals, T, will become sick again with the disease at a given RECURRENCE RATE, and again pass on the virus to the susceptible population.

The effect of temporary immunity is an overall larger stock of infective individuals. The results are plotted for a RECURRENCE RATE ranging from 0.02 to 0.1 (Fig. 20.7). The level of the initial outbreak of the disease is different under each of these rates, but the remaining dynamics show convergence toward the same steady-state level.

Change the duration of temporary immunity and observe the results. Introduce an incubation period for the disease and vaccination program as you did in the first section of this chapter. Disaggregate your model to deal with the case of a virus that affects different age cohort in the population differently. An example for such a disease is chicken pox. Chicken pox is a highly infectious childhood disease, caused by the varicella zoster virus. This virus can be spread either through direct contact with infected individuals or through the air. After exposure to the virus, the incubation period

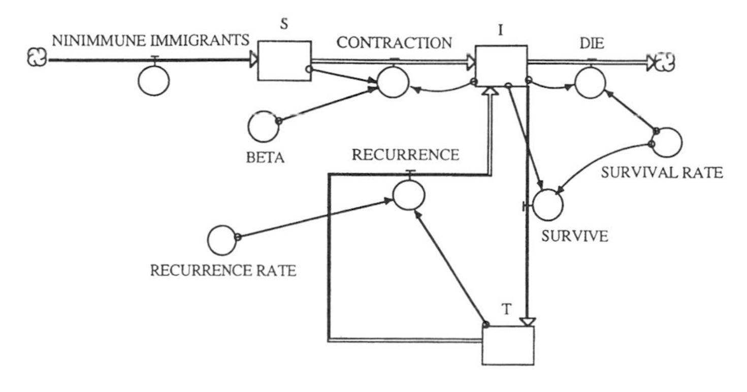

FIGURE 20.6

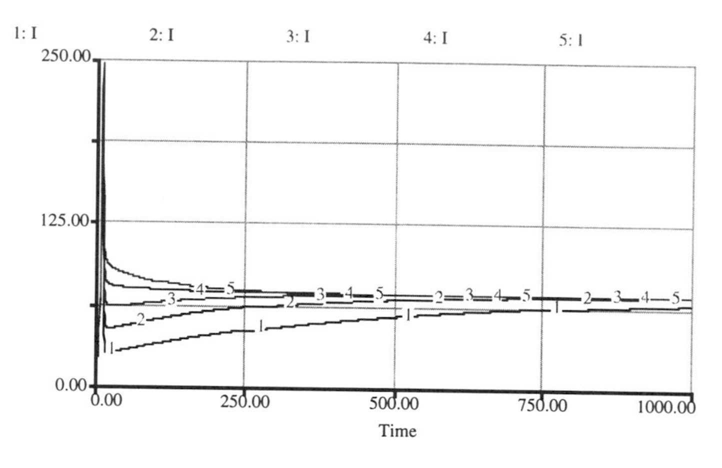

FIGURE 20.7

before an individual becomes contagious is approximately 7 days. Individuals are contagious for about 7 days, during which symptoms including fever and blisters appear, and then remain sick for an additional 14 days. Once an individual recovers, he or she develops a natural immunity and is unlikely to get the disease again.

Later in life, the varicella zoster virus manifests itself in the form of shingles in about 15% of the population who contracted chicken pox. Shingles has symptoms that are similar to chicken pox but strikes mostly individuals over the age of 50 who are fatigued or under stress. It takes approximately 10 days to recover from shingles, and during this time susceptible individuals can contract chicken pox from those suffering from shingles.

Recently, a vaccine has been approved to immunize people against chicken pox. The target population for immunization is children, because they comprise the highest infectious class. What are the effects of chicken pox on a given population, and how do those effects change as immunization takes place? What are the effects of immunization during childhood on the occurrence of shingles? How does an immunization policy affect the average age at which a person contracts the disease? Set up a model to provide answers to these questions.[2]

---

[2]To achieve some realism with your model, consult the following literature for parameter values:

L. Edelstein-Keshet, *Mathematical Models in Biology.* New York: Random House, 1988.

R. Finger, J. Hughes, B.J. Meade, A.R. Pelletier, and C.T. Palmer. Age-Specific Incidence of Chickenpox. *Public Health Reports* Nov/Dec:750–755, 1994.

H.W. Hethcote, Qualitative Analyses of Communicable Disease Models. *Mathematical Biosciences* 28:335–356, 1976.

R.M. May, Parasitic Infections as Regulator of Animal Populations. *American Scientist* 71:36-45, 1983.

Reader's Digest Association, *The American Medical Association Family Medical Guide.* New York: Random House, 1982.

```
                    TEMPORARY IMMUNITY

I(t) = I(t - dt) + (CONTRACTION + RECURRENCE - DIE -
SURVIVE) * dt
INIT I = 20
INFLOWS:
CONTRACTION = BETA*S*I
RECURRENCE = RECURRENCE_RATE*T
OUTFLOWS:
DIE = (1-SURVIVAL_RATE)*I
SURVIVE = SURVIVAL_RATE*I

S(t) = S(t - dt) + (NINIMMUNE_IMMIGRANTS - CONTRACTION)
* dt
INIT S = 1000
INFLOWS:
NINIMMUNE_IMMIGRANTS = 7
OUTFLOWS:
CONTRACTION = BETA*S*I

T(t) = T(t - dt) + (SURVIVE - RECURRENCE) * dt
INIT T = 10
INFLOWS:
SURVIVE = SURVIVAL_RATE*I
OUTFLOWS:
RECURRENCE = RECURRENCE_RATE*T

BETA = .002
RECURRENCE_RATE = .1
SURVIVAL_RATE = .9
```

20.4 Epidemic with Vaccination

Let us further expand on the models of the previous sections and introduce a number of features that make those models more meaningful. Among these features are

- The explicit inclusion of birth rates
- Death rates that are not only influenced by the disease but that result also from natural mortality
- A vaccination program that allows the population to become immune to the disease without having to first be sick

- Mutations in the disease that result in immune people not staying immune forever
- Ignorance of a fixed portion of the contagious population. These people are assumed not to know that they carry the disease. Consequently, we assume that ignorance would increase the rate at which the disease gets passed on from the infective to the susceptiple population.

The birth and natural death rates are specified graphically in this model as shown in Figure 20.8.

The CONTRACTION is a function of susceptible, healthy people coming into contact with people who are aware that they are contagious, people who are unaware that they are contagious, or people who come into contact with sick people:

CONTRACTION = BETA*((CONTAGIOUS-UNAWARE CONTAGIOUS)*3/5 +UNAWARE CONTAGIOUS+SICK/2)*NONIMMUNE (1)

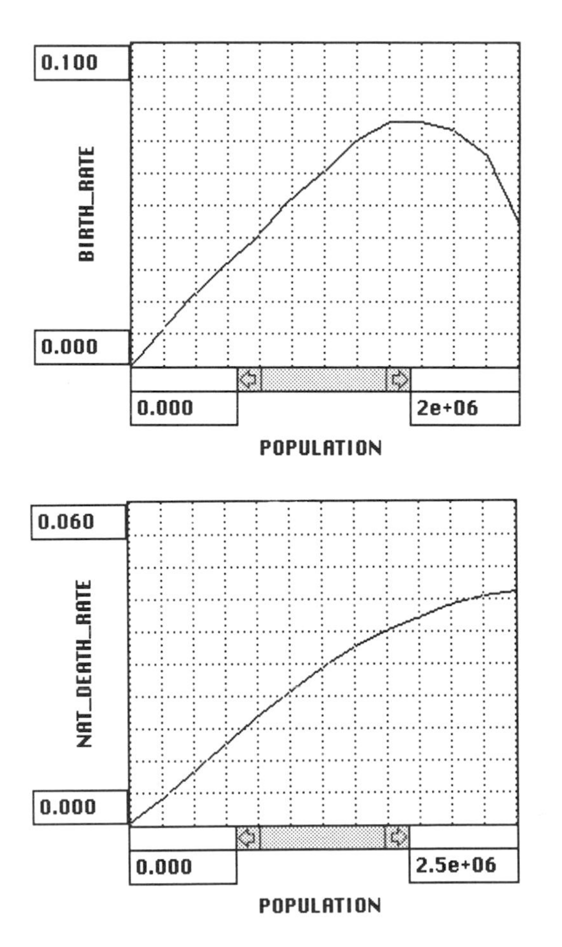

FIGURE 20.8

The VACCINE flow is specified as

$$VACCINE = NONIMMUNE*(1-CONTRACTION) \quad (2)$$

and contains the number of people vaccinated. A flow from IMMUNE to NONIMMUNE captures 1.5% of immune people who lose their immunity.

Consistent with the model of the previous section, there are a series of epidemic outbreaks. Due to the additional features of this model (Fig. 20.9), however, the numbers of immune and nonimmune people tends to stabilize, and so does the number of the sick. Even though an effective immunization program is in place, there are always some sick people present because the disease is assumed to undergo mutations (Fig. 20.10).

Introduce a spatial component into the model by considering two regions with different contact rates and different vaccination programs. Investigate the implications of travel restrictions imposed by one of the regions on people originating in the other region.

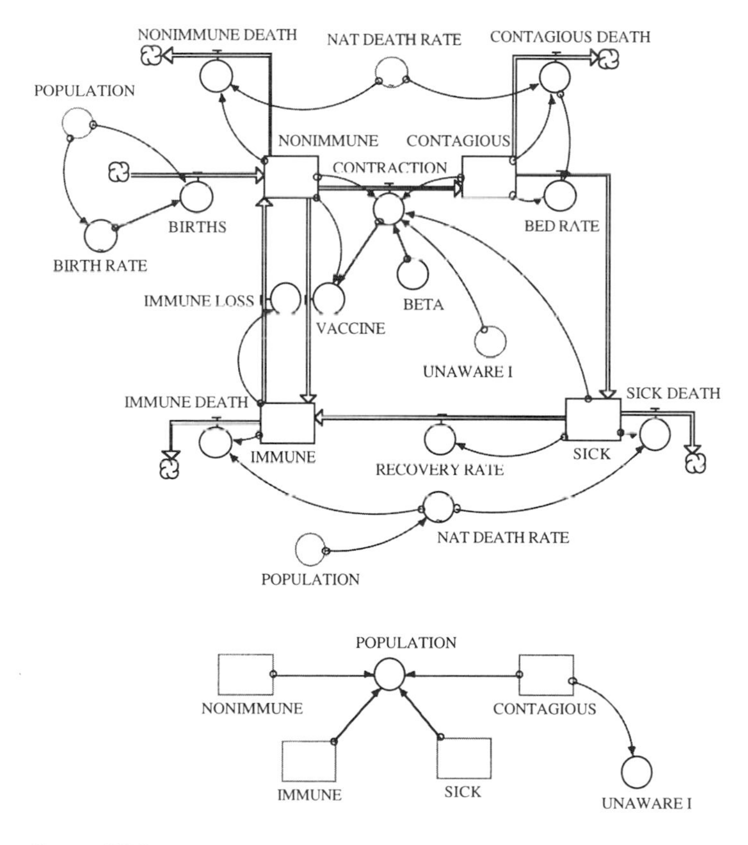

FIGURE 20.9

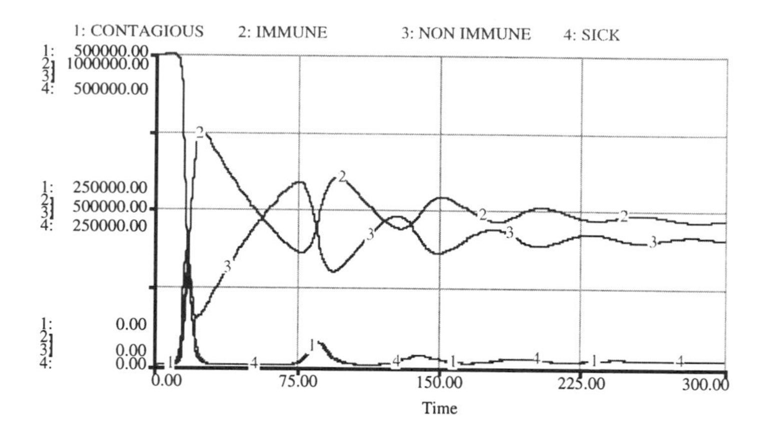

FIGURE 20.10

EPIDEMIC WITH VACCINATION

```
CONTAGIOUS(t) = CONTAGIOUS(t - dt) + (CONTRACTION -
BED_RATE - CONTAGIOUS_DEATH) * dt
INIT CONTAGIOUS = 1 {People}
INFLOWS:
CONTRACTION = BETA*((CONTAGIOUS-
UNAWARE_CONTAGIOUS)*3/5+UNAWARE_CONTAGIOUS+SICK/2)*NONI
MMUNE {People per Week}
OUTFLOWS:
BED_RATE = CONTAGIOUS-CONTAGIOUS_DEATH {People per
Week}

CONTAGIOUS_DEATH = CONTAGIOUS*NAT_DEATH_RATE/52 {People
per Week}
IMMUNE(t) = IMMUNE(t - dt) + (RECOVERY_RATE + VACCINE -
IMMUNE_DEATH - IMMUNE_LOSS) * dt
INIT IMMUNE = 0 {People}
INFLOWS:
RECOVERY_RATE = 0.9*SICK {People per Week}
VACCINE = NONIMMUNE*(1-CONTRACTION) {People per Week}
OUTFLOWS:
IMMUNE_DEATH = IMMUNE*NAT_DEATH_RATE/52 {People per
Week}
IMMUNE_LOSS = .015*IMMUNE {People per Week}
```

```
NONIMMUNE(t) = NONIMMUNE(t - dt) + (BIRTHS +
IMMUNE_LOSS - CONTRACTION -

NONIMMUNE_DEATH - VACCINE) * dt
INIT NONIMMUNE = 1000000 {People}
INFLOWS:
BIRTHS = POPULATION*BIRTH_RATE/52 {People per Week}
IMMUNE_LOSS = .015*IMMUNE {People per Week}
OUTFLOWS:
CONTRACTION = BETA*((CONTAGIOUS-
UNAWARE_CONTAGIOUS)*3/5+UNAWARE_CONTAGIOUS+SICK/2)*NONI
MMUNE {People per Week}
NONIMMUNE_DEATH = NAT_DEATH_RATE/52*NONIMMUNE {People
per Week}
VACCINE = NONIMMUNE*(1-CONTRACTION) {People per Week}

SICK(t) = SICK(t - dt) + (BED_RATE - RECOVERY_RATE -
SICK_DEATH) * dt
INIT SICK = 0 {People}
INFLOWS:
BED_RATE = CONTAGIOUS-CONTAGIOUS_DEATH {People per
Week}
OUTFLOWS:
RECOVERY_RATE = 0.9*SICK {People per Week}
SICK_DEATH = (.1*SICK)+(NAT_DEATH_RATE/52*SICK) {People
per Week}

BETA = 0.000002 {+SINWAVE(0.0000005,52 ) }
POPULATION = NONIMMUNE+CONTAGIOUS+IMMUNE+SICK {People}
UNAWARE_CONTAGIOUS = CONTAGIOUS/3 {A third of the
contagious persons are unaware of being contagious;
People}
BIRTH_RATE = GRAPH(POPULATION)
(0.00, 0.00), (166667, 0.0115), (333333, 0.023),
(500000, 0.0325), (666667, 0.041), (833333, 0.0525),
(1000000, 0.061), (1.2e+06, 0.0705), (1.3e+06, 0.076),
(1.5e+06, 0.076), (1.7e+06, 0.0735), (1.8e+06, 0.066),
(2e+06, 0.045)
NAT_DEATH_RATE = GRAPH(POPULATION)
(0.00, 0.0005), (208333, 0.005), (416667, 0.0102),
(625000, 0.0153), (833333, 0.0204), (1e+06, 0.0249),
(1.2e+06, 0.0294), (1.5e+06, 0.033), (1.7e+06, 0.0363),
(1.9e+06, 0.0387), (2.1e+06, 0.0411), (2.3e+06,
0.0426), (2.5e+06, 0.0432)
```

Part 5

Single Population Models

21

Adaptive Population Control

> To make increased population the cause of improved agriculture, is to commit the absurd blunder of confounding cause and effect.
>
> Rogers, *Political Economy*, 1868

In this chapter we model the action of a population that is collectively trying to control their population size by imperfectly recalling much of what they have done in terms of birth rate control over the recent past. They assess the gap between their current population size and that size dictated by their physical environment. It takes time to gain the information about these two population sizes. Once the population knows these levels, we assume they react by changing their birth rate. The new birth rate is an average of the ones remembered over the recent past. The death rate for this population and the level of the desired population size are dependent on the population density. The part of the model illustrated in Figure 21.1 captures these relationships among area, population density, death rates, and desired population level.

Three graphical functions are used to express these relationships. They are shown in Figures 21.2 to 21.4. One of these functions shows arable land changing over time. By defining population density as

$$\text{POP DENSITY} = \text{POPULATION}/(\text{VARIABLE AREA}*0 + \text{FIXED AREA}*1) \qquad (1)$$

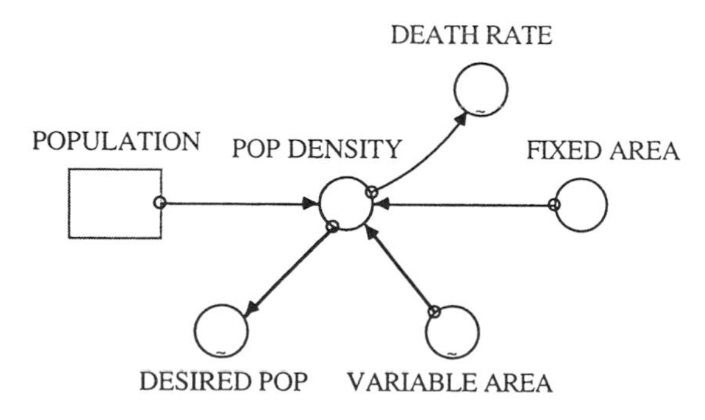

FIGURE 21.1

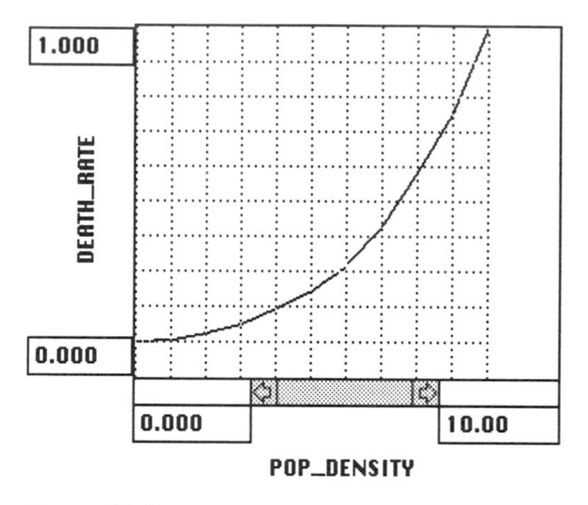

FIGURE 21.2

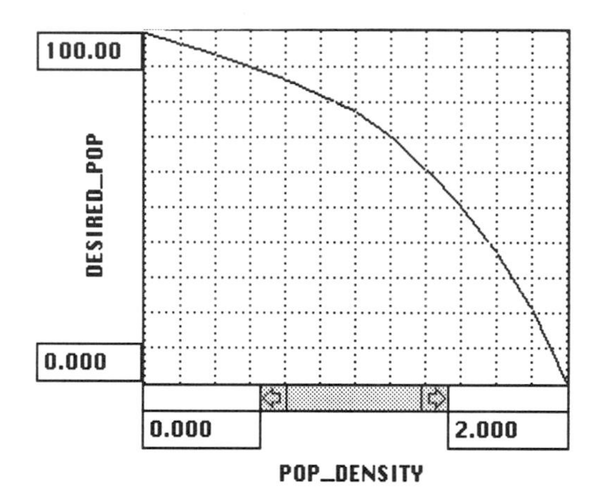

FIGURE 21.3

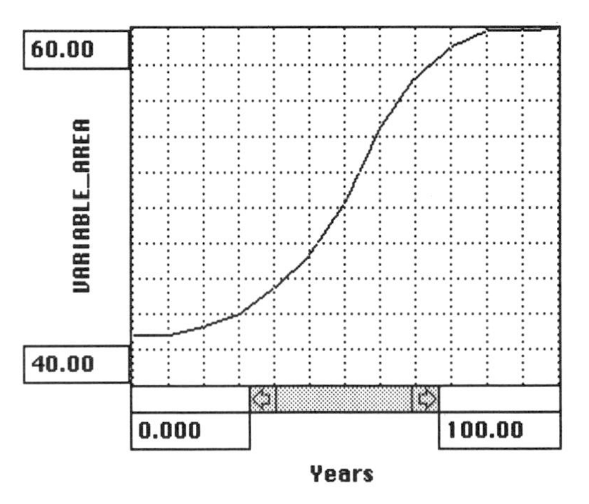

FIGURE 21.4

we can make use of either a fixed area or a variable area in our calculation of the population density simply by multiplying one of them by 1 and the other by 0. With this setup, the model can easily be run under alternative assumptions without having to redo parts of the diagram.

The desired level of the population is compared with the actual level and a GAP or difference is calculated as follows (Fig. 21.5):

$$\text{GAP} = \text{DESIRED POP} - \text{DELAY(POPULATION,2)} \qquad (2)$$

Note how the model is structured with a two-period DELAY of the POPULATION level to account for the time it takes to obtain a population census.

The GAP is then normalized by the current population level and multiplied by a reaction sensitivity to get the fractional birth rate (FRACTIONAL BR). We now wish to keep track of these FRACTIONAL BRs for a length of time called the YEARS RECALL. This part of the model captures the *adaptive control*, the influence of the remembered composite history of the system on its current and future performance.

To keep track of the FRACTIONAL BRs, we use a stock that accumulates the birth rates. We call this stock CUM BIRTH RATE. However, we do not specify this stock as a reservoir but as a Conveyor. Conveyors are stocks that function very much like conveyor belts. Objects, in our case, the FRACTIONAL BRs of a particular time period, are placed on a slat on the belt and with time move along this belt. To generate a conveyor, place a stock in the STELLA diagram, open it, and click in the dialog box on Conveyor. A new window appears. Give the conveyor an initial value, say, 0.05, then click on OK. Next, draw an outflow from the conveyor into a cloud to specify when objects leave the conveyor. Define a transforming variable YEARS RECALL, set it to 20, and connect it to the outflow from the conveyor. With this setup, the conveyor dumps off the oldest values on the conveyor, those that have been on the conveyor for YEARS RECALL. Finally, draw a second outflow from the conveyor and call it FORGET. The leakage, or exponential forgetting of the collected birth rate signals stored in the conveyor CUM BIRTH RATE, is controlled by a FORGETTING RATE. We set it here to 0.05, that is, 5% of the information is lost from each slat on the conveyor. Had we wanted to reserve this forgetting process to some portion of the conveying process, we could set the No Leak Zone of the conveyor at something less than the full length in the FORGET variable.

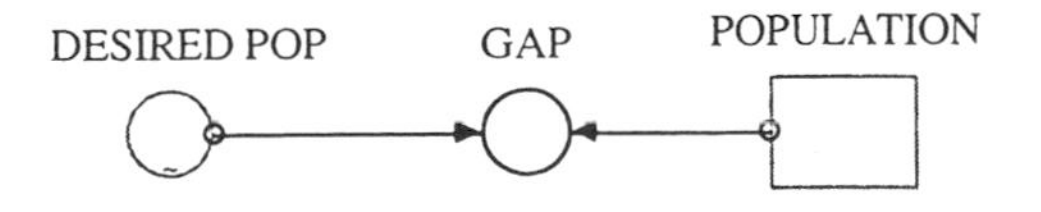

FIGURE 21.5

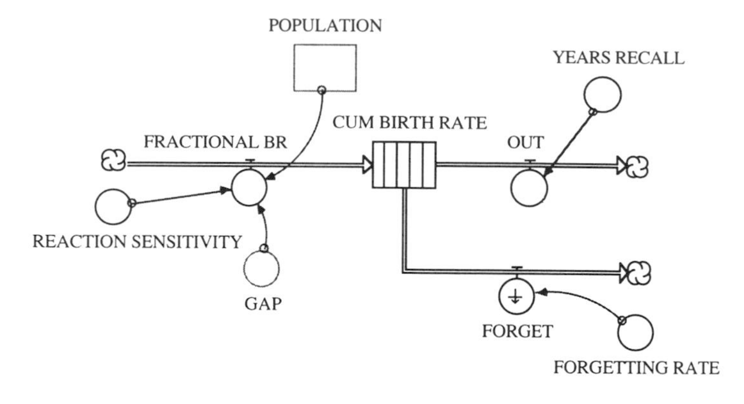

FIGURE 21.6

The module that captures the adaptive control is illustrated in Figure 21.6. The CUM BIRTH RATE, the sum of all the birth rates on the various conveyor slats, is sent to NET BIRTHS, where it is divided by the YEARS RECALL to get an average rate, and then it is compared with the death rate to calculate a net birth rate. This net birth rate is then multiplied by the population level to find the addition or subtraction to the total POPULATION (Fig. 21.7).

Curve 1 in Figure 21.8 refers to the case of a fixed area, while curve 2 depicts the population dynamics under the assumption of a variable area. In both cases, the result is a population that rises sharply from its initial low level and then is damped back to a steady state. Of course, part of the oscillation is due to the fact that there is no initial history of the birth rates in

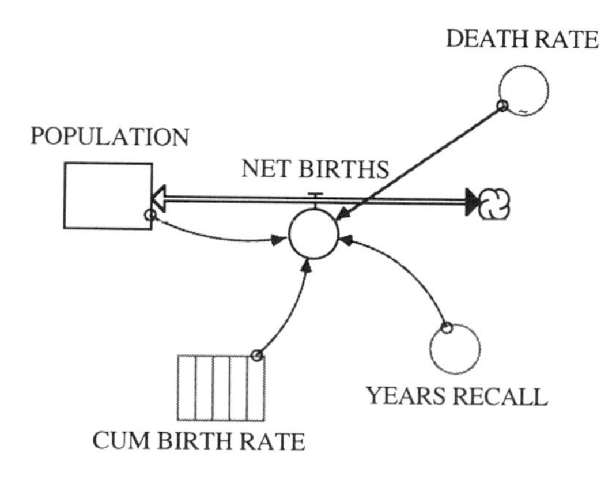

FIGURE 21.7

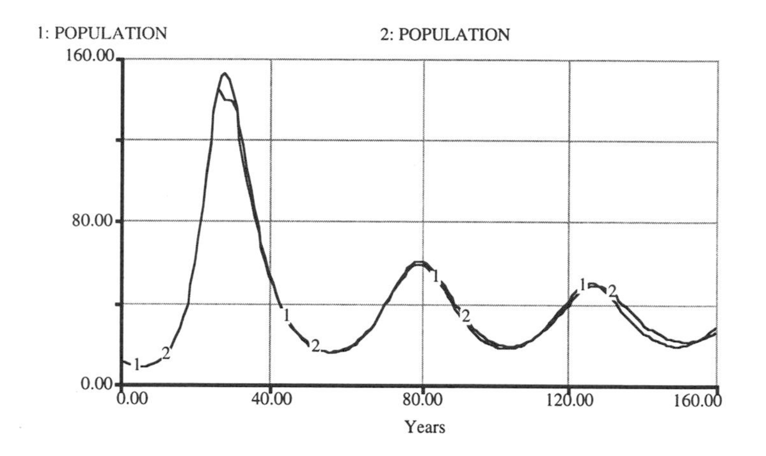

FIGURE 21.8

this population. To this extent, the model might be thought unrealistic. Yet suppose that we had a population that was initially stabilized at 10 for many years, and then due to a technological change or a conquest of new lands was suddenly exposed to the given amounts of land. In this case, the model is more realistic. Such a population would have no remembrance of any adaptation process to sudden new resource availability. But we show that as long as they are trying to adapt to the available resources (close the GAP), they will eventually reach a new steady state. In essence, this population is remembering a process of adaptation, but not the details. A more sophisticated model would allow them to remember how much social trauma the cycling produced, and the next period of sudden resource increase would be met with less very high peaks and lows.

In general we are trying to capture here the idea that a population reacts to an average remembered history of efforts to achieve some goal, in this case, closing the gap between the actual and desired levels of the population. Our approach to the final steady state might be framed as an optimality problem: What controls (birth rates) do we use to proceed to the steady state within a certain time, perhaps with certain constraints, such as avoiding population decline while rising to the ultimate steady state level? This constraint may be appropriate given the trauma of reducing populations. The problem with the optimality proposal above though is that we don't know a priori what the final steady state is going to be, given all these parameters. The final steady-state population is a function of the very parameters that we wish to change in order to control the ascent to the steady state. We could use as an optimality goal, the rise (only) in a prescribed time to *a* steady state and not be partial to the actual value of that steady state. In that case we would want to know which of the controllable parameters are accessible; in our case here, only the forgetting rate, the number of years of recall, and perhaps the reaction sensitivity. Then we wish to know which of these cause damping in the system. We would next set those that cause damping such that there is no oscillation in the population. This slope constraint is probably the best goal to use: When that value gets sufficiently low, we have reached the goal—we are sufficiently close to the steady state. Try out this idea and ones of your own to achieve the "no oscillation" goal.

ADAPTIVE POPULATION CONTROL

```
CUM_BIRTH_RATE(t) = CUM_BIRTH_RATE(t - dt) +
(FRACTIONAL_BR - OUT - FORGET) * dt
INIT CUM_BIRTH_RATE = .05
      TRANSIT TIME = varies
      INFLOW LIMIT = ∞
      CAPACITY = ∞
INFLOWS:
FRACTIONAL_BR = REACTION_SENSITIVITY*GAP/POPULATION
OUTFLOWS:
OUT = CONVEYOR OUTFLOW
      TRANSIT TIME = YEARS_RECALL
FORGET = LEAKAGE OUTFLOW
      LEAKAGE FRACTION = FORGETTING_RATE
      NO-LEAK ZONE = 0%

POPULATION(t) = POPULATION(t - dt) + (NET_BIRTHS) * dt
INIT POPULATION = 10
INFLOWS:
NET_BIRTHS = (CUM_BIRTH_RATE/YEARS_RECALL-
DEATH_RATE)*POPULATION

FIXED_AREA = 50+RANDOM(-20,20)*1
FORGETTING_RATE = .05
DOCUMENT: Acts like a damping coefficient.
GAP = DESIRED_POP-DELAY(POPULATION,2)
POP_DENSITY = POPULATION/(VARIABLE_AREA+FIXED_AREA*0)
REACTION_SENSITIVITY = .05
DOCUMENT: The inverse of this number is like a damping
coefficient.
YEARS_RECALL = 20
DEATH_RATE = GRAPH(POP_DENSITY)
(0.00, 0.1), (1.00, 0.105), (2.00, 0.125), (3.00,
0.15), (4.00, 0.195), (5.00, 0.245), (6.00, 0.315),
(7.00, 0.425), (8.00, 0.58), (9.00, 0.75), (10.0, 1.00)
```

```
DESIRED_POP = GRAPH(POP_DENSITY)
(0.00, 99.5), (0.167, 96.5), (0.333, 93.5), (0.5,
90.0), (0.667, 86.5), (0.833, 82.0), (1, 77.5), (1.17,
70.5), (1.33, 61.0), (1.50, 50.0), (1.67, 37.0), (1.83,
21.0), (2.00, 0.00)
VARIABLE_AREA = GRAPH(TIME)
(0.00, 42.8), (8.33, 42.8), (16.7, 43.2), (25.0, 44.0),
(33.3, 45.5), (41.7, 47.2), (50.0, 50.1), (58.3, 54.3),
(66.7, 57.2), (75.0, 59.0), (83.3, 59.8), (91.7, 59.9),
(100, 60.0)
```

22

Roan Herds

We descended into a valley, bent upon the destruction of a roan antelope.

W.C. Harris,
The Wild Sports of Southern Africa, xxii, 194.
Cape Town, C. Struik 1839

In Chapter 2 we captured environmental influences on population dynamics through the concept of carrying capacity. As parameters of the physical environment change, so does the carrying capacity of the ecosystem. Let us model environmental influences in more detail and focus on seasonal fluctuations and spatial differences in environmental parameters. We develop the model of this chapter for roans.

The roan is an antelope-like animal in Africa. Our problem is to model its population cycles given a weather-grass availability pattern on two different habitats called *prime* and *marginal* ground. The converter KP controls the birth and death rate of any roan that live on the prime ground; the converter KM controls the same for those roan living on the marginal ground. These converters are shown in Figures 22.1 and 22.2.

For these converters, the variable YEAR is defined by

$$\text{YEAR} = \text{INT(MOD(TIME,12))} + 1 \qquad (1)$$

which utilizes the built-in function INT, which returns the largest integer less than or equal to its argument. Thus, YEAR counts the numbers 1, 2, . . . , 12 corresponding to the year of the 12 year rainfall cycle. KP and KM reflect that rainfall pattern on births.

Ideally, one should imagine a spatial pattern of prime and marginal grounds and provide an initial population to each. This is the subject of further model building. For now let us say that there are only two places in the model, prime ground and marginal ground. If the population in either place drops below 3, the herd is lost from that area, that is, the outflows DUMP PGP and DUMP MGP empty the population stocks on the respective ground type.

The herd on the prime ground has a tendency to split to the marginal ground on a random basis. The probability that a split occurs is

$$\text{SPLIT PROB} = \text{INT(2*RANDOM(0,1))}, \qquad (2)$$

making the SPLIT flow either 0 or 1 according to the following rule:

$$\begin{aligned}\text{SPLIT} = {} & \text{IF PRIME GRND POP} > 7 \text{ THEN} \\ & \text{INT(PRIME GRND POP/3)*SPLIT PROB ELSE 0}\end{aligned} \qquad (3)$$

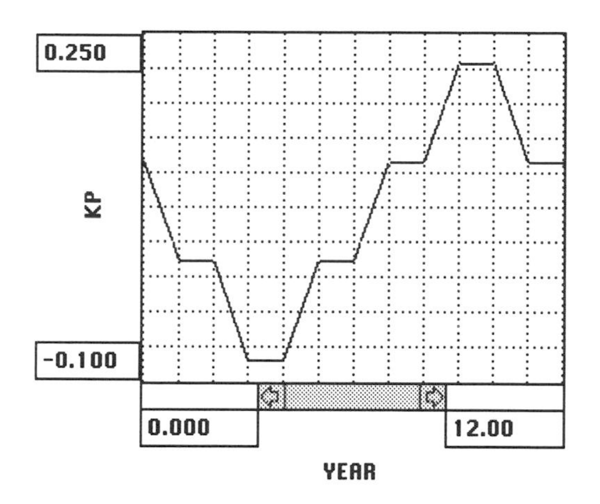

FIGURE 22.1

The nature of the split function is taken from Starfield and Blelock.[1] These authors defined the problem in much more detail than we have done here. In fact, they define a simplified approach and this model is simpler yet (Fig. 22.3).

The results of this simplified approach seem to agree with those in the more elaborate model of Starfield and Blelock, reconfirming our earlier statements that you should start with simple models. These are often the most powerful ones.

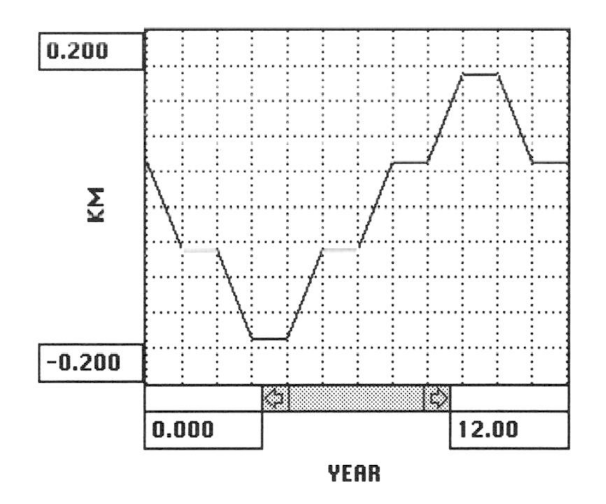

FIGURE 22.2

[1]A.M. Starfield, and A.L. Bleloch, *Building Models for Conservation and Wildlife*, New York: MacMillan Publishing, 1986, Chap. 2.

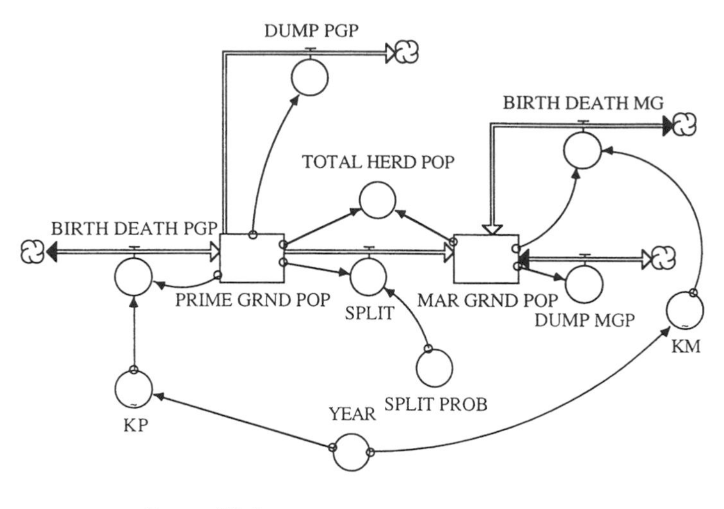

FIGURE 22.3

As the graph shows (Fig. 22.4), the population cycles higher in the first few decades and then proceeds to cycle to a steady-state population of 6 on the prime ground and 0 on the marginal ground. However, we should not expect to accurately model anything more than a couple of weather cycles at most. In the first 25 years, the cycle seems normal enough, even though the long-term effect is quite different.

Population cycles that are more pronounced than the ones found in this model are typical for animals that produce more rapidly than roans do. Voles and lemmings are two prominent examples, and we will model their population dynamics in the following two chapters.

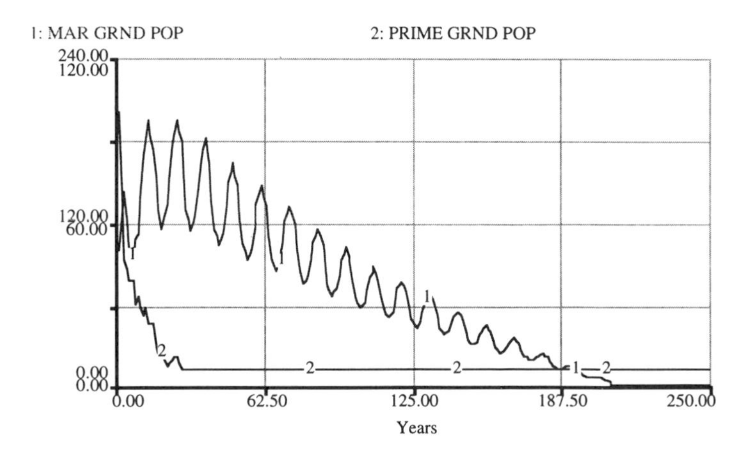

FIGURE 22.4

Try adding another marginal land unit. Double the initial prime ground herd size. The newer marginal unit, call it MARGINAL 2, receives splits from the prime ground with half of the split probability of the first unit. The marginal units are connected and can transfer roan back and forth with the following rule: If one of the marginal units is larger by 3 or more roan than the other for more than 1 year, then that unit transfers roan to the less populated unit in groups of 3. Run your model for 24 years and interpret your results.

ROAN HERDS

```
MAR_GRND_POP(t) = MAR_GRND_POP(t - dt) + (SPLIT +
BIRTH_DEATH_MG - DUMP_MGP) * dt
INIT MAR_GRND_POP = 100 {Individuals}
INFLOWS:
SPLIT = IF PRIME_GRND_POP > 7 THEN
INT(PRIME_GRND_POP/3)*SPLIT_PROB ELSE 0 {Individuals
per Year}
BIRTH_DEATH_MG = INT(KM*MAR_GRND_POP) {Individuals per
Year}
OUTFLOWS:
DUMP_MGP = IF MAR_GRND POP < 3 THEN MAR_GRND_POP ELSE
0 {Individuals per Year}
PRIME_GRND_POP(t) = PRIME_GRND_POP(t - dt) +
(BIRTH_DEATH_PGP - SPLIT - DUMP_PGP) * dt

INIT PRIME_GRND_POP = 100 {Individuals}
INFLOWS:
BIRTH_DEATH_PGP = IF PRIME_GRND_POP ≥ 7 THEN
INT((KP*PRIME_GRND_POP)) ELSE 0 {Individuals per Year}
OUTFLOWS:
SPLIT = IF PRIME_GRND_POP > 7 THEN
INT(PRIME_GRND_POP/3)*SPLIT_PROB ELSE 0 {Individuals
per Year}
DUMP_PGP = IF PRIME_GRND_POP < 3 THEN PRIME_GRND_POP
ELSE 0 {Individuals per Year}

SPLIT_PROB = INT(2*RANDOM(0,1))
TOTAL_HERD_POP = PRIME_GRND_POP+MAR_GRND_POP
{Individuals}
YEAR = INT(MOD(time,12)) + 1
KM = GRAPH(YEAR)
(0.00, 0.05), (1.00, -0.05), (2.00, -0.05), (3.00, -
0.15), (4.00, -0.15), (5.00, -0.05), (6.00, -0.05),
```

```
(7.00, 0.05), (8.00, 0.05), (9.00, 0.15), (10.0, 0.15),
(11.0, 0.05), (12.0, 0.05)
KP = GRAPH(YEAR)
(0.00, 0.12), (1.00, 0.02), (2.00, 0.02), (3.00, -
0.08), (4.00, -0.08), (5.00, 0.02), (6.00, 0.02),
(7.00, 0.12), (8.00, 0.12), (9.00, 0.22), (10.0, 0.22),
(11.0, 0.12), (12.0, 0.12)
```

23

Population Dynamics of Voles[1]

The true Voles . . . number about fifty known species.

Cassell's Natural History, III, 115
edited by Martin Duncan, 1877

23.1 Basic Model

Microtine rodents, such as voles and lemmings, have been the subject of intense interest in population biology for over 50 years. Much of this interest stems from the dramatic fluctuations in density observed in many populations. These fluctuations are often cyclic in nature, with large-scale irruptions occurring every 2 to 4 years. Voles and other small rodents are also of great economic importance due to their potential as agricultural pests and vectors of disease. Voles may cause substantial damage to a wide variety of crops and cause severe damage to fruit orchards by girdling trees. Renewed attention is also being given to the population dynamics of small rodents due to their prospective role in outbreaks of Lyme disease and the Hanta virus. Understanding the factors that regulate their population densities is the first step in controlling future outbreaks.

The periodic oscillations these animals experience, often referred to as *vole cycles,* have generated a tremendous number of experimental studies. Both field and laboratory approaches have been utilized to determine how voles respond to environmental changes. Although no consensus has been reached concerning the causes of vole cycles, it is clear that no single extrinsic factor, such as weather or food availability, can be directly responsible for this phenomenon. Could intrinsic factors be invoked to explain the occurrence of cyclical population changes in voles?

The logistic model of population growth (Fig. 23.1) has been previously used as a starting point for simulating vole dynamics but has failed to generate cycles using parameters that are commonly observed. One attribute that may have been missing from previous models is some form of time lag. It may be that vole populations behave in a logistic fashion, but with a delayed response to changes in density. In this chapter we construct a simple logistic model of population density to determine whether time-delayed components can be used to generate multiyear population cycles in voles.

[1]This chapter was contributed by Gary Fortier.

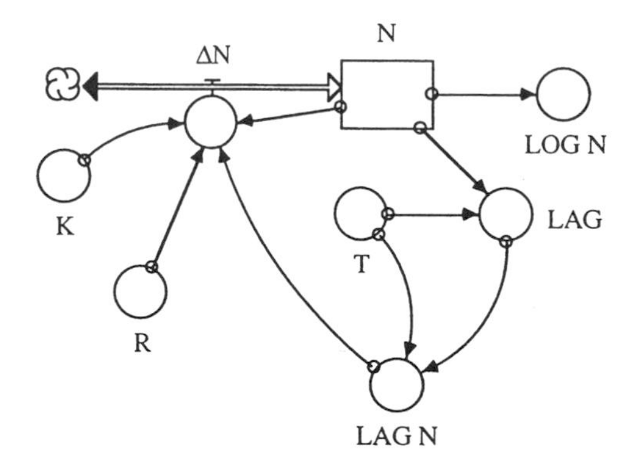

FIGURE 23.1

The basic logistic equation for population growth is

$$\Delta N = \frac{dN}{dt} = R * N * \left(1 - \frac{N}{K}\right), \tag{1}$$

where R is the intrinsic rate of increase, K is the environmental carrying capacity, and N is the population size or, in the case of a fixed area, the population density. In real populations of animals, there may often be a delay between a change in total population size and the animals' response to that change. For example, there may be an increase in birth rates due to increasing food availability, but if reproduction is limited by competition, the new animals entering the population may not create any appreciable impact until they reach adult size several months later. This may mean that current density is dependent upon the density at some time period in the past, a phenomenon known as *delayed-density dependence.* Delayed-density dependence can be incorporated into our logistic equation as

$$\Delta N = \frac{dN}{dt} = R * N * \left(1 - \frac{N_T}{K}\right), \tag{2}$$

where N_T represents the population size at some earlier time period T.

Using the standard logistic growth equation, the incremental change in population size, ΔN, would decrease as the population approaches the environmental carrying capacity, K. You can verify this by setting the time lag, T, equal to zero. When a nonzero time lag is added to the model (Fig. 23.2), the population may no longer reach a steady state. Run the model using K = 78, R = .15, and an initial population of 2. Rerun your model with T = 12 and T = 18. The values for R and K were measured from a cycling population of meadow voles (*Microtus pennsylvanicus*) in Massachusetts.[2] N = 2 represents a founding pair of voles in a new population, and T = 12, T = 18

---

[2]R.H. Tamarin, Demography of the Beach Vole (*Microtus breweri*) and the Meadow Vole (*Microtus pennsylvanicus*) in Southern Massachusetts, *Ecology,* 58: 1310–1321, 1977.

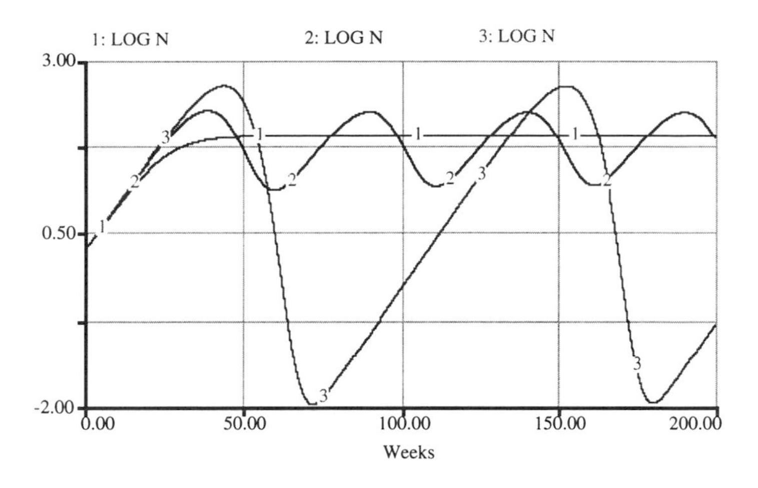

FIGURE 23.2

represent time lags of approximately one and one and a half generations, respectively. You should find that at T = 12 the population oscillates around the carrying capacity, with a period of approximately 1 year. Increases in the time lag lead to oscillations of higher amplitude and lower periodicity. The results in Figure 23.2 were derived for T = 0, 12 and 18, respectively.

Now experiment with each of the parameters to determine their potential effects on the population. Predict the qualitative changes you would expect before running the model with modified parameters. The carrying capacity K only raises or lowers the amplitude of the cycle without altering the periodicity. Run the model with different initial population sizes, N. Any $N \leq K$ alters the starting point of the population without affecting the shape of the oscillation itself. For N much larger than K, extreme density fluctuations result that eventually settle down into the standard 1-year oscillation (Fig. 23.3).

The model is quite sensitive to modest changes in the population's intrinsic rate of increase, R. A small increase in this value produces tremendous changes in the amplitude of the oscillation and also increases the period of the oscillation. As R increases, the population will overshoot the carrying capacity to a larger extent before the damping effects of the carrying capacity pull the population back into a decline. We will retain our R = 0.15, as it has been empirically determined from a natural population. K and N have no effect on periodicity; thus, if we are to obtain multiyear cycles from our model we must turn our attention to the effects of the time lag.

Experiment with a large range of values to determine the sensitivity of the model to changes in T. Values of $T > 30$ weeks generate very unstable populations that eventually crash to extinction. Very small time lags reduce the period and amplitude of the oscillation; $T < 4$ weeks produces only temporary cycles that ultimately dampen out on the environmental carrying capacity, K. Conversely, increasing $T > 12$ increases both the amplitude and period of our cycle.

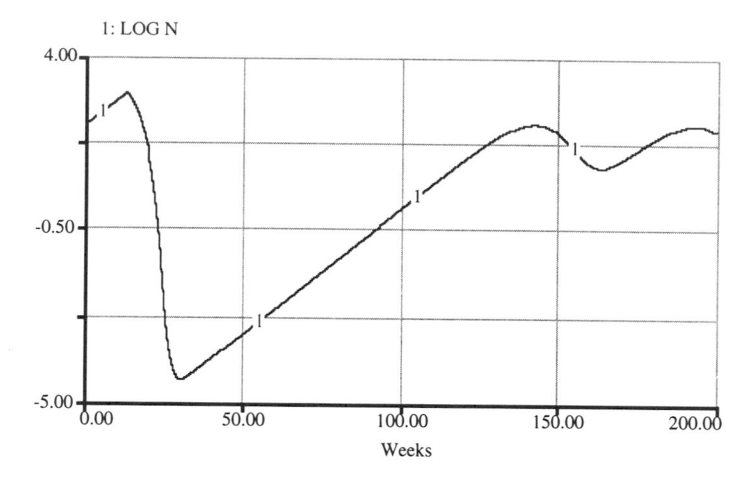

FIGURE 23.3

A time lag of 18 weeks yields a "correct" population cycle with a period of approximately 2 years. At first glance it appears that we have now achieved our task, simulating a vole cycle from field parameters. A closer look at the oscillation, however, reveals a serious flaw in the model. If we plot the simulation using the logarithm of population density, as we have done earlier, we see that the density undergoes changes of over four orders of magnitude during the course of a cycle. Such changes are not observed in North American vole populations and are not biologically meaningful. The densities of one animal per 100 hectares implied by the graph would certainly lead to extinction! The model must be re-examined to determine whether multiyear cycles can be generated without experiencing density changes of over three orders of magnitude. This is done in the following section.

BASIC VOLE MODEL

```
N(t) = N(t - dt) + (ΔN) * dt
INIT N = 2 {Individuals}
INFLOWS:
ΔN = R*N*(1-(LAG_N/K)) {Individuals per Week}

K = 78 {Individuals}
LAG = DELAY(N,T) {Individuals}
LAG_N = IF (TIME>T) THEN LAG ELSE 0 {Individuals}
LOG_N = LOG10(N) {Individuals}
R = .15 {Individuals per Individuals per Week}
T = 12 {Weeks}
```

23.2 Vole Population Dynamics with Seasonality

One shortcoming of our previous model may be its failure to account for seasonal shifts in behavior. Seasonal changes in the demographics of voles have been well documented. Let us add seasonality to our delayed-logistic model by varying R, the intrinsic rate of increase, as a function of time. This modification has a solid foundation in field data as these rodents are seasonally reproductive; breeding is sharply curtailed during the winter months. If we assume that limited winter reproduction is roughly equal to winter mortality, we can allow R to alternate between zero and its maximum value during the breeding season.

We will now modify our model to set R = 0 for 16 weeks of each 52 week period and set R = 0.15 for the remainder of the year. This is done with VAR R (Fig. 23.4).

WEEK serves as a counter; once an entire year has elapsed, the counter is reset to one. The parameter VAR R is set to zero during the 16 weeks of winter, then reverts back to its original value for the rest of the year. Run the model and observe the results on a logarithmic density plot, using a value of T = 18 weeks (Fig. 23.5).

Initially the population cycles over a period of approximately 2 years, within a fairly reasonable amplitude. After several years elapse, however, the amplitude fluctuations become more irregular and we encounter a familiar problem: The range of fluctuations in population density in our model is unreasonably large.

Another shortcoming of this model may be its boolean-style approach to seasonality. A graph of VAR R reveals an on-off or boolean mode in the intrinsic rate of increase: The value is reset immediately from one value to another without any transition. It would be more realistic to alter the seasonal

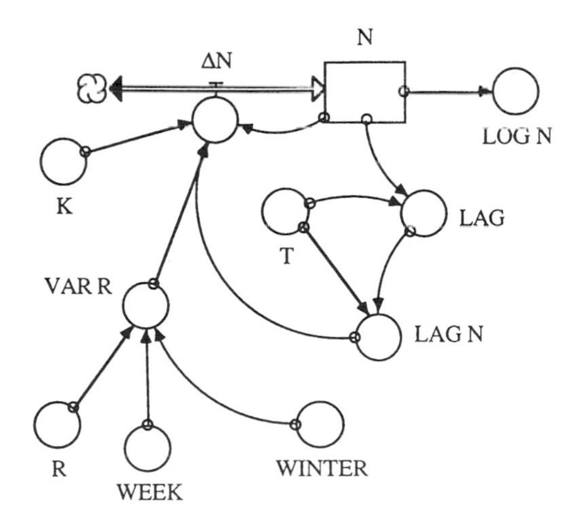

FIGURE 23.4

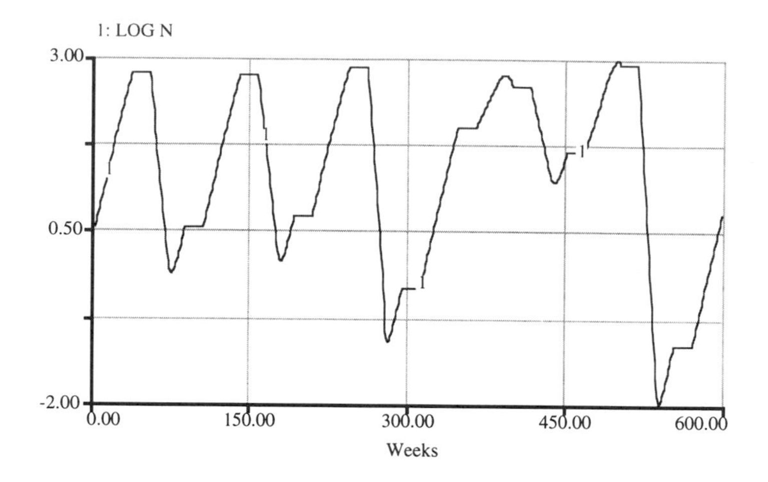

FIGURE 23.5

component of R in a more continuous fashion. This type of change is modeled in the following section.

```
VOLE POPULATION DYNAMICS WITH SEASONALITY

N(t) = N(t - dt) + (ΔN) * dt
INIT N = 2 {Individuals}
INFLOWS:
ΔN =  VAR_R*N*(1-(LAG_N/K)) {Individuals per Week}

K = 78 {Individuals}
LAG = DELAY(N,T) {Individuals}
LAG_N = IF (TIME>T) THEN LAG ELSE 0 {Individuals}
LOG_N = LOG10(N)
R = 0.15 {Individuals per Individuals per Week}
T = 18 {Weeks}
VAR_R = IF (week ≥ (52 - WINTER) and week ≤ 52) THEN 0
ELSE  R {Individuals per Individual per Week}
WEEK = MOD(TIME,52)+1
WINTER = 16 {weeks}
```

23.3 Sinusoidal Seasonal Change

Allowing the seasonal component of R to change in a continuous fashion is a better approximation of reality as the population is not perfectly synchronous in its behavioral response to seasonal changes. This form of seasonal-

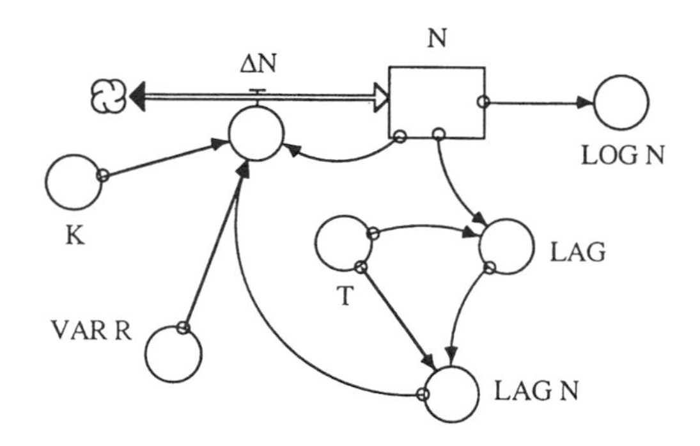

FIGURE 23.6

ity may be achieved in our model by allowing the intrinsic rate of growth to oscillate between zero and R in a sinusoidal fashion. In our sinusoidal model, a new variable, SIN R, is simply a sine wave function with a 52 week period oscillation with an amplitude of –0.75 to +0.75. By setting

$$\text{VAR R} = 0.075 + \text{SINWAVE}(0.075, 52), \tag{1}$$

we create an R value varying continuously between 0.15 and 0 over the course of 1 year (Fig. 23.6). The simulation that results from this modification provides a very satisfying result—multiyear cycles within biologically reasonable amplitudes of density (Fig. 23.7).

What implications do these simulations have for our understanding of vole biology? First, they demonstrate that a simple delayed-logistic equation, with a single seasonal component, is capable of generating multiyear fluctuations in population density. This was accomplished without resorting to external

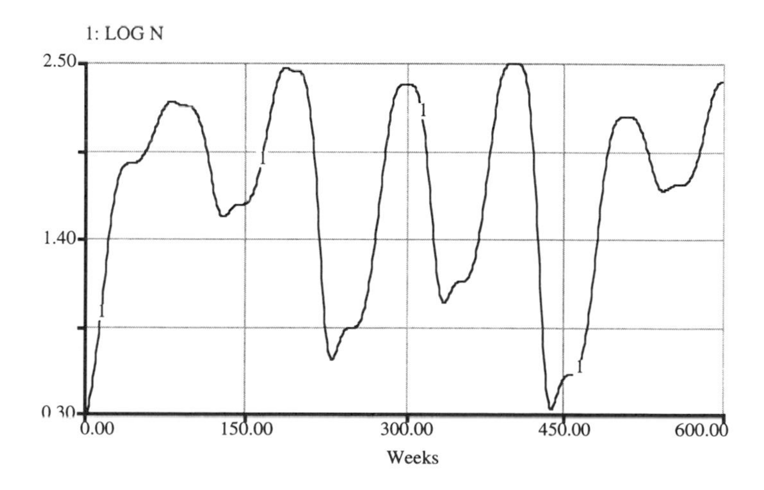

FIGURE 23.7

factors such as predation or climate change. These simulations were completed using actual field data for the intrinsic rate of increase (R) and the environmental carrying capacity (K). Second, the lag component was a mere 18 weeks, as opposed to the considerably longer lags used in previous models[3]. The 18 week delay used in this simulation roughly approximates the generation time observed in many species of vole. This implies that roughly one generation elapses between any change in density and a corresponding change in population growth rate. In our simple model, the time lag was of paramount importance in generating population cycles. Interestingly, regular periodic oscillations in density do not occur in populations where dispersal is prevented, such as on island populations or populations that are enclosed within vole-tight fences. In light of our model results, this suggests that blocking dispersal may influence the time-delayed component of vole cycles. A greater understanding of the relationship between dispersal and its potential effects on time-delayed responses may shed new light on the field of microtine population dynamics. Third, modeling of population dynamics, combined with data from field studies, can provide insight into the mechanisms of population change that not only enhance our understanding of the driving forces of the system but further sharpen the focus of subsequent studies of these populations.

---

[3]R.M. May (ed.), *Theoretical Ecology*, 2nd ed., Oxford: Blackwell Scientific Publishers, 1981, pp. 5–29.

SINUSOIDAL SEASONAL CHANGE

```
N(t) = N(t - dt) + (ΔN) * dt
INIT N = 2 {Individuals}
INFLOWS:
ΔN = VAR_R*N*(1-(LAG_N/K)) {Individuals per Week}

K = 78 {Individuals}
LAG = DELAY(N,T) {Individuals}
LAG_N = IF (TIME>T) THEN LAG ELSE 0 {Individuals}
LOG_N = LOG10(N)
T = 18 {Weeks}
VAR_R = SINWAVE(.075,52) + .075 {Individuals per
Individual per Week}
```

24

Lemming Model

> A kind of Mice, (they call Lemming . . .) in Norway, which eat up every green thing. They come in such prodigious Numbers, that they fancy them to fall from the Clouds.
>
> William Derham,
> *Physico-Theology,* 56 (note) London, 1713

Similar to the vole population modeled in the previous chapter, lemming populations can experience significant population fluctuations from year to year. Many legends surround the seemingly erratic population dynamics of lemmings.[1]

To model lemming population dynamics it seems advisable[2] to distinguish two types of lemmings. Type 1 reproduces rapidly and migrates in response to overcrowding. Type 2 has a lower reproductive capacity but is less sensitive to high population densities. The change in the densitities of each type are given, respectively, by

$$\Delta N1 = \text{IF } N1 > 0 \text{ THEN } N1 * (A1 - (B1 - C1) * N2 - C1 * (N1 + N2)) \text{ ELSE } 0 \quad (1)$$

$$\Delta N2 = \text{IF } N2 > 0 \text{ THEN } N2*(-A2 + B2*N1) \text{ ELSE } 0 \quad (2)$$

The parameters A1, A2, B1, B2, and C1 capture the density dependence of birth and death rates of the two types of lemmings (Fig. 24.1). Appropriate choice of these parameters yields oscillatory behavior noted in the lemming populations. For the results illustrated in Figures 24.2 and 24.3, we have set A1 = 0.9 (1/time period), A2 = 0.5 (1/ time period), B1 = 0.9 (1/area/time period), B2 = 0.2(1/individuals/area/time period), and C1 = 0.0043 (1/individuals/area/time period).

As in the previous chapter, introduce seasonality with random variations into the model. Then find the parameters that yield oscillation of the lemming population.

---

[1] H. Decker, A Simple Mathematical Model of Rodent Population Cycles, *Journal of Mathematical Biology,* 2:57–67, 1975.

[2] HJ.H. Myers, and C.J. Krebs, Genetic, Behavioral and Reproductive Attributes of Dispersing Field Voles *Microtus pennsylvanicus* and *Microtus ochragaster, Ecological Mongraphs,* 41:53–78, 1971.

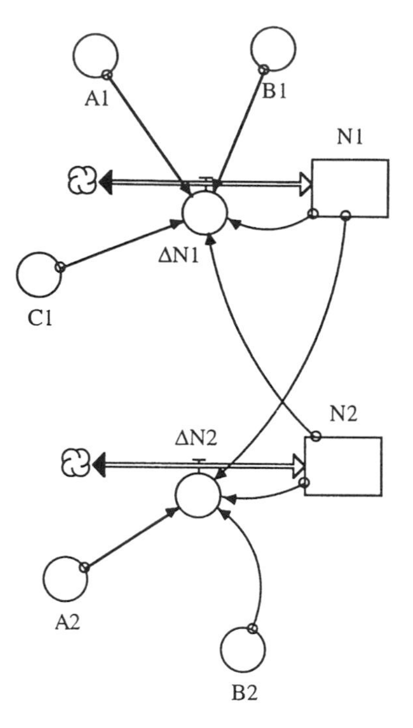

Figure 24.1

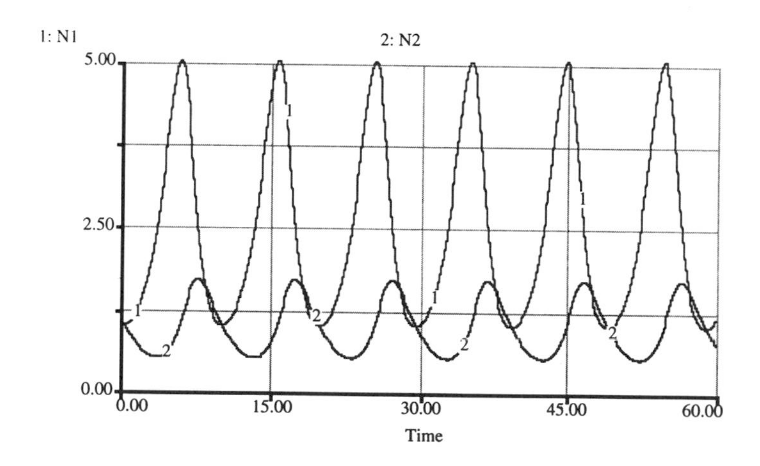

Figure 24.2

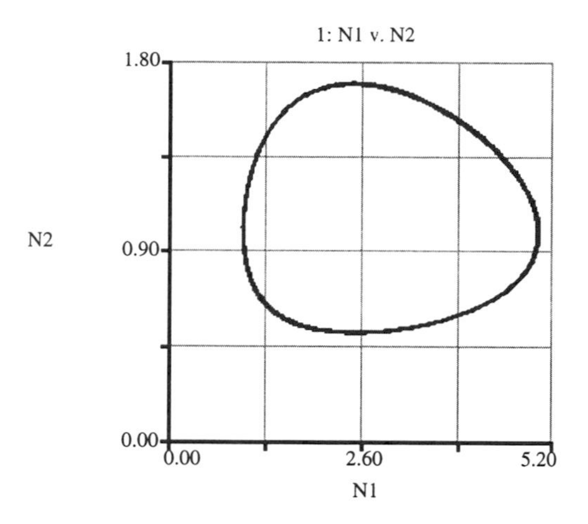

FIGURE 24.3

LEMMING MODEL

```
N1(t) = N1(t - dt) + (ΔN1) * dt
INIT N1 = 1 {Individuals per Unit Area}
INFLOWS:
ΔN1 = IF N1>0 THEN N1^(A1-(B1-C1)^N2-C1^(N1+N2)) ELSE 0
{Individuals per Unit Area per Time Period}

N2(t) = N2(t - dt) + (ΔN2) * dt
INIT N2 = 1 {Individuals per Unit Area}
INFLOWS:
ΔN2 = IF N2>0 THEN N2*(-A2+B2*N1) ELSE 0 {Individuals
per Unit Area per Time Period}

A1 = .9 {1/Time Period}
A2 = .5 {1/Time Period}
B1 = .9 {1/Individuals per Unit Area per Time Period}
B2 = .2 {1/Individuals per Unit Area per Time Period}
C1 = .0043 {1/Individuals per Unit Area per Time
Period}
```

25

Multistage Insect Models

> We may define insects to be little animals without red blood, bones or cartilages, furnished with a trunk or else a mouth, opening lengthwise, with eyes which they are incapable of covering, and with lungs which have their openings in the sides.
>
> Oliver Goldsmith,
> *Natural History,* IV, 13, 1774

25.1 Matching Experiments and Models of Insect Life Cycles

One of the most advanced areas of dynamic modeling at the organism level is found in entomology. Insects are economic creatures; they cause billions of dollars of damage to food supplies around the world every year. We try to control their population levels, having long ago realized they multiply and evolve too fast for elimination.

To better understand the dynamics of insect populations, we model the life cycle of an insect, simplified into two stages, egg and adult. Typically, the data used in understanding insect population dynamics come from laboratory experiments in which one watches each egg and notes that at the end of the maturation period, only a fraction of the original eggs have survived. Eggs are only pronounced dead after they fail to mature. This observation gives us two numbers, the experimental survival fraction, EXP SURV FRAC, and the experimental maturation time, EXP MATURE TIME. How can we use such data to parameterize a model when the model time step is dramatically different from the maturation time?

We develop a new concept, the model survival fraction, MOD SURV FRAC. In ecological experiments, the instantaneous survival rate cannot be measured. But the survival rate can be measured over some real time period, the maturation time, by counting the number of eggs surviving to maturation. A problem arises when we wish to model the system at a shorter time step than the real one. We need to model at these shorter times because the characteristic time of the system may be shorter than the shortest feasible measurement time of some particular part of the system. So we have the experimental time step (the maturation time) and the model time step (DT), and we must devise a conversion from experiment to model.

That conversion is based on the assumption that the survival fraction is a declining exponential. First, we simply assume that

$$\text{EXP SURV FRAC(t)} = \text{N(t)/N(t} - \text{DT)} = \text{EXP}(-\text{m*t})$$
$$= \text{the dimensionless survival fraction.} \quad (1)$$

N is the population level. Later, when we attempt to verify the experimental data with our model, we will shut off the birth and hatch rate and observe the (necessarily exponential) decline in egg population due to death. Then we will realize that the exponential form assumption was necessary. The resulting instantaneous survival fraction is designated by some constant m. Using the experimental data to solve this equation for –m gives

$$-\text{m} = \ln(\text{EXP SURV FRAC})/\text{T}. \quad (2)$$

The *model* survival fraction (based on our choice of the time step DT), via the same exponential assumption, is

$$\text{MOD SURV FRAC} = \text{EXP}(-\text{m*DT}), \quad (3)$$

and when the expression for –m is substituted

$$\text{MOD SURV FRAC} = \text{EXP}(\ln(\text{EXP SURV FRAC*DT/T}), \quad (4)$$

which is the basic equation for the model survival fraction. We now have the instantaneous survival fraction and those surviving will mature or hatch at the modified maturation rate,

$$\text{EGGS/T*MOD SURV FRAC}, \quad (5)$$

that is, the survivors mature at the experimentally found maturation rate, EGGS/T. Remember, eggs don't have to hatch or die; they may simply wait. But survivors do hatch.

To use this fraction, let us look first at the ADULT DEATH rate and the model survival fraction, MOD SURV FRAC, divided by DT. This step converts the experimental survival fraction, EXP SURV FRAC, to a daily model survival rate (when DT = 1.0 it is equal to a 1 day step). From this we can readily calculate the death rate as 1 – MOD SURV FRAC, divided by DT. A similar calculation is done to yield the egg death rate, adjusting for the fact that the experimental period is 5 days, not 1 day as in the adult death rate case (Fig. 25.1).

The model is run for initial stocks of adults and eggs equal to 0 and 50, respectively, EXP ADULT SURV FRAC = 0.8, EXP MATURE RATE = 0.2, EXP SURV FRAC = 0.7, and EXP ADULT SURV RATE = 1. The results are shown in Figure 25.2.

Suppose we are uncertain about the exact egg experimental fraction. We may suspect that using literature data is not good enough, and think that this number is within ±10%. Do an experiment to find this number if the total number of adults in 24 days is within ±10%. Insert a larval stage into this model with a larval survival fraction of 0.8 in 3 days maturation time. Why doesn't the stock of adults in this model grow as is did in the first version?

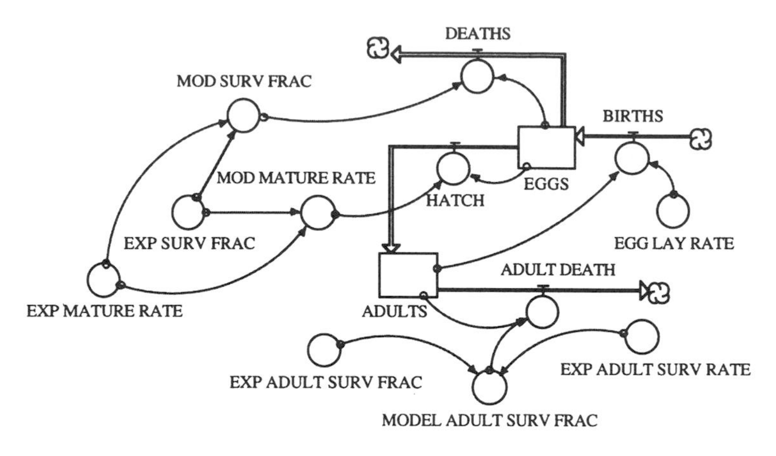

FIGURE 25.1

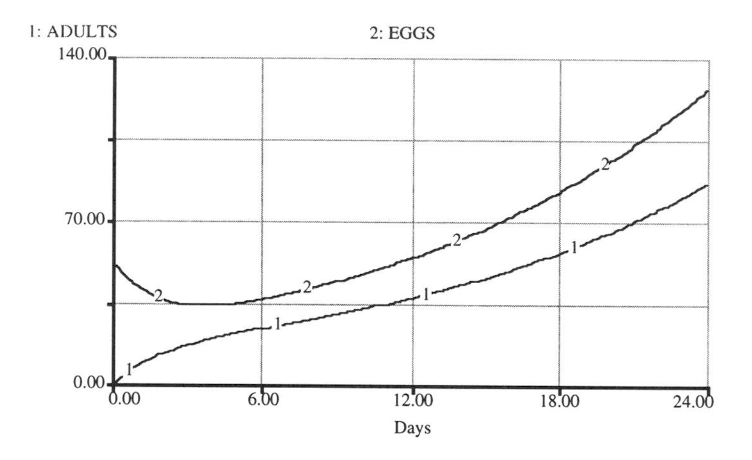

FIGURE 25.2

TWO-STAGE INSECT POPULATION MODEL

```
ADULTS(t) = ADULTS(t - dt) + (HATCH - ADULT_DEATH) * dt
INIT ADULTS = 0
INFLOWS:
HATCH = MOD_MATURE_RATE*EGGS  {Some of the eggs
neither die nor mature.}
OUTFLOWS:
ADULT_DEATH = ADULTS*MODEL_ADULT_SURV_FRAC

EGGS(t) = EGGS(t - dt) + (BIRTHS - DEATHS - HATCH) * dt
```

```
INIT EGGS = 50
INFLOWS:
BIRTHS =  EGG_LAY_RATE*ADULTS
OUTFLOWS:
DEATHS =  MOD_SURV_FRAC*EGGS
HATCH =  MOD_MATURE_RATE*EGGS  {Some of the eggs
neither die nor mature.}

EGG_LAY_RATE =  .5  {Experimental laying rate. EGGS PER
ADULT PER DAY.}
EXP_ADULT_SURV_FRAC =  .8 {Experimental daily adult
survival fraction per stage, dimensionless.}
EXP_ADULT_SURV_RATE = 1
DOCUMENT:  One day = experimental period for which
adult mortality is measured.
EXP_MATURE_RATE =  .2  {Experimental maturation rate,
1/DAY, i.e., 20 eggs per 100 eggs mature each day, as
noted in the experiment. In other words, a surviving
egg matures on the average in 5 days under the
experimental conditions. T = 5 days.}
EXP_SURV_FRAC = .7     {Experimental egg survival
fraction, dimensionless, per stage.  Stage = 1/EXP
MATURE RATE, i.e., 70 eggs per 100 eggs survive each
1/EXP MATURE RATE days, as noted in the experiment.}
MODEL_ADULT_SURV_FRAC =  (1-
EXP(LOGN(EXP_ADULT_SURV_FRAC)*DT/EXP_ADULT_SURV_RATE))/
DT  {Adult mortality rate, 1/day.   Instantaneous sur-
vival fraction + instantaneous mortality fraction - 1.}
MOD_MATURE_RATE =
EXP_MATURE_RATE*EXP(LOGN(EXP_SURV_FRAC)*DT*EXP_MATURE_
RATE)    {Model maturation rate for survivors, 1/DAY.
Hatch rate = instantaneous survival fraction*model
maturation rate.}
MOD_SURV_FRAC =  (1 -
EXP(LOGN(EXP_SURV_FRAC)*EXP_MATURE_RATE*DT))/DT
{Egg mortality rate,
1/DAY.  Instantaneous survival fraction + instantaneous
mortality fraction = 1.}
```

25.2 Two-Stage Insect Model with a Degree-Day Calculation Controlling the Maturation Rate

Let us take the model of the previous section and specify the maturation rate as a function of the temperature. In this model, time and temperature are now the two independent variables. We have assumed a sine function for the mean daily temperature and assumed ± 10 degrees + the mean daily to get the high and low temperature for the day:

$$\text{DALY MEAN TEMP} = 47 + 27*\text{SIN}(2*\text{PI}/365*\text{TIME}). \qquad (1)$$

If the high temperature is less than the threshold or base temperature, no degree-days for that day are calculated. If the minimum temperature for that day is less than the threshold temperature but the maximum temperature is greater than the threshold, the degree-days, DD 1, are calculated as the maximum temperature minus the threshold temperature, divided by 2:

$$\begin{aligned}\text{DD} = {} & \text{IF DAILY MEAN TEMP+10} < \text{BASE TEMP THEN 0 ELSE} \\ & \text{(DAILY MEAN TEMP+10-BASE TEMP)/2} \qquad (2)\end{aligned}$$

The calculation is illustrated in Figure 25.3.

For this model we needed to assume a relation between the degree-days and the maturation time, which we took to be linear. Degree-days DD are not accumulated over time but are determined for each time step (1 day). We have also set up a similar effect on the birth or oviposition rate in order to prevent the population from growing too large (Figs. 25.4 to 25.6).

Together these additions make the population rise and fall. But the peak declines exponentially. Do you know why?

The graph in Figure 25.7 shows the result. Note how the adults die off and the egg number levels off when the threshold temperature is reached. Run this model for 2 years. When will it begin to repeat itself and thus be clear of the initial conditions? Try different initial conditions and find the same sort of independence. The insects are gradually dying out of this sys-

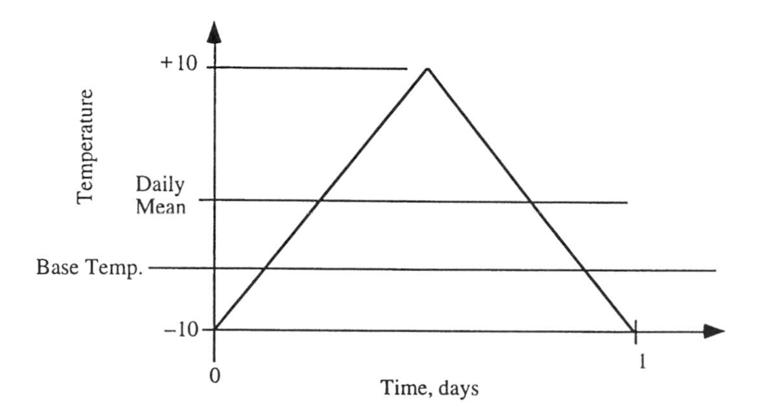

FIGURE 25.3

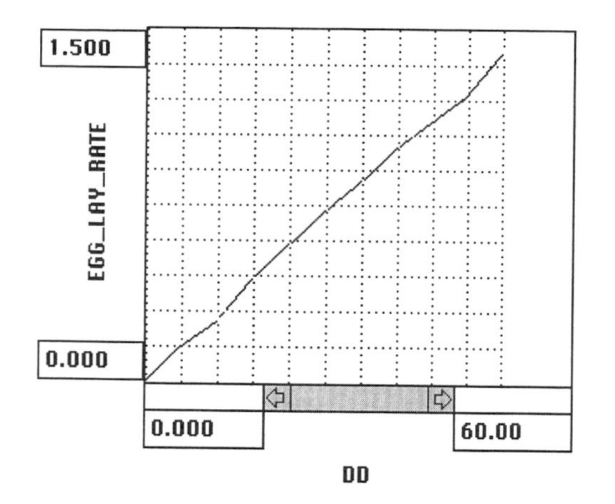

FIGURE 25.4

tem. What are the logical changes that you could make to stabilize their annual peaks?

Assume that this model represents the pattern of a needed predator insect. Try to pulse in the fewest number of eggs at the best time to keep this species between 30 and 50 adults at the most throughout the indefinite future.

As a modification to the model illustrated in Figure 25.6, we could assume that the egg maturation rate was a function of time and the *accumulated* degree-days. We would have to represent oviposition by pulsing in the eggs over a very short period. Can you devise a feasible model

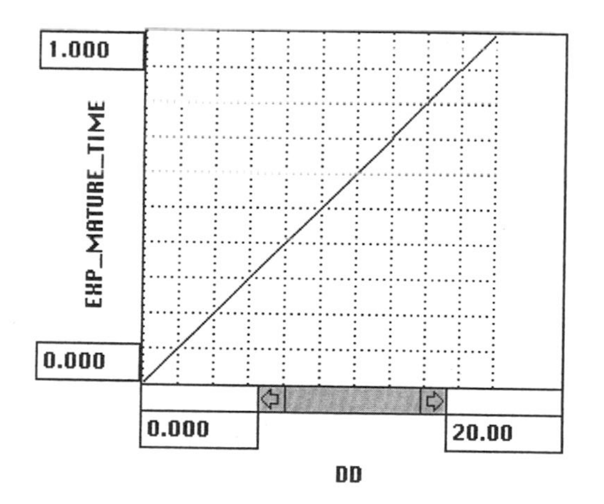

FIGURE 25.5

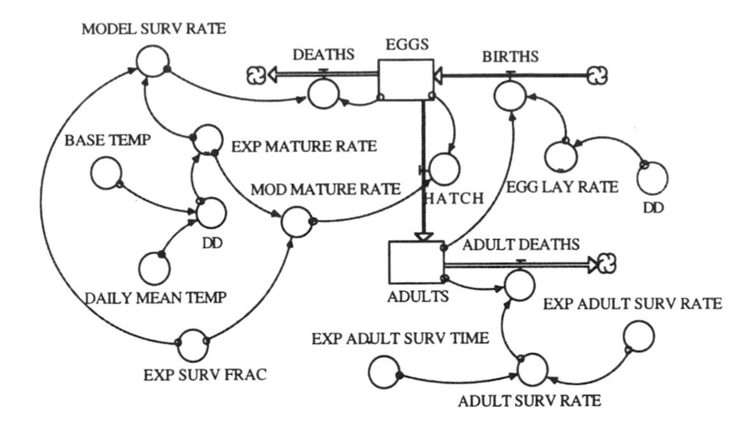

FIGURE 25.6

of this sort? Use a normal distribution of the daily temperature and assume a standard deviation, and then use the base temperature as the basis for calculating the accumulated degree-days. Take the difference between the daily mean as now calculated, add the normal deviation for the day, and subtract the base temperature to determine the number of degree-days for each day. Then have the maturation rate a function of the cumulative DD and time.

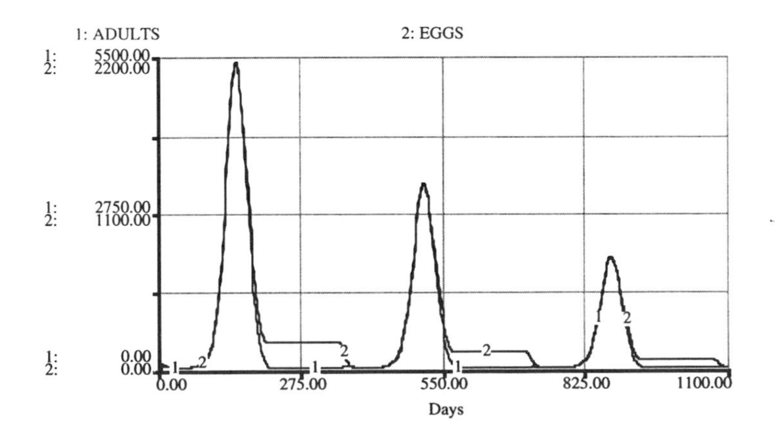

FIGURE 25.7

TWO-STAGE INSECT MODEL WITH DEGREE-DAY CALCULATION CONTROLLING THE MATURATION RATE

```
ADULTS(t) = ADULTS(t - dt) + (HATCH - ADULT_DEATHS) *
dt
INIT ADULTS = 100
INFLOWS:
HATCH = MOD_MATURE_RATE*EGGS
OUTFLOW:
ADULT_DEATHS =  ADULT_SURV_RATE*ADULTS

EGGS(t) = EGGS(t - dt) + (BIRTHS - DEATHS - HATCH) * dt
INIT EGGS = 0
INFLOWS:
BIRTHS =  EGG_LAY_RATE*ADULTS
OUTFLOW:
DEATHS =  MODEL_SURV_RATE*EGGS
HATCH = MOD_MATURE_RATE*EGGS

ADULT_SURV_RATE = (1-
EXP(LOGN(EXP_ADULT_SURV_RATE)*DT/EXP_ADULT_SURV_TIME))/
DT
BASE_TEMP =  48 {Fahrenheit; below this temperature egg
maturation and oviposition rates stop.}
DAILY_MEAN_TEMP = 47 + 27*SIN(2*PI/365*TIME)
DD = if DAILY_MEAN_TEMP+10 < BASE_TEMP then 0 else
(DAILY_MEAN_TEMP+10-BASE_TEMP)/2     {This equation
estimates the number of degree-days per day for egg
maturation rate or oviposition rate. Tmax, Tmin = +,-
10 = Tavg. }
EXP_ADULT_SURV_RATE =  .8  {The adult survival rate.}
EXP_ADULT_SURV_TIME = 1
EXP_SURV_FRAC =  .7      {Survival Rate: fraction of
those eggs surviving to maturity, per adult.}
MODEL_SURV_RATE =  (1-
EXP(LOGN(EXP_SURV_FRAC)*EXP_MATURE_RATE*DT))/DT {egg
mort. rate per day, 1/day}
MOD_MATURE_RATE =
EXP_MATURE_RATE*EXP(LOGN(EXP_SURV_FRAC)*EXP_MATURE_RATE
*DT)
EGG_LAY_RATE = GRAPH(DD)
(0.00, 0.00), (6.00, 0.158), (12.0, 0.263), (18.0,
0.443), (24.0, 0.593), (30.0, 0.735), (36.0, 0.863),
(42.0, 1.00), (48.0, 1.11), (54.0, 1.23), (60.0, 1.41)
```

```
EXP_MATURE_RATE = GRAPH(DD)
(0.00, 0.00), (2.00, 0.1), (4.00, 0.2), (6.00, 0.3),
(8.00, 0.4), (10.0, 0.5), (12.0, 0.6), (14.0, 0.7),
(16.0, 0.8), (18.0, 0.9), (20.0, 1.00)
```

25.3 Three-Stage Insect Model

In this section we return again to the basic two-stage insect model but introduce now a larval stage. The survival rate for larvae is given by

$$\text{LARV INSTANT SURV} = \text{EXP(LOGN(S2)*U2*DT)} \tag{1}$$

Unlike in the previous models, we must now distinguish the experimental egg maturation rate from the experimental larval maturation rate. The resulting model is shown in Figure 25.8. The result is plotted in Figure 25.9.

Perform a sensitivity analysis for the parameters A, U1, and U2, and interpret your results. Then, extend your model to incorporate the degree-day influence on maturation rates of eggs discussed earlier as well as a

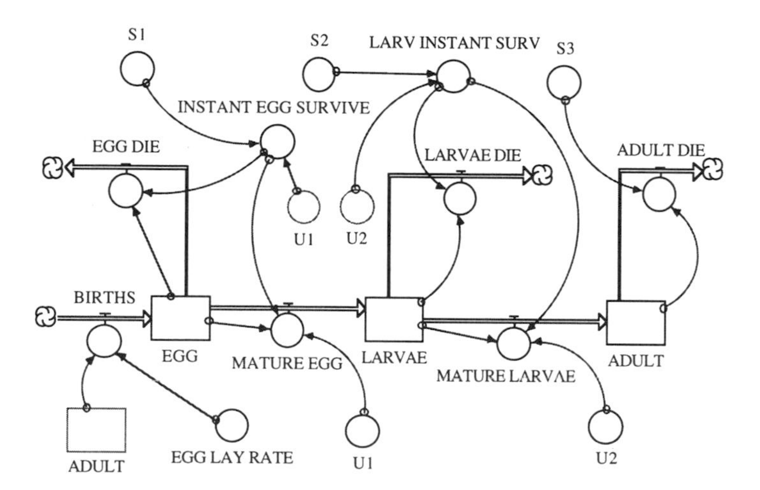

U = Model Maturation Rate; B = Model Death Rate; F = Experimental Maturation Rate; S = Experimental Survival Rate.

Figure 25.8

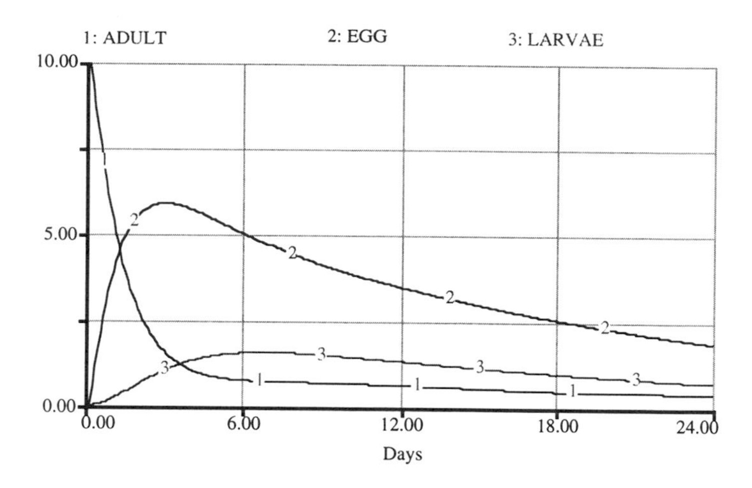

FIGURE 25.9

degree-day influence on maturation rates of larvae. How does the introduction of the larval stage affect your results?

In the following chapter and Chapter 30 we will combine our insight into the spread of a disease, modeled in Chapter 20, with the knowledge we gained here on insect population dynamics and the ways in which laboratory experiments can be used to set up dynamic models of biological systems.

THREE-STAGE INSECT MODEL

```
ADULT(t) = ADULT(t - dt) + (MATURE_LARVAE - ADULT_DIE)
* dt
INIT ADULT = 10 {Number of Adults}
INFLOWS:
MATURE_LARVAE = U2*LARV_INSTANT_SURV*LARVAE {Adults per
Time Period}
OUTFLOWS:
ADULT_DIE = (1 - S3^DT)*ADULT/DT

EGG(t) = EGG(t - dt) + (BIRTHS - MATURE_EGG - EGG_DIE)
* dt
INIT EGG = 0 {Number of Eggs}
INFLOWS:
BIRTHS = A*ADULT {Eggs per Time Period}
```

```
OUTFLOWS:
MATURE_EGG = U1*INSTANT_EGG_SURVIVE*EGG {Larvae per
Time Period}
EGG_DIE = (1-INSTANT_EGG_SURVIVE)*EGG/DT {Eggs per Time
Period}

LARVAE(t) = LARVAE(t - dt) + (MATURE_EGG -
MATURE_LARVAE - LARVAE_DIE) * dt
INIT LARVAE = 0 {Number of Larvae}
INFLOWS:
MATURE_EGG = U1*INSTANT_EGG_SURVIVE*EGG {Larvae per
Time Period}
OUTFLOWS:
MATURE_LARVAE = U2*LARV_INSTANT_SURV*LARVAE {Adults per
Time Period}
LARVAE_DIE = (1 - LARV_INSTANT_SURV)*LARVAE/DT {Larvae
per Time Period}

EGG LAY RATE = 0.6 {Eggs/adult/day. Try 1.112}
INSTANT_EGG_SURVIVE = EXP(LOGN(S1)*U1*DT)
LARV_INSTANT_SURV = EXP(LOGN(S2)*U2*DT)
S1 = .7 {Experimental egg-larvae survival rate,
1/stage.}
S2 = .8 {Experimental larvae-adult survival rate,
1/stage.}
S3 = .5 {Experimental adult survival per time step.}
U1 = .12 {1/Day, 1/Days per Stage}
U2 = .3 {1/Day, 1/Days per Stage}
```

26

Two Age-Class Parasite Model

Man is infested with internal . . . and is plagued by external parasites.
Charles Darwin,
Decent of Man, I, i. i. 12, 1871

This model shows how a disease spreads in an insect population, such as asexually reproducing aphids, consisting of two life stages—nymphs and adults. The model has two parts, one for the healthy population and one for the diseased population. Diseased nymphs can infect healthy nymphs and become diseased adults. Diseased adults cannot infect healthy adults or nymphs but can produce infected nymphs. Note how these features are expressed in the model by the appropriate flows and links.

Unlike the models of chapter 20, the infection coefficient is based on an exponential model.

$$\text{INFECTION COEF} = 1\text{-EXP}(-.3 * \text{NYMPHS} * \text{NYMPHS D}) \quad (1)$$

NYMPHS and NYMPHS D refer to the population sizes of healthy and infected nymph populations, respectively. The INFECTION RATE is calculated as the product of the INFECTION COEF, the number of healthy nymphs, and a model maturation rate for survivors, U1

$$\text{INFECTION RATE} = \text{U1*INFECTION COEF*NYMPHS} \quad (2)$$

with

$$\text{U1} = \text{F1*EXP(LOGN(S1)*DT*F1)} \quad (3)$$

where F1 is the experimental maturation rate and S1 is the dimensionless experimental egg survival fraction. When there are no sick nymphs or healthy nymphs, the probability of becoming infected equals zero. The specification of the INFECTION COEF translation variable is a purely empirical formulation, but it gives the correct value at the extremes, 0 when the number of diseased nymphs is zero or when the number of healthy nymphs is zero, and near 1 when either at least one of the stocks NYMPHS or NYMPHS D is very large.

Note well the specification of the MATURE function in the model that ensures the total rate of change from nymphs to adults here is still U1*NYMPHS,

$$\text{MATURE} = \text{U1*NYMPHS*(1–INFECTION COEF)} \quad (4)$$

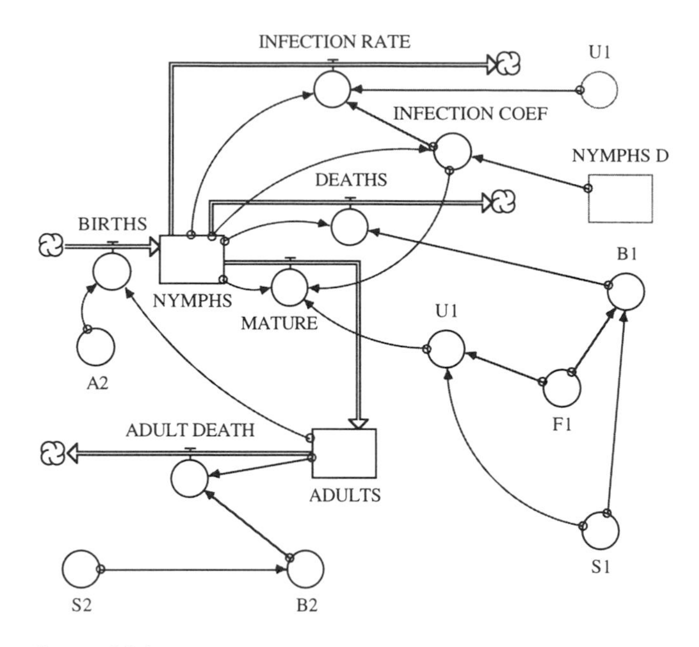

FIGURE 26.1

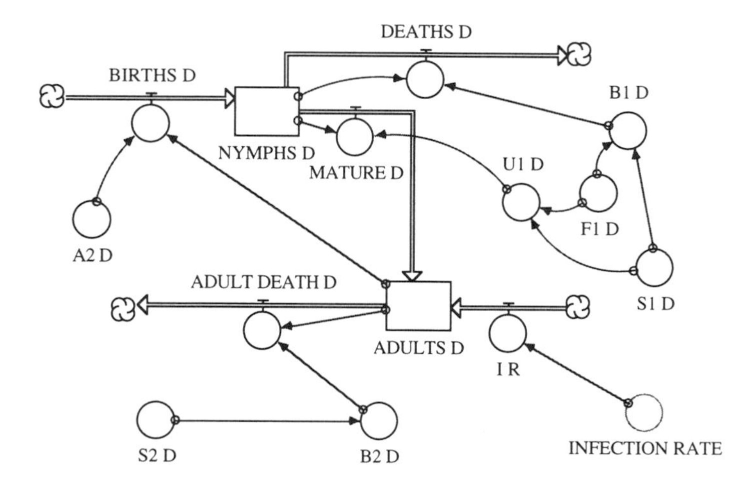

A2, A2 D = EXPERIMENTAL EGG LAYING FRACTION; B1, B1 D = EXPERIMENTAL MORTALITY FRACTION; B2, B2 D = EXPERIMENTAL ADULT MORTALITY FRACTION; F1, F1 D = EXPERIMENTAL MATURATION FRACTION; S1, S1 D = EXPERIMENTAL EGG SURVIVAL FRACTION; S2, S2 D = EXPERIMENTAL ADULT SURVIVAL FRACTION; U1, U1 D = MODEL MATURATION FRACTION.

FIGURE 26.2

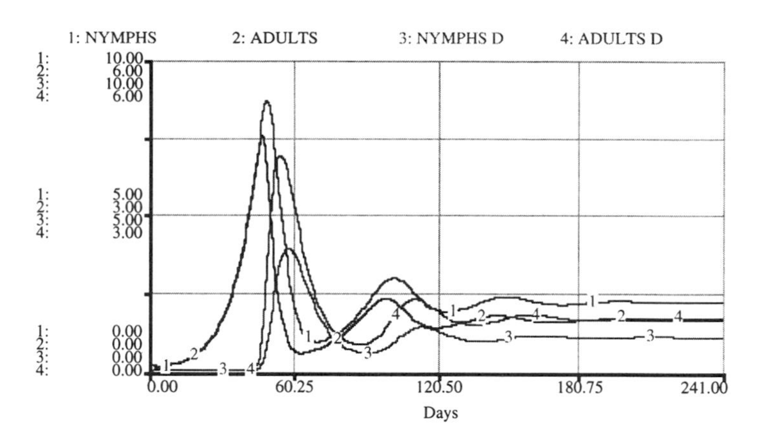

FIGURE 26.3

These newly diseased nymphs are converted to diseased adults rather than directly into diseased nymphs in a effort to reflect the fact that these nymphs, who contract rather than acquire the disease, have the normal nymph survival rate and, they are not able to convey the disease to other healthy nymphs (Figs. 26.1 and 26.2).

The graph shows in Figure 26.3 the proportions of healthy nymphs and adults, and the number of diseased nymphs and adults. Similar to the previous section of this chapter, we find distinct phases for the outbreak of a disease. Try varying the separate birth and death rates and the infection coefficient, and note the effect on the relative size of the healthy and diseased portions of the populations.

TWO-AGE CLASS PARASITE MODEL

```
ADULTS(t) = ADULTS(t - dt) + (MATURE - ADULT_DEATH) *
dt
INIT ADULTS = .10 {Individuals}
INFLOWS:
MATURE = U1*NYMPHS*(1-INFECTION_COEF) {Individuals per
Day}
OUTFLOWS:
ADULT_DEATH = ADULTS*B2 {Individuals per Day}

ADULTS_D(t) = ADULTS_D(t - dt) + (MATURE_D + I_R -
ADULT_DEATH_D) * dt
INIT ADULTS_D = .10 {Individuals}
```

```
INFLOWS:
MATURE_D = U1_D*NYMPHS_D {Individuals per Day}
I_R = INFECTION_RATE {Individuals per Day}
OUTFLOWS:
ADULT_DEATH_D = ADULTS_D*B2_D {Individuals per Day}

NYMPHS(t) = NYMPHS(t - dt) + (BIRTHS - DEATHS - MATURE
- INFECTION_RATE) * dt
INIT NYMPHS = 0 {Individuals}
INFLOWS:
BIRTHS = A2*ADULTS {Individuals per Day}
OUTFLOWS:
DEATHS = B1*NYMPHS {Individuals per Day}
MATURE = U1*NYMPHS*(1-INFECTION_COEF) {Individuals per
Day}
INFECTION_RATE = U1*INFECTION_COEF*NYMPHS {Individuals
per Day}

NYMPHS_D(t) = NYMPHS_D(t - dt) + (BIRTHS_D - DEATHS_D -
MATURE_D) * dt
INIT NYMPHS_D = 0 {Individuals}
INFLOWS:
BIRTHS_D = A2_D*ADULTS_D {Individuals per Day}
OUTFLOWS:
DEATHS_D = B1_D*NYMPHS_D {Individuals per Day}
MATURE_D = U1_D*NYMPHS_D {Individuals per Day}

A2 = .6 {Experimental laying fraction. Eggs per Adult
per Day}
A2_D = .35 {Experimental laying fraction. Eggs per
Adult per Day}
B1 = (1 - EXP(LOGN(S1)*F1*DT))/DT {Egg mortality
fraction, 1/DAY. Instantaneous survival fraction +
instantaneous mortality fraction = 1.}
B1_D = (1 - EXP(LOGN(S1_D)*F1_D*DT))/DT {Egg mortality
fraction, 1/DAY. Instantaneous survival fraction +
instantaneous mortality fraction = 1.}
B2 = (1-EXP(LOGN(S2)*DT/1))/DT {Adult mortality fraction,
1/day. One day = T = 1 = experimental period for which
adult mortality is measured. Instantaneous survival
fraction + instantaneous mortality fraction = 1.}
B2_D = (1-EXP(LOGN(S2_D)*DT/1))/DT {Adult mortality
fraction, 1/day. One day = T = 1 = experimental period
```

```
for which adult mortality is measured. Instantaneous
survival fraction + instantaneous mortality fraction =
1.}
F1 = .2 {Experimental maturation fraction, 1/DAY, i.e.,
20 eggs per 100 eggs mature each day, as noted in the
experiment. In other words, a surviving egg matures on
the average in 5 days under the experimental
conditions. T = 5 days.}
F1_D = .2 {Experimental maturation fraction, 1/DAY,
i.e., 20 eggs per 100 eggs mature each day, as noted in
the experiment. In other words, a surviving egg matures
on average in 5 days under the experimental conditions.
T = 5 days.}
INFECTION_COEF = 1-EXP(-.3*NYMPHS*NYMPHS_D)
S1 = .7 {Experimental egg survival fraction,
dimensionless, per stage. Stage = 1/F1, i.e., 70 eggs
per 100 eggs survive each 1/F1 days, as noted in the
experiment.}
S1_D = .5 {Experimental egg survival fraction,
dimensionless, per stage. Stage = 1/F1, i.e., 70 eggs
per 100 eggs survive each 1/F1 days, as noted in the
experiment.}
S2 = .8 {Experimental daily adult survival fraction per
stage, dimensionless.}
S2_D = .65 {Experimental daily adult survival fraction
per stage, dimensionless.}
U1 = F1*EXP(LOGN(S1)*DT*F1) {Model maturation fraction
for survivors, 1/DAY. Hatch rate = instantaneous
survival fraction*model maturation rate.}
U1_D = F1_D*EXP(LOGN(S1_D)*DT*F1_D) {Model maturation
fraction for survivors, 1/DAY. Hatch rate =
instantaneous survival fraction*model maturation rate.}
```

27

Monkey Travels[1]

> Well, little monkeys mine,
> I must go write;
> and so good-night.
>
> Jonathon Swift, Journal to Stella, 2 Nov. 171.
> edited by Harold Williams, Clarendon Press, Oxford 1948

In the rain forests of Peru, a small monkey, a tamarin, no larger than a squirrel, lives in groups of about a dozen and spends most of its time in the canopy. It eats, travels, and sleeps in the canopy, apparently fearing predators on the ground and in the air above the canopy. The canopy, for the most part, is so thick that vision is limited to a few meters at most. Tamarins travel on the average about 90 meters before finding a sufficient quantity of fruit to stop for a feeding bout. They have about 5 to 10 such bouts each day, before stopping to sleep in the largest local tree. An interesting question arises as to how these tamarins find food. One possibility is that they use their noses. This is the central assumption for the model of this chapter.[2]

Let us lay out a horizontal plane, which runs through the canopy of all the trees, and is divided into cells of unit width and height. The symbol X measures the distance from the origin in the center of the space, and Y measures the distance perpendicular to X. The odor strength is measured vertically, perpendicular to the X-Y plane. The tamarin troop is considered a point in this X-Y plane and moves in a straight line for the visual sight distance in the canopy. We assumed that this distance, called VISIBILITY in our model, was a constant 2.0 meters. At the end of this STEP DISTANCE, a new assessment is made and a new direction is chosen. This process is continued until the fruited tree is in sight of the troop. We assume that the typical tree could be seen at a distance of 17 meters from the tree. We also assume that the troop cannot travel exactly in the direction it desires to go because of the lack of suitable branches. So the chosen direction is modified in the model by a small random angle (Fig. 27.1).

---

[1]Based on P. Garber, and B. Hannon, Modeling Monkeys: A Comparison of Computer Generated and Empirical Measures, *International Journal of Primatology*, 14:827–852, 1993.

[2]We thank Prof. Paul Garber, an anthropologist at the University of Illinois who studies these animals in the field, for contributing to this model.

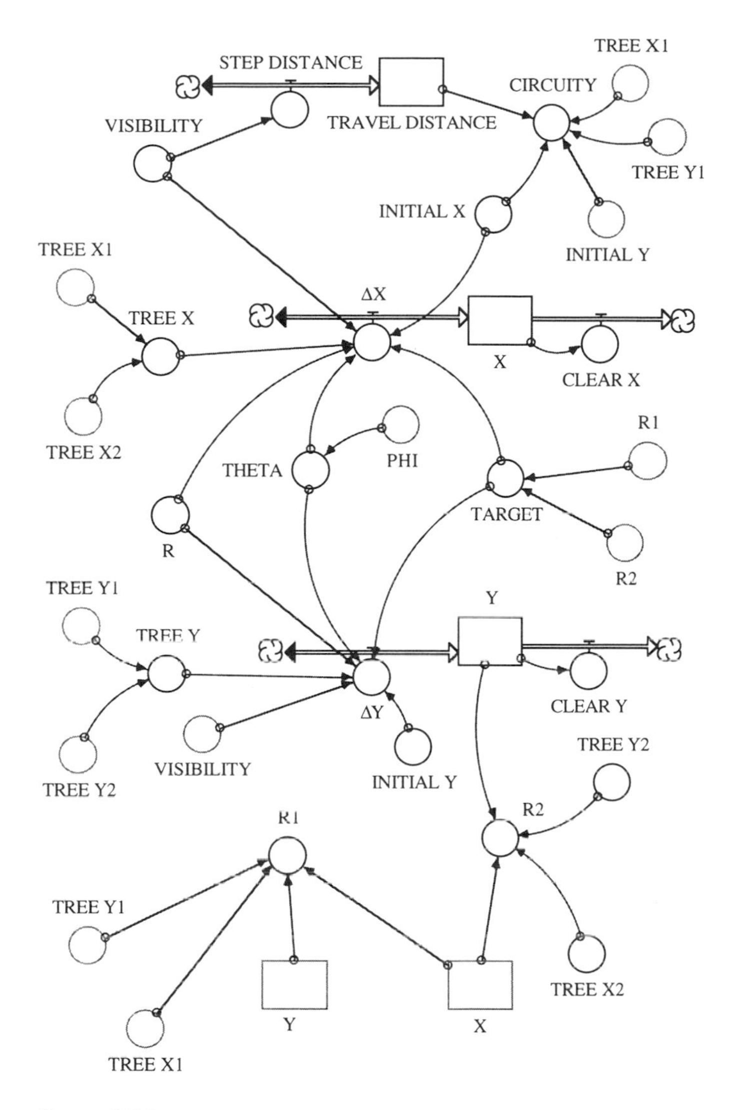

FIGURE 27.1

The location of the troop is determined in the diagram in Figure 27.1. Here, the state variables are the coordinates X and Y at which the troop is located. For example, the X variable is initialized as a specific position, such as the origin of the grid, and updated by ΔX.

The variables TREE X1, TREE Y1, and TREE X2 TREE Y2 draw the outline of each of the trees present in the plane. VISIBILITY is the specified step distance, and it is used, after the trees are drawn, to calculate the total travel distance from the starting point to the sighting point of the tree. The controller TARGET calculates the relation of the troop to the tree circle of a visual influence of 17 meter radius. When this circle is reached, it is assumed that the troop can proceed by sight:

$$\text{TARGET} = \text{IF (TIME} > 16) \text{ AND ((R1} \le 17) \text{ OR (R2} \le 17)\text{) THEN 0 ELSE 1,} \quad (1)$$

with R1 and R2 as the straight line distance from the troop to the center of tree 1 and tree 2, respectively. R1 and R2 are defined by

$$\text{R1} = \text{SQRT((X} - \text{TREE X1)\^{}2} + \text{(Y}-\text{TREE Y1)\^{}2)} \quad (2)$$

$$\text{R2} = \text{SQRT((X} - \text{TREE X2)\^{}2} + \text{(Y}-\text{TREE Y2)\^{}2).} \quad (3)$$

The straight line distances, R1 and R2 from the troop to the center of each tree are used for the calculation of the odor strength at the troop location. This calculation is performed in the STELLA modules in Figure 27.2. It is based on information on the WIND ANGLE DEGREES from which we derive the WIND ANGLE in radians, counterclockwise from the positive X axis (Fig 27.2). The WIND ANGLE is then used to calculate the troop location coordinate relative to the tree 1 in terms of the plume:

$$\text{X1} = \text{(X}-\text{TREE X1)*COS(WIND ANGLE)} + \text{(Y}-\text{TREE Y1)*SIN(WIND ANGLE)} \quad (4)$$

$$\text{Y1} = \text{(Y} - \text{TREE Y1)*COS(WIND ANGLE)} - \text{(X} - \text{TREE X1)*SIN(WIND ANGLE)} \quad (5)$$

X1 is measured parallel to wind, and Y1 perpendicular to X1.

To determine troop movement we need to further specify variables for the fruit ripenesses, tree height, and the wind speed and direction. The

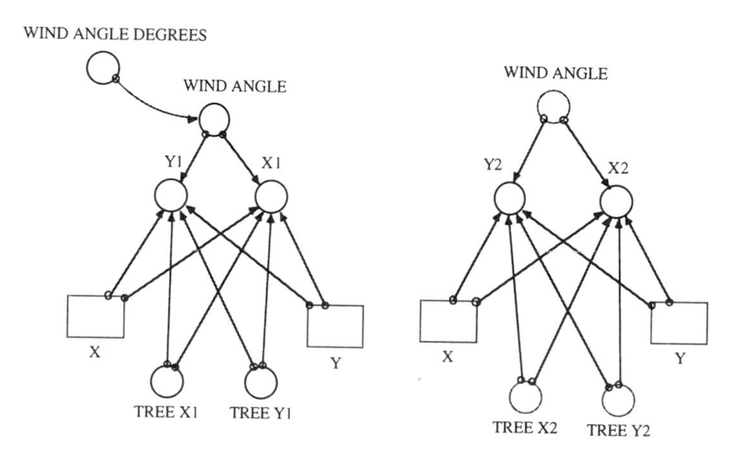

FIGURE 27.2

concentration of the odors caused by each tree is calculated separately (CONC 1 and CONC 2). The basis for the calculation are the air pollution stack emission equations. These are empirically derived equations that give the dispersion of a pollutant, such as sulfur dioxide, emitted from a chimney, depending on the chimney height, wind speed, and emission rate. The variables that begin with SIG_, X1, X2, Y1, Y2, RY, RZ, and R Y, R Z are the elements of these equations. The vertical emission velocity is set at zero (Fig. 27.3).

The behavioral part of the model is shown in the module in Figure 27.4. First, we add the odor concentrations from the two trees to form the total concentration at the troop location, CONCENTRATION. The problem here is to calculate the direction that optimally points up the odor "mountain," the direction of maximum odor increase. The direction is to be calculated on the minimum necessary information: the location and odor strength at three points. These three points are the current location and the last two locations of the troop. The monkeys must remember the locations and the odor concentrations at the last two points and compare them with their current situation. They then have the minimum necessary information.

Exactly how the monkeys would figure which direction is the maximum ascent up the odor mountain is not known, of course, nor do we know how we would do it if we were in the monkeys place. Perhaps each individual has a sort of stereo olfactory system and with their two nostrils can detect odor gradients, or perhaps they sense these gradients by just moving

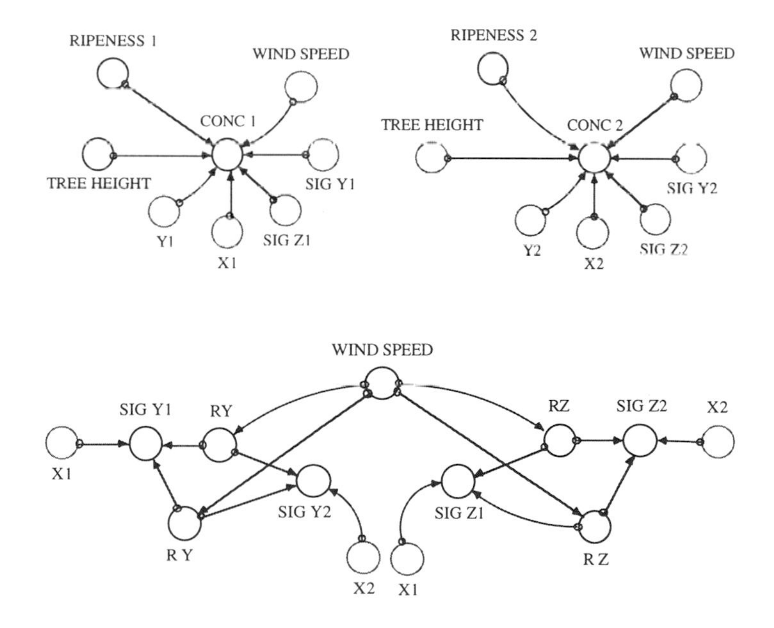

FIGURE 27.3

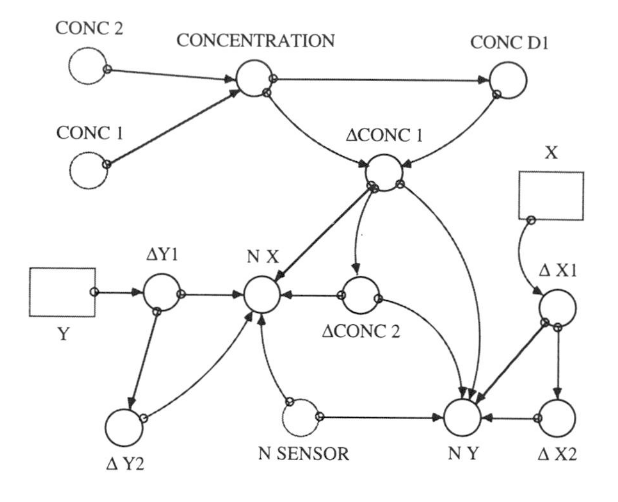

FIGURE 27.4

their heads? Here we assume that they develop a sense of the odor gradient by moving their entire body and refiguring the gradient periodically. If we were to use odor to find its source, we could ascend the odor mountain in the direction of steepest ascent but delete and not solve any computer programs. Yet to mimic the monkey pattern, we assume that they do solve the problem and we proceed with a mathematical solution. That solution requires that we calculate the normal to the plane formed by the last two locations of the troop and the current location. The projection of this normal vector onto the horizontal plane, components N X and N Y, gives the direction of the steepest descent. The opposite direction to this projection is the direction of steepest ascent, or, more precisely, a local approximation to the true direction of steepest ascent (Fig. 27.4).

The plane formed by these three points is approximately a plane tangent to the odor mountain in the area of the three points. It is possible for the normal to point down rather than up on certain occasions. This possibility requires the calculation of a detector for the normal vector's orientation, the N SENSOR. It is calculated from knowledge of the change in the directions taken in the past, PHI D1 and PHI D2, the two previous directions that the troop has taken. Still another possibility exists. The directions could be so accurate that the plane formed is perpendicular to the X-Y plane. In this case, the direction of the normal vector is ambiguous. Fortunately, we need to add a small random variation (RAND) to the angle chosen for the next direction (PHI), and that variation nearly always prevents the ambiguous case (Fig. 27.5). On the rare occasion when the ambiguous case does form, the troop will take a 90 degree turn to the right or left, randomly.

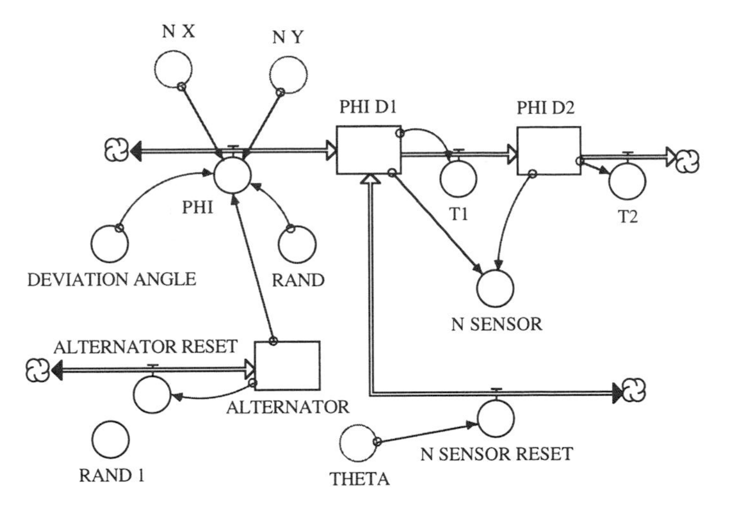

FIGURE 27.5

The model was calibrated to the best extent possible by comparing the averaged field data to the trip distances traveled in the model. We devised a CIRCULARITY measure that compares the distance actually traveled to the straight line distance between the starting point and the tree. We adjusted the random angle and visibility variables until the model circularity agreed with the values found in the field. Unfortunately, we have no data on the wind speed or direction in the canopy in the rain forest. We do have a reference on canopy wind speeds in another rain forest, and we used an approximate average daily value of 0.5 meters per second at the upper canopy level for our model.

Sometimes, the monkeys in the field will be stopped by a wide stream as they proceed in their search for food. In such cases they seem to abandon that search path and start out on another. Such a situation is not covered by the present model, nor is the model sensitive to absolute odor levels. It is assumed in the model that they can sense the slightest of odors and only relative odor strengths can influence their path. After these brief calibrations, we are as ready as possible to test the model.

First we set up the troop at the center, lower edge of the X-Y plane and turn the emissions from the left tree off by setting its "ripeness" to zero. We set the wind at 0.5 m/s from the north. The results are shown in Figure 27.6.

Next we turn the left tree on with the same odor emission rate as the right tree, and run again (Fig. 27.7).

Finally, we move the troop to a point nearly between the trees and run again (Fig. 27.8).

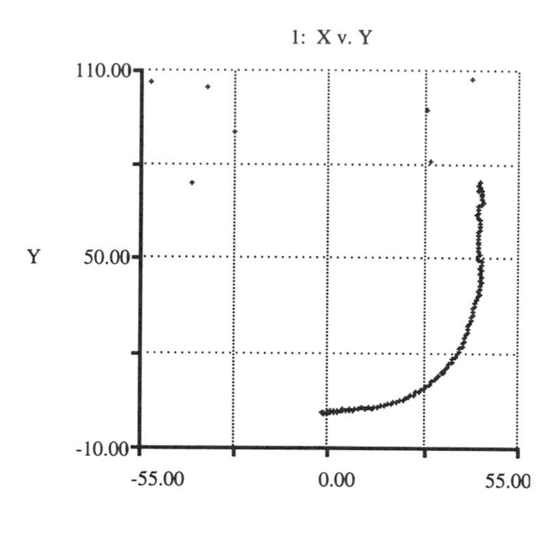

FIGURE 27.6

In the first case, the troop starts out heading to the right of the tree but gradually curves into the goal. With both trees emitting evenly, the troop starts out headed between the trees and rather suddenly turns up the odor "ridge" formed by the left tree and then proceeds along the ridge to that tree. The ridge effect is pronounced when the troop starts out between the trees. First, they head nearly downwind before circling back into the right tree.

One other interesting result seems worth mentioning. If we plot the odor concentration directly downwind of the tree, we find that, due to ground reflection of the odor, the peak odor is at the canopy level only near the emit-

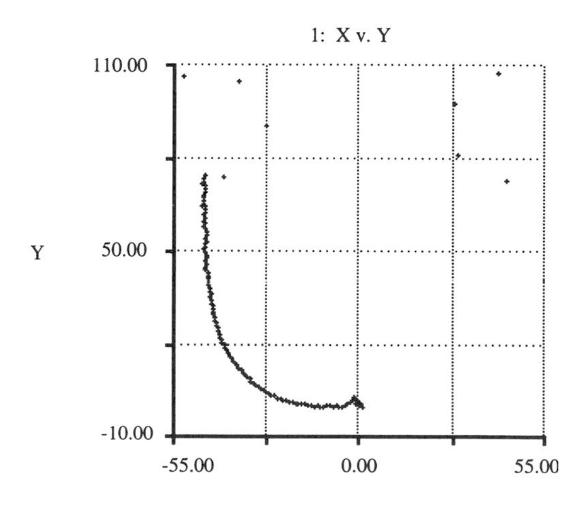

FIGURE 27.7

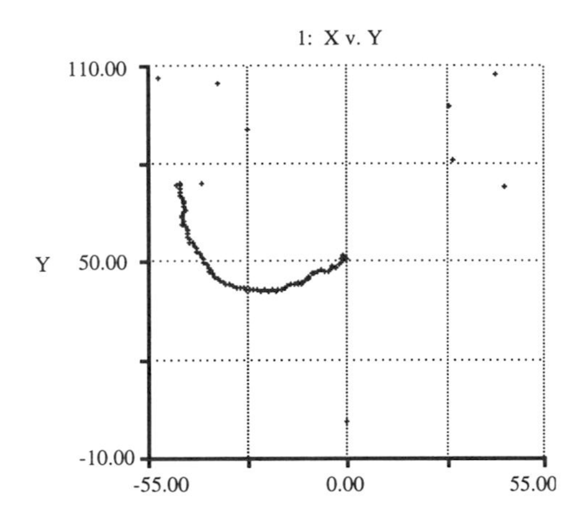

FIGURE 27.8

ting tree. As one moves away from that tree, the peak concentration falls both in intensity of course, but also in height. So, for very distant fruit trees the best place to be is not in the canopy but down the trunk, perhaps halfway to the ground. There is another monkey species that travels with the tamarin and that stays at about the mid-height level of the tree until the fruit tree is reached. These lower monkeys are in a better position to direct the combined troop in the early stages of the search for a distant tree with ripe fruit.

Although quite elaborate, our model has several shortcomings. The diffusion equations are for a uniform medium, usually air. In the actual forest, we do not have such a medium. The leaves and branches of the trees no doubt cause much more rapid mixing than an air-only medium. The leaves are not uniformly distributed vertically in the rain forest. Above the canopy, the model has a single medium (clear air). How would this affect the readings in the forest? What about temperature inversions caused by the forest? How would the model include such effects? Perhaps the monkeys use a mixture of odor tracking and memory. How would the memory effects be included? It is assumed that the monkeys can detect the difference between under-ripe and over-ripe fruit. How can this be modeled?

MONKEY TRAVELS

```
ALTERNATOR(t) = ALTERNATOR(t - dt) + (ALTERNATOR_RESET)
* dt
INIT ALTERNATOR = 1*RAND_1
INFLOWS:
ALTERNATOR_RESET = -2*ALTERNATOR
```

```
PHI_D1(t) = PHI_D1(t - dt) + (PHI + N_SENSOR_RESET -
T1) * dt
INIT PHI_D1 = 0 {Phi delayed. }
INFLOWS:
PHI = IF (TIME > 16) AND (N_X≥0) THEN ARCTAN(N_Y/N_X) +
ALTERNATOR*(RAND+DEVIATION_ANGLE) ELSE IF (TIME>16) AND
(N_X<0) THEN ARCTAN(N_Y/N_X) + PI +
ALTERNATOR*(RAND+DEVIATION_ANGLE) ELSE 0 {The chosen
angle for the next step.}
N_SENSOR_RESET = IF TIME =15 THEN THETA ELSE 0
DOCUMENT: (TIME ≥ 14) AND (TIME ≤ 15)
OUTFLOWS:
T1 = PHI_D1

PHI_D2(t) = PHI_D2(t - dt) + (T1 - T2) * dt
INIT PHI_D2 = 0
INFLOWS:
T1 = PHI_D1
OUTFLOWS:
T2 = PHI_D2

TRAVEL_DISTANCE(t) = TRAVEL_DISTANCE(t - dt) +
(STEP_DISTANCE) * dt
INIT TRAVEL_DISTANCE = 0 {The distance traveled from
the starting point of the troop to the circle of
influence of the tree.}
INFLOWS:
STEP_DISTANCE = IF TIME ≥ 15 THEN VISIBILITY*DT ELSE 0

X(t) = X(t - dt) + (ΔX - CLEAR_X) * dt
INIT X = 0 {The X horizontal distance to the troop from
the origin of the designated space.}
INFLOWS:
ΔX = IF TIME > 14 THEN VISIBILITY*COS(THETA)*TARGET
ELSE IF TIME = 13 THEN INITIAL_X ELSE IF TIME ≤ 11 THEN
(R*COS(TIME) + TREE_X)/DT ELSE 0
{Changing the X location of the troop of monkeys,
looking for trees with ripe fruit.}
OUTFLOWS:
CLEAR_X = IF TIME ≤ 12 THEN X/DT ELSE 0

Y(t) = Y(t - dt) + (ΔY - CLEAR_Y) * dt
INIT Y = 0 {The Y horizontal distance to the troop from
the origin of the designated space.}
```

```
INFLOWS:
ΔY = IF TIME > 14 THEN VISIBILITY*SIN(THETA)*TARGET
ELSE IF TIME = 13 THEN INITIAL_Y ELSE IF TIME ≤ 11 THEN
(R*SIN(TIME) + TREE_Y)/DT ELSE 0
{Changing the y location of the troop of monkeys,
looking for trees with ripe fruit.}
OUTFLOWS:
CLEAR_Y = IF TIME ≤ 12 THEN Y/DT ELSE 0

CIRCUITY = (TRAVEL_DISTANCE+9)/(SQRT((TREE_X1-
INITIAL_X)^2+
(TREE_Y1-INITIAL_Y)^2)-7.5)
CONCENTRATION = CONC_1+CONC_2 {The combined odor
concentration at the canopy top at the current troop
location.}
CONC_1 = IF X1 > 0 THEN EXP(-.5*(Y1/SIG_Y1)^2)*EXP(-
.5*(2*TREE_HEIGHT/SIG_Z1)^2)/(2*PI*SIG_Y1*SIG_Z1)*RIPEN
ESS_1/WIND_SPEED ELSE 0 {Plume and ground reflection
conc at top of tree; ug/m3}
CONC_2 = IF X2 > 0 THEN EXP(-.5*(Y2/SIG_Y2)^2)*EXP(-
.5*(2*TREE_HEIGHT/SIG_Z2)^2)/(2*PI*SIG_Y2*SIG_Z2)*RIPEN
ESS_2/WIND_SPEED ELSE 0 {See note in CONC 1}
CONC_D1 = DELAY(CONCENTRATION,1)
DEVIATION_ANGLE = PI/16
INITIAL_X = 0 {Specified X coordinate of starting point
for the troop in the designated space.}
INITIAL_Y = 0
N_SENSOR = IF (PHI_D2<PI) AND ((PHI_D1>PHI_D2) AND
(PHI_D1<(PHI_D2+PI))) THEN 1 ELSE IF (PHI_D2≥PI) AND
((PHI_D1>PHI_D2) OR (PHI_D1<(PHI_D2-PI))) THEN 1 ELSE
-1 {Rel. angle sizes sets normal up or down.}
N_X = (-Δ_Y2*ΔCONC_1 + ΔCONC_2*ΔY1)*N_SENSOR {The
projection of the normal vector to the plane by the
last two steps, on the X-Y plane. This is the X
component of the max. rise vector.}
N_Y = (-ΔCONC_2*Δ_X1 + Δ_X2*ΔCONC_1)*N_SENSOR {The y
component of the projection of the normal vector to the
plane by the last two steps, onto the X-Y plane. This
is the y component of the max. rise vector}
R = 17.5 {At this radius (meters), the fruit can be
seen by the troop.}
R1 = SQRT((X-TREE_X1)^2 + (Y-TREE_Y1)^2) {The
straight line distance from the troop to the center
of tree 1.}
```

```
R2 = SQRT((X-TREE_X2)^2 + (Y-TREE_Y2)^2) {The straight
line distance from the troop to the center of tree 2.}
RAND = 2*PI*RANDOM(1,0)*.05*1
RAND_1 = IF RANDOM(0,2) > 1 THEN 1 ELSE -1
RIPENESS_1 = 0 {Specified fruit odor emission rate,
ug/sec, from tree 1.}
RIPENESS_2 = 7000 {Specified fruit odor emission rate,
ug/sec, from tree 2.}
RY = IF WIND_SPEED < 2 THEN .4 ELSE IF (WIND_SPEED ≥ 2)
AND (WIND_SPEED ≤ 5) THEN .36 ELSE .32
RZ = IF WIND_SPEED < 2 THEN .4 ELSE IF (WIND_SPEED ≥ 2)
AND (WIND_SPEED ≤ 5) THEN .33 ELSE .22 {For the odor
dispersion equation. See SIG Y.}
R_Y = IF WIND_SPEED < 2 THEN .9 ELSE IF (WIND_SPEED ≥
2) AND (WIND_SPEED ≤ 5) THEN .86 ELSE .78
R_Z = IF WIND_SPEED < 2 THEN 2 ELSE IF (WIND_SPEED ≥ 2)
AND (WIND_SPEED ≤ 5) THEN .86 ELSE .78
SIG_Y1 = IF X1 > 0 THEN RY*X1^R_Y ELSE 0
SIG_Y2 = IF X2 > 0 THEN RY*X2^R_Y ELSE 0
SIG_Z1 = IF X1 > 0 THEN RZ*X1^R_Z ELSE 0
SIG_Z2 = IF X2 > 0 THEN RZ*X2^R_Z ELSE 0
TARGET = IF (TIME > 16) AND ((R1 ≤ 17) OR (R2 ≤ 17) )
THEN 0 ELSE 1 {Senses relation of troop to tree circle
of visual influence (17 m. radius). When this circle is
reached, it is assumed that the troop can proceed by
sight.}
THETA = IF (TIME ≥ 14) AND (TIME ≤ 15) THEN -
2*PI*RANDOM(1,0) ELSE PHI
TREE_HEIGHT = 20 {Specified height of the ripe fruit
and canopy of tree, meters.}
TREE_X = IF TIME < 6 THEN TREE_X1 ELSE IF TIME < 12
THEN TREE_X2 ELSE 0
TREE_X1 = -45 {X location of the first tree with ripe
fruit in the general coordinate system for the
designated space.}
TREE_X2 = 45 {X coordinate of tree 2 in the designated
space.}
TREE_Y = IF TIME < 6 THEN TREE_Y1 ELSE IF TIME < 12
THEN TREE_Y2 ELSE 0
TREE_Y1 = 90 {Y location of the first tree w/ ripe
fruit in the general coord. system for the designated
space.}
TREE_Y2 = 90 {Y coordinate of tree 2 in the designated
space.}
```

```
VISIBILITY = IF TIME ≤16 THEN .1 ELSE 1 {Specified step
rate for the troop. 2 meters per minute. Actual tamarin
data from P. Garber: 300 meter max. distance or 86
minute max. travel time. Avg. travel time = 26 min.}
WIND_ANGLE = WIND_ANGLE_DEGREES*PI/180 {Wind angle in
radians, counterclockwise from + x axis.}
WIND_ANGLE_DEGREES = -90 {Specified wind angle,
degrees; counterclockwise from the + x axis, from the
origin.}
WIND_SPEED = .5
X1 = (X-TREE_X1)*COS(WIND_ANGLE) + (Y-
TREE_Y1)*SIN(WIND_ANGLE) {The troop location
coordinate, relative to tree 1, in terms of the plume;
X1 parallel to wind, Y1 perpendicular to X1.}
X2 = (X-TREE_X2)*COS(WIND_ANGLE) + (Y-
TREE_Y2)*SIN(WIND_ANGLE) {The troop location
coordinate, relative to tree 2, in terms of the plume;
X2 parallel to wind, Y2 perpendicuar to X2.}
Y1 = (Y - TREE_Y1)*COS(WIND_ANGLE) - (X -
TREE_X1)*SIN(WIND_ANGLE) {The troop location
coordinate, relative to the tree 1, in terms of the
plume; X1 parallel to wind, Y1 perpendicular to X1.}
Y2 = (Y-TREE_Y2)*COS(WIND_ANGLE) - (X-
TREE_X2)*SIN(WIND_ANGLE) {The troop location
coordinate, relative to tree 2, in terms of the plume;
X2 parallel to wind, Y2 perpendicular to X2.}
ΔCONC_1 = CONCENTRATION-CONC_D1
ΔCONC_2 = DELAY(ΔCONC_1,1)
ΔY1 = Y-DELAY(Y,1)
Δ_X1 = X-DELAY(X,1) {See note ΔY2.}
Δ_X2 = DELAY(Δ_X1,1) {See note ΔY2.}
Δ_Y2 = DELAY(ΔY1,1) {The Y location of the troop 2
steps ago. We now have 3 odor readings at 3 known
locations. This forms a plane, the normal to which
gives us the local approximation of the direction of
steepest ascent up the odor hill.}
```

28

Biosynchronicity

With Heav'nly touch of instrumental sounds
In full harmonic number joind.

1667 Milton,
Paradise Lost, iv. 687

We know from the observation of fireflies in India[1] that whole trees containing tens of thousands of these insects begin to blink in unison shortly after dusk. Casual observation of the sounds of night-time insects around the common suburban home shows us that audio-synchronous behavior occurs. We assume some group reproduction advantage is conferred by such synchrony. The pacemakers in the heart of every mammal are really the synchronous pulsing of thousands of special cells, yielding sufficient signal to cause a muscle action. What process allows such synchronization? How can these organisms, and even cells, conform to each other's signal?

Apparently, Charles Peskin of New York University first successfully formulated a model of this process. Nearly any electrical engineer would understand the process immediately, as he began with an electrical analogy: a resistance and a capacitor in parallel, subjected to a steady electrical current input. The voltage builds on the capacitor to a limit, when it suddenly discharges and the voltage drops quickly to zero, only to repeat the process. This system is analogous to a weight hanging on a damper, subjected to a constant extending force. When the damper reaches its limiting extension, the velocity of the weight become zero.

In the model (Fig. 28.1), we represent four fireflies by four cells of a spatial model. The state variable for each cell is the voltage, V, or brightness, of the cell. We choose different starting values for each cell in order to get the cells initially out of phase. When the TOTAL V, total voltage, or in our case the brightness of the combined flashes, exceeds a given level, each cell boosts its own voltage by a small amount (BOOST STRENGTH). That boost voltage hastens the time for the receiver to reach its peak and discharge. The result of this process, providing the parameters are within certain limits, is that the cells gradually merge into synchronous firing.

Even if the cells are significantly different, they can still be driven into synchrony, but the reaction times of the cells (DT) must be shorter, the

[1]S. Strogatz, and I. Stewart, Coupled Oscillators and Biological Synchronization, *Scientific American,* Dec:102–109, 1993

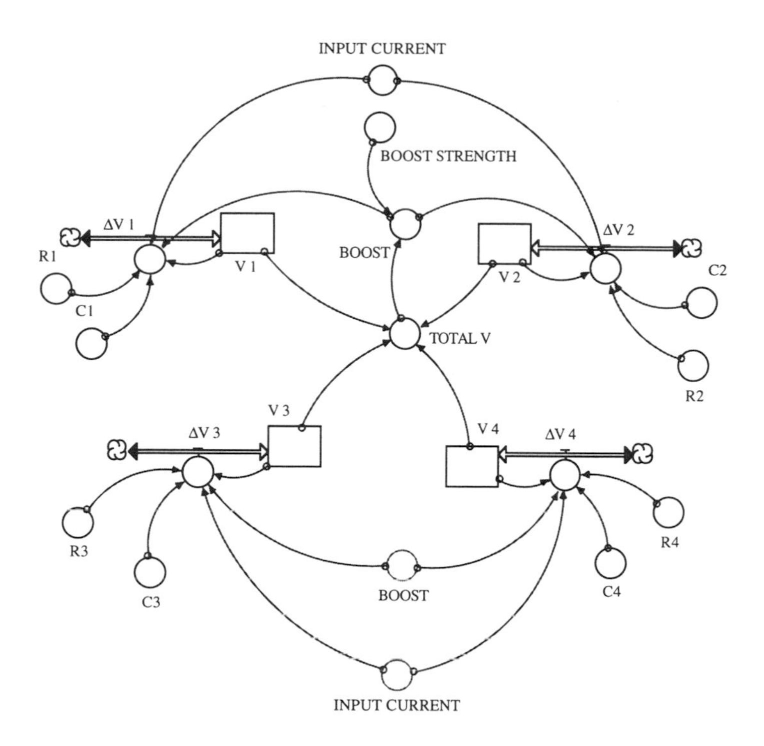

FIGURE 28.1

booster signal strength must be larger, and the resulting synchronous frequency is greater. These requirements probably mean that the organisms must be more highly developed and more energy demanding if they are not identical, and, therefore, the organic forms of these cells would tend to become genetically more similar through time, at least with regard to their flashing mechanisms. Such an evolutionary direction would tend to reduce or even avoid the biological cost of faster reaction times and high strength booster signals.

In this model we represent four such systems subjected to a common steady INPUT CURRENT. Reactance (R) is the value of the resistance constant and capacitance (C) is the measure of the capacitor constant.

The governing differential equation for each firefly is a simple linear one:

$$C^*dV/dt + V/R = \text{INPUT CURRENT} = 0.15$$

that is solved for each of the four fireflies.

As shown in the two graphs in Figure 28.2, TOTAL V is erratic at first but grows in size and fluctuation as the cells become synchronized.

Experiment with this model. See what the first DT and BOOSTER STRENGTH settings give when the cells are not the same. Change the input

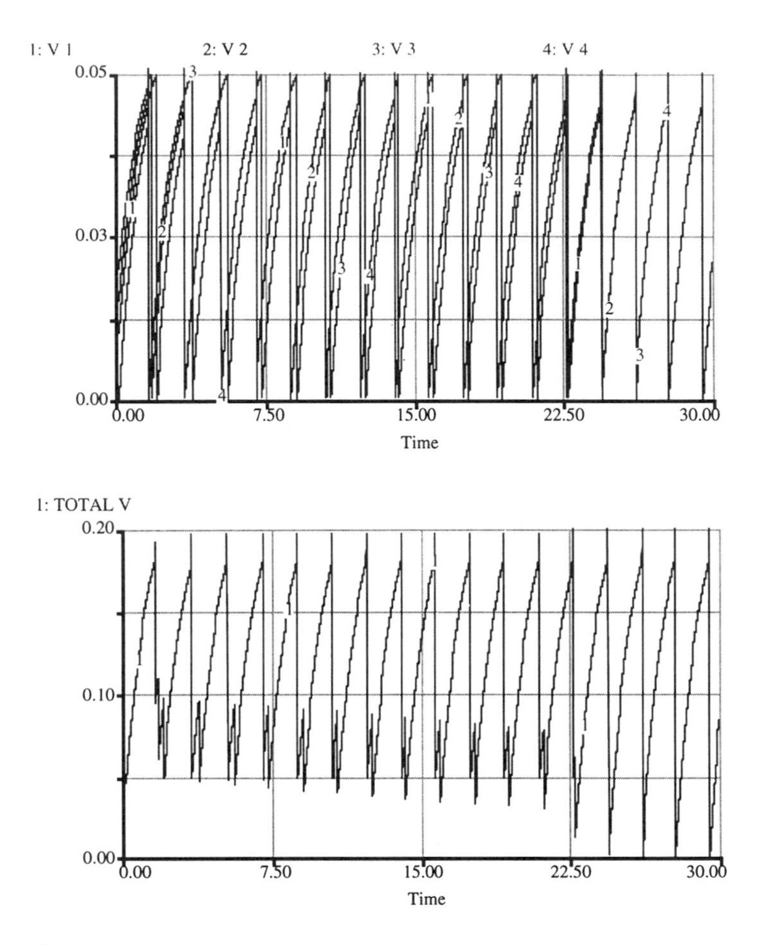

FIGURE 28.2

current level. All sorts of interesting results can occur. You should even find, as Strogatz and Stewart report, that there are a variety of steady conditions in which the peaks are not synchronous.

Think of those thousands of fireflies. Is each interconnected to only one other? Or does each connect to only its nearest neighbors, in a kind of regional association, with the regions eventually acting as a single unit that must swing somehow into synchrony with other regional units, a kind of hierarchy of synchronous behavior? Or does each somehow average the peak of the signal from the whole and adjust its own flash initiation? Add more cells and try out these, and perhaps other ideas. You will find no doubt that this glorious process of nature is not as complex a process as you might have thought.

We may imagine that the group flashing is the behavior of individuals whose reproductive chances are enhanced by synchronous behavior, to attract distant mates into the proximity from a long distance. But what happens when the attracted mate is close? The appeal of belonging to a group is lost, the act of mating is not a many-to-one relationship, it is the ultimate in one-to-one behavior. Perhaps those in the vicinity of attractee stop flashing once they realize the situation. Perhaps once those locals notice the newcomer of the opposite sex, only these locals begin to flash synchronously with that of the attractee, while the bulk of the group continues to flash in synchrony. With enough attractees the group synchrony falls apart, destroying the potential for success.

BIOSYNCHRONICITY

```
V_1(t) = V_1(t - dt) + (ΔV_1) * dt
INIT V_1 = .01
INFLOWS:
ΔV_1 = if V_1≤.05 then (INPUT_CURRENT-V_1/R1)/C1+BOOST
else -V_1/dt

V_2(t) = V_2(t - dt) + (ΔV_2) * dt
INIT V_2 = .015
INFLOWS:
ΔV_2 = if (V_2 ≤ .05) then (INPUT_CURRENT-
V_2/R2)/C2+BOOST else -V_2/dt

V_3(t) = V_3(t - dt) + (ΔV_3) * dt
INIT V_3 = 0
INFLOWS:
ΔV_3 = if V_3≤.05 then (INPUT_CURRENT-V_3/R3)/C3+BOOST
else -V_3/dt

V_4(t) = V_4(t - dt) + (ΔV_4) * dt
INIT V_4 = .02
INFLOWS:
ΔV_4 = if (V_4 ≤ .05) then (INPUT_CURRENT-
V_4/R4)/C4+BOOST else -V_4/dt

BOOST = IF (TOTAL_V > .18) THEN BOOST_STRENGTH ELSE 0
BOOST_STRENGTH = .001/dt
C1 = 3
C2 = 3
C3 = 3
```

```
C4 = 3
INPUT_CURRENT = .15
R1 = .4
R2 = .4
R3 = .4
R4 = .4
TOTAL_V = V_1+V_2+V_3+V_4
```

Part 6

Multiple Population Models

29

Wildebeest Model

There is another species of wild ox, called by the natives gnoo.

Johan Forster,
A Voyage Around the World, I. 83, 1777

The wildebeest model is developed for a wildlife park on an African grassland. Wildebeest are eaten by lions and are shot by park rangers attempting to manage the ecosystem. The data on the number of lions in the park is thought to be between 400 and 600. We choose 500 for the initial runs of the model. The rate at which wildebeest are killed by the lions varies between 4.5 and 3.8 wildebeest per lion in the first 6 years of the data (Fig. 29.1)

The wildebeest were shot by the park rangers during the first 4 years. The calf survival rate varies from 0.35 to 0.48 during the first 6 years. Females constitute 70% of the total population. For the first 6 years there are population census estimates for the calves, yearlings, 2 year olds, and adults (Fig. 29.2).

The lion population of our model should depend on the availability of food, which in our case is mostly the noncalf population of wildebeest. This dependency is given as a graph in Figure 29.3.

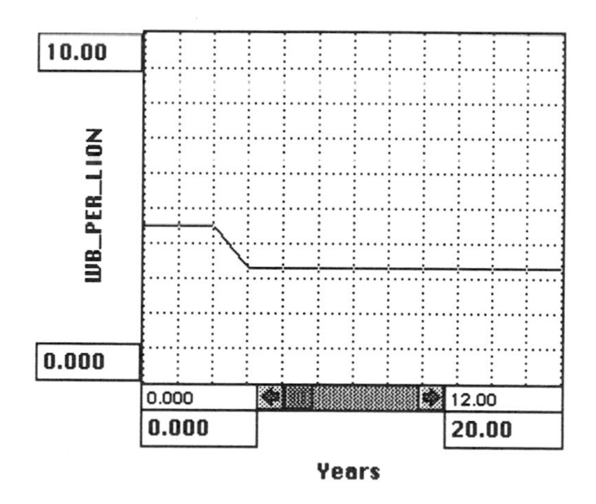

FIGURE 29.1

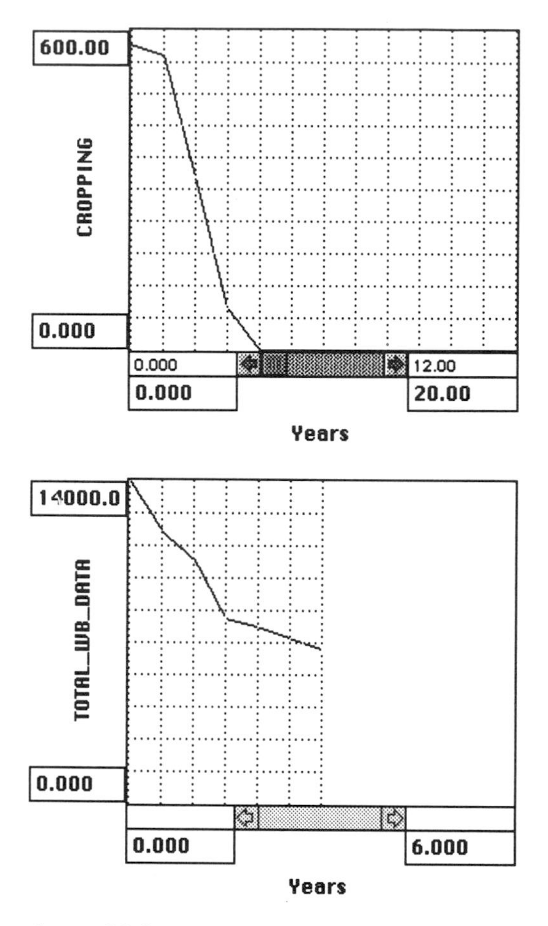

FIGURE 29.2

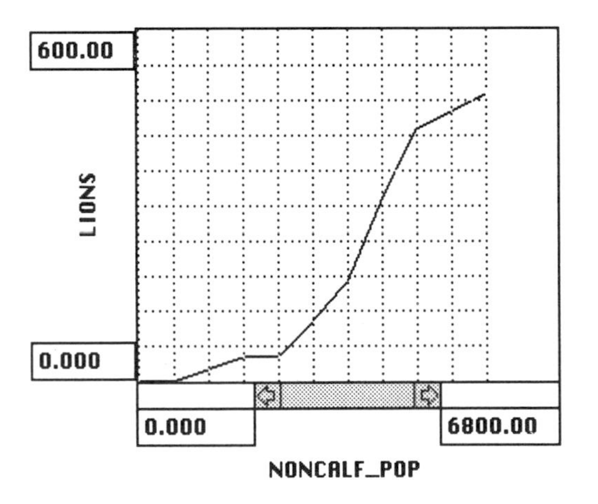

FIGURE 29.3

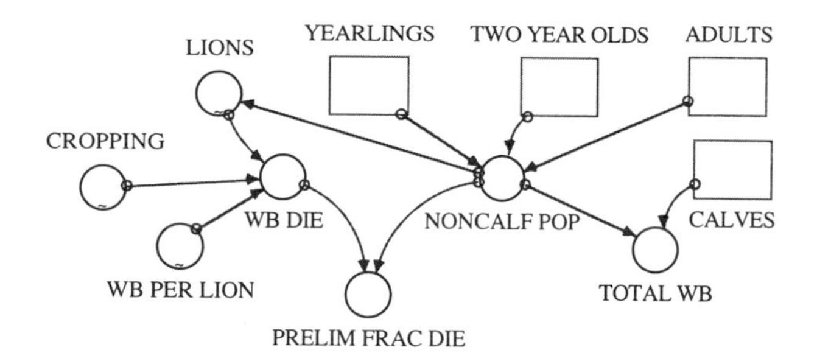

FIGURE 29.4

Given the population of YEARLINGS, TWO YEAR OLDS, and ADULTS, we can calculate an average death rate assumed to apply to these "cohorts." This death rate is assumed to be distinct from that for calves (Fig. 29.4).

The death rate for calves is calculated from data on calf survival rates Q as (1 – Q). We fit the data[1] by adjusting the calf survival rate Q downward in the early years for the given estimates. This is done in the graph in Figure 29.5.

We are now ready to put the pieces of our model together to assess changes in the wildebeest population given the influences of lions and rangers (Fig. 29.6).

One can now change all of the parameters in turn to find the sensitivity of each, that is, the kill rate, the number of lions, the calf survival rate, the

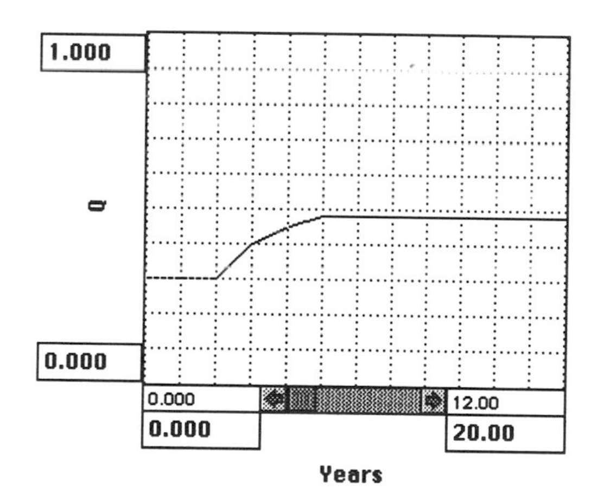

FIGURE 29.5

[1]See A.M. Starfield, and A.L. Bleloch, *Building Models for Conservation and Wildlife*, New York: MacMillan, 1986, Chap. 2.

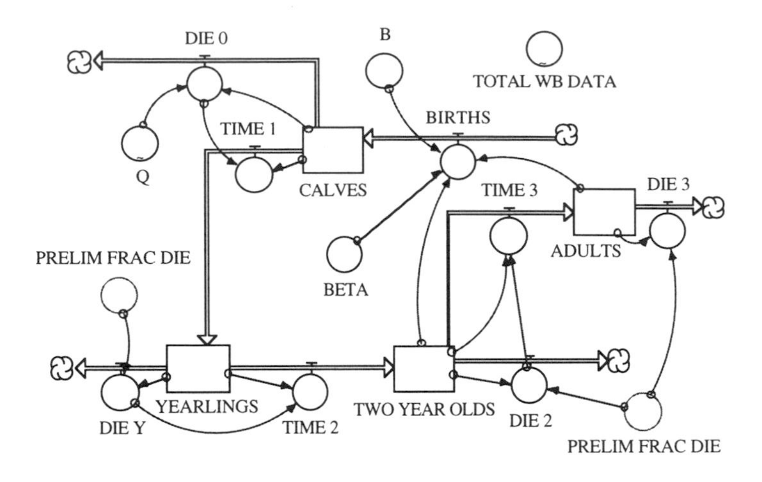

FIGURE 29.6

fecundity coefficients BETA and B for the 2 year olds and for adults, respectively, and the female fraction (Fig. 29.7).

Run the model yourself and test for the sensitivity of the parameters. How would you modify this model to correct the lion population and eating rate to bring the wildebeest herd to a 5000 animal steady state?

Lions are just one factor in the system affecting the wildebeest population. The amount of rainfall influences grass height during the calving season, which in turn influences the predation rate on calves. Making predictions about the rainfall variations can determine the expected appropriate cropping rates for wildebeests and lions. Introduce seasonal rainfall into your model and find a (non-zero) lion population and eating rate that prevent the wildebeest population from crashing.

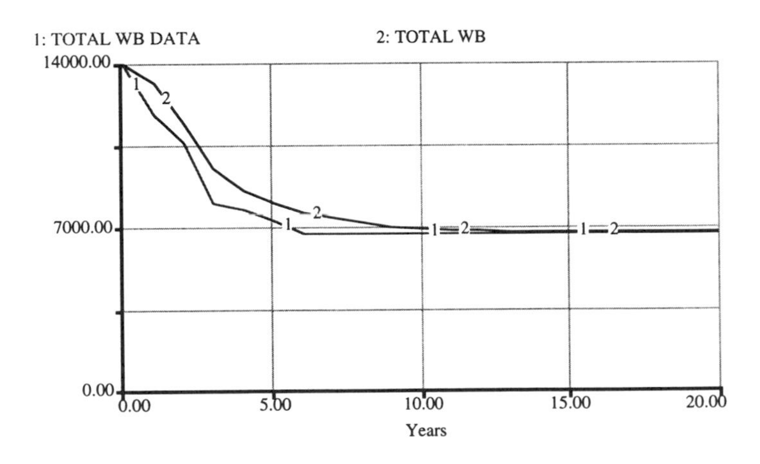

FIGURE 29.7

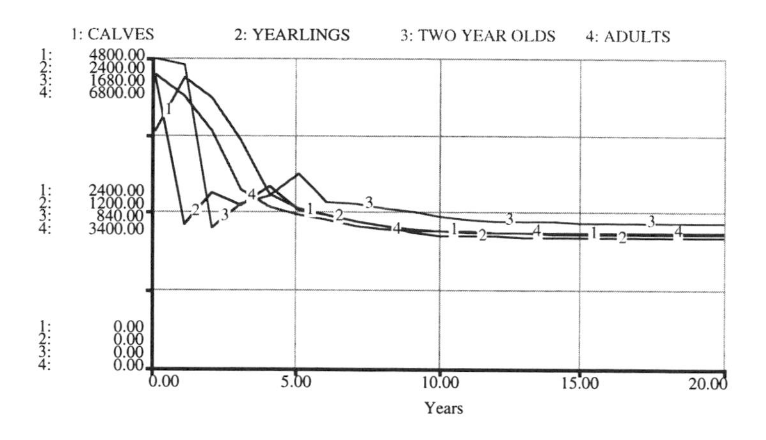

FIGURE 29.7 Continued

WILDEBEEST MODEL

```
ADULTS(t) = ADULTS(t - dt) + (TIME_3 - DIE_3) * dt
INIT ADULTS = 6440 {Individuals}
INFLOWS:
TIME_3 = TWO_YEAR_OLDS-DIE_2 {Net graduation of the 2
year olds to adulthood. Individuals per Time Period}
OUTFLOWS:
DIE_3 = ADULTS*PRELIM_FRAC_DIE {Individuals per Time
Period}

CALVES(t) = CALVES(t - dt) + (BIRTHS - TIME_1 - DIE_0)
* dt
INIT CALVES = 3640 {Individuals}
INFLOWS:
BIRTHS = (BETA*TWO_YEAR_OLDS + B*ADULTS)*.7 {.7 is the
fraction of the adult population that is female.
Fecundity is greater among the adults. Individuals per
Time Period}
OUTFLOWS:
TIME_1 = CALVES - DIE_0 {Individuals per Time Period}
DIE_0 = (1 - Q)*(CALVES) {The survival rate q improves
with time, reducing the death rate of the calves.
Individuals per Time Period}

TWO_YEAR_OLDS(t) = TWO_YEAR_OLDS(t - dt) + (TIME_2 -
TIME_3 - DIE_2) * dt
INIT TWO_YEAR_OLDS = 1680 {Individuals}
```

```
INFLOWS:
TIME_2 = YEARLINGS-DIE_Y {The net number of yearlings
graduating to 2 year olds. Individuals per Time Period}
OUTFLOWS:
TIME_3 = TWO_YEAR_OLDS-DIE_2 {Net graduation of the 2
year olds to adulthood. Individuals per Time Period}
DIE_2 = TWO_YEAR_OLDS*PRELIM_FRAC_DIE {Individuals per
Time Period}

YEARLINGS(t) = YEARLINGS(t - dt) + (TIME_1 - TIME_2 -
DIE_Y) * dt
INIT YEARLINGS = 2240 {Individuals}
INFLOWS:
TIME_1 = CALVES - DIE_0 {Individuals per Time Period}
OUTFLOWS:
TIME_2 = YEARLINGS-DIE_Y {The net number of yearlings
graduating to 2 year olds. Individuals per Time Period}
DIE_Y = YEARLINGS*PRELIM_FRAC_DIE {Individuals per Time
Period}

B = .92 {Given adult female fecundity. Births per
Female}
BETA = .3 {Given 2 year old female fecundity. Births
per Female}
NONCALF_POP = YEARLINGS + TWO_YEAR_OLDS + ADULTS
PRELIM_FRAC_DIE = WB_DIE/NONCALF_POP {A death rate
assumed to apply to yearlings, 2 year olds and adults.
Individuals per Individuals per Time Period}
TOTAL_WB = CALVES+NONCALF_POP {Individuals}
WB_DIE = ( WB_PER_LION*LIONS + CROPPING*1) {The number
of wildebeest eaten and shot in the game park per
year.}
CROPPING = GRAPH(TIME)
(0.00, 572), (1.00, 550), (2.00, 320), (3.00, 78.0),
(4.00, 0.00), (5.00, 0.00), (6.00, 0.00), (7.00, 0.00),
(8.00, 0.00), (9.00, 0.00), (10.0, 0.00), (11.0, 0.00),
(12.0, 0.00), (13.0, 0.00), (14.0, 0.00), (15.0, 0.00),
(16.0, 0.00), (17.0, 0.00), (18.0, 0.00), (19.0, 0.00),
(20.0, 0.00)
LIONS = GRAPH(NONCALF_POP)
(0.00, 0.00), (680, 0.00), (1360, 20.0), (2040, 40.0),
(2720, 40.0), (3400, 100), (4080, 170), (4760, 310),
(5440, 430), (6120, 460), (6800, 490)
```

```
Q = GRAPH(TIME)
(0.00, 0.3), (1.00, 0.3), (2.00, 0.3), (3.00, 0.4),
(4.00, 0.45), (5.00, 0.48), (6.00, 0.48), (7.00, 0.48),
(8.00, 0.48), (9.00, 0.48), (10.0, 0.48), (11.0, 0.48),
(12.0, 0.48), (13.0, 0.48), (14.0, 0.48), (15.0, 0.48),
(16.0, 0.48), (17.0, 0.48), (18.0, 0.48), (19.0, 0.48),
(20.0, 0.48)
TOTAL_WB_DATA = GRAPH(time)
(0.00, 14000), (1.00, 11800), (2.00, 10600), (3.00,
8000), (4.00, 7700), (5.00, 7200), (6.00, 6700)
WB_PER_LION = GRAPH(TIME)
(0.00, 4.50), (1.00, 4.50), (2.00, 4.50), (3.00, 3.30),
(4.00, 3.30), (5.00, 3.30), (6.00, 3.30), (7.00, 3.30),
(8.00, 3.30), (9.00, 3.30), (10.0, 3.30), (11.0, 3.30),
(12.0, 3.30), (13.0, 3.30), (14.0, 3.30), (15.0, 3.30),
(16.0, 3.30), (17.0, 3.30), (18.0, 3.30), (19.0, 3.30),
(20.0, 3.30)
```

30

Nicholson–Bailey Host–Parasitoid Model[1]

You knot of Mouth-Friends: . . . Most smiling, smooth, detested Parasites.
William Shakespeare,
Timon of Athens, iii. vi. 104, 1607

In Chapter 26 we modeled the spread of a parasitic infection in an insect population of two life stages. The focus of that model was the spread of the infection. Therefore, we did not pay any explicit attention to the fate of the parasitoid. In this chapter, however, we model explicitly the interactions between the host and the parasitoid populations. Rather than setting up our model in terms of population sizes, we specify host–parasitoid interactions in terms of population densities.

In order to model the host–parasitoid interactions, we abstract away from the fact that only specific life cycle stages exhibit those interactions. After you work through this chapter, you may want to refine the model to account for the fact that, for example, adult parasitoids lay their eggs in the pupae of hosts, but not in the eggs of their hosts or with the larvae or adults.

Denote, respectively, H(t) and P(t) as the host and parasitoid densities in time period t, and F(H(t), P(t)) as the fraction of hosts that is not parasitized. Then

$$H(t + 1) = \lambda(H(t)) * F(H(t), P(t)) \quad (1)$$

$$P(t + 1) = C * H(t) * [1 - F(H(t), P(t))], \quad (2)$$

where $\lambda(H(t))$ is the host growth rate and C is the parasitoid fecundity.

Let us assume that the fraction of hosts that become parasitized depends on the density-dependent rate of encounter of parasitoids and hosts. Encounters occur randomly, allowing us to invoke the law of mass action that we already discussed in Chapters 6 and 20. Accordingly, the number of encounters of hosts HE with parasitoids is

$$HE(t) = A*H(t)*P(t), \quad (3)$$

where A is the searching efficiency of the parasitoids.

[1]This chapter follows L. Ederstein-Keshet, *Mathematical Models in Biology,* New York: Random House, 1988, pp. 79–85, and D. Brown and P. Rothery, *Models in Biology: Mathematics, Statistics and Computing,* New York: Wiley, 1993, pp. 399–406.

Unlike the models of the spread of a disease from an infected to a non-immune population, subsequent encounters of individuals in the two populations do not alter the rate at which parasitoids are propagated. Therefore, we need to modify the law of mass action to account for the fact that only the first encounter of hosts and parasitoids is significant in propagating the parasitoid. Once a host carries the parasitoid's eggs, subsequent encounters with parasitoids will not change the number of parasitoid progeny that hatch from the host. We need only to distinguish between hosts that had no encounter and hosts that had at least one encounter with parasitoids.

The Poisson distribution describes the occurrence of such discrete, random events as encounters of hosts and parasitoids. We can make use of the Poisson probability distribution to calculate the probability that there is no attack of parasitoids on a host within a certain time period. In general, therefore,

$$P(X) = \frac{EXP\left(-\frac{HE(t)}{P(t)}\right)\left(\frac{HE(t)}{P(t)}\right)^X}{X!} \tag{4}$$

is the probability of X attacks. This probability depends on the average number of attacks in the given time interval, HE/P. From equation (3) we know

$$HE(t)/P(t) = A^*P(t). \tag{5}$$

Thus, for zero attacks by the parasitoids, equation (4) yields

$$P(0) = \frac{EXP(-A * P(t))(A * P(t))^0}{0!} = \frac{EXP(-A * P(t)) * 1}{1} = EXP(-A * P(t)). \tag{6}$$

Equations (1) and (2) can therefore be rewritten as

$$H(t + 1) = H(t) * \lambda(N(t)) * EXP(-A*P(t)) \tag{7}$$

$$P(t + 1) = H(t) * [1 - EXP(-A*P(t))]. \tag{8}$$

Let us also assume that without parasitoids, the hosts will grow toward a carrying capacity K set by the environment. To capture growth of the host population up to a density H(t) = K and decline of the host population for H(t) > K, we replace in equation (7) the growth rate l(H(t)) with

$$\lambda(H(t)) = EXP\left(R * \left(1 - \frac{H(t)}{K}\right)\right), \tag{9}$$

where R is the maximum host growth rate. Thus, the equation governing the size of the host population in time t+1 becomes

$$H(t + 1) = H(t) * EXP\left(R * \left(1 - \frac{H(t)}{K}\right) - A * P(t)\right), \tag{10}$$

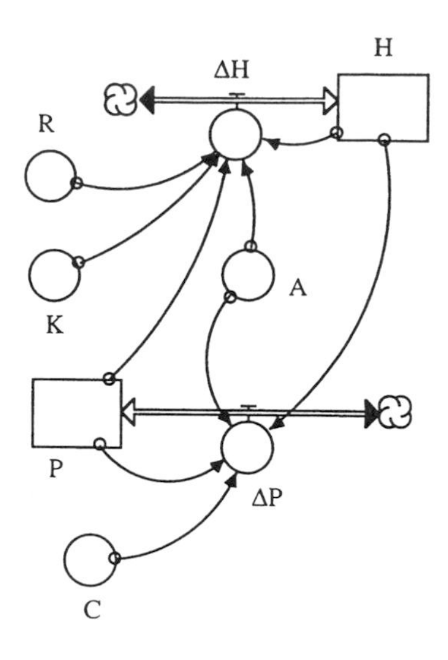

FIGURE 30.1

and after subtracting the respective state variables in time period t from equations (8) and (10), we have a set of differential equations that capture the change of host and parasitoid densities from time period t to t + 1:

$$\Delta H(t) = H(t) * EXP\left(R * \left(1 - \frac{H(t)}{K}\right) - A * P(t)\right) - H(t) \tag{11}$$

$$\Delta P(t) = H(t) * [1 - EXP(-A * P(t))] - P(t). \tag{12}$$

We can now see the dynamics exhibited by this model (Fig. 30.1). These equations describing changes in the host and parasitoid densities can yield a variety of results, from the production of steady-state conditions for the host and parasitoid, to their lock in a limit cycle, to chaos.

The graphs in Figures 30.2 to 30.4 result from the parameters and initial conditions in Table 30.1 and a DT = 1.

Try reducing the DT. For the graphs in Figures 30.2 to 30.4 it is set at one. A smaller DT fetches a completely different answer. What is going on here? Is the DT of 1.00 required by the host or the parasitoid? The Nicholson–Bailey model views t = 1 as one generation, and all the dynamics for 1 DT go on inside that time period of 1. It is as though the

TABLE 30.1 Model parameters and initial conditions

Graph	Description	R	A	K	H(t=0)	P(t=0)
1	Steady state	0.50	0.20	14.5	10.00	1.00
2	Limit cycle	2.00	0.20	21.5	10.00	1.00
3	Chaos	2.65	0.20	25.0	10.00	1.00

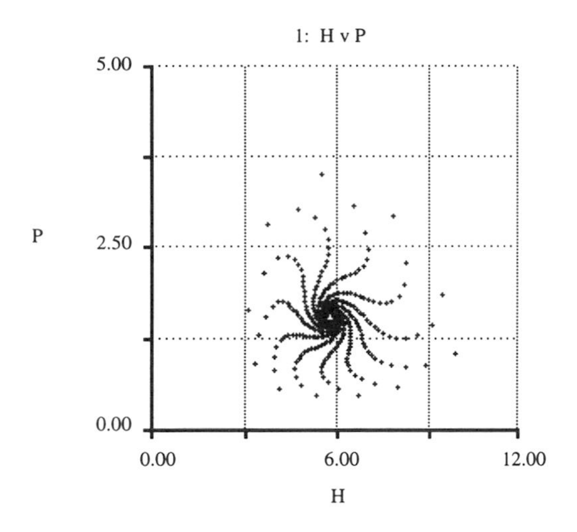

FIGURE 30.2 The approach of the populations to the steady state.

whole new generation of the two populations is formed just before the beginning of that generation. So, in the sense of this model, a DT less than 1 has no meaning. Do a sensitivity analysis on the initial values of H and P, on R, A, and K.

We have modeled in this chapter one type of species interaction that is almost exclusively found among insects. Typically, both the parasitoid and host have a number of life cycle stages—eggs, larvae, pupae, and adults—

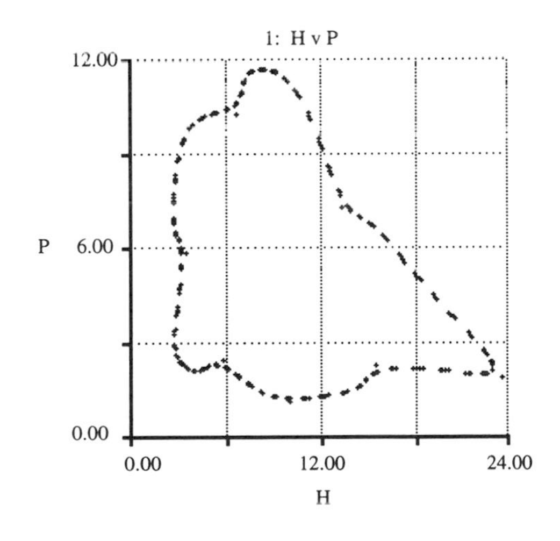

FIGURE 30.3 The populations in a limit cycle.

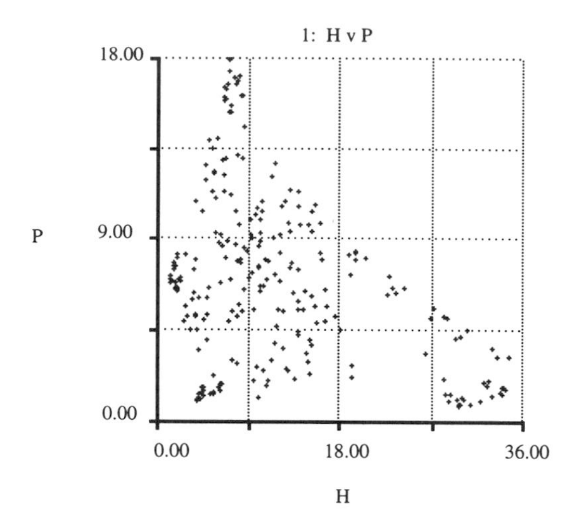

FIGURE 30.4 The populations in chaos.

and their interaction is limited to a subset of these. Can you modify the model to account for the fact that it is typically only the larvae of the host that get parasitized by adult parasitoids? How does this disaggregation of the parasitoid and host population affect your results? Can you find parameters and initial values that generate alternatively steady state, limit cycles, or chaos? What is the appropriate DT to use here and how are the results affected by its choice?

NICHOLSON-BAILEY HOST-PARASITOID MODEL

```
H(t) = H(t - dt) + (ΔH) * dt
INIT H = 10
INFLOWS:
ΔH = H*EXP(R*(1-H/K)-A*P)-H

P(t) = P(t - dt) + (ΔP) * dt
INIT P = 1
INFLOWS:
ΔP = C*H*(1-EXP(-A*P))-P

A = .2
C = 1
K = 14.5
R = .5
```

31

Diseased and Healthy Immigrating Insects

And migrant tribes these fruitful shorelands hail.

Joel Barlow, 1807
The Columbiad, ii. 178.

This chapter builds on the models developed in Chapter 26 by distinguishing two cohorts of a population infected with a disease. The two populations modeled here are insects that suffer from a disease that increases mortality for the infected nymphs and adults, and also decreases their egg-laying rate. Unlike the previous chapters we assume here two populations of insects, living in two fields. One of the fields has generally better living conditions than the other, although the current year's carrying capacities are randomly generated, and there is some overlap in the ranges within which the carrying capacity fluctuates. Carrying capacity has a direct effect on birth rates.

The carrying capacities of the two fields are defined as

$$\begin{aligned} K1 = {} & \text{IF CARRY R1} > .666 \text{ THEN } 2 \\ & \text{ELSE IF CARRY R1} < .333 \text{ THEN } .5 \\ & \text{ELSE } 1 \end{aligned} \tag{1}$$

and

$$\begin{aligned} K2 = {} & \text{IF CARRY R2} > .666 \text{ THEN } 4 \\ & \text{ELSE IF CARRY R2} < .333 \text{ THEN } 1 \\ & \text{ELSE } 2, \end{aligned} \tag{2}$$

respectively, with CARRY R1 and CARRY R2 as random numbers between 0 and 1. These random numbers are calculated in the module in Figure 31.1 with

$$\begin{aligned} \text{R COUNT1} = {} & \text{IF MOD(TIME,52)} \\ & = 0 \text{ THEN RANDOM(0,1)/DT ELSE } 0 \end{aligned} \tag{3}$$

and

$$\begin{aligned} \text{R COUNT2} = {} & \text{IF MOD(TIME,52)} \\ & = 0 \text{ THEN RANDOM(0,1)/DT ELSE } 0. \end{aligned} \tag{4}$$

where MOD is the built in function that computes the remainder (modulo) of a division—in our case the division of TIME by 52. We make use of the

MOD function here to set up a recurring counter. Note that with some DT values, whose fractional representation does not have n^2 in the denominator, STELLA rounds the remainder in the MOD function; so the re-starting values of R COUNT1 and R COUNT2 for each new year are not exactly zero.

When over-crowding develops, healthy adult insects leave their home field and join the other population. Furthermore, it is assumed that 10% of healthy adults migrate under all circumstances. Changes in population sizes are no longer only dependent on births and on deaths but additionally on migration.

The model is composed of the following additional modules. The first captures the population dynamics of healthy insects in the first field. The structure and workings of this module are analogous to the ones outlined in Chapter 26 with the additional feature of migration from and to that region.

The second module is set up to calculate the change in nymph and adult population in field 1 that are affected by the disease (Fig. 31.2).

A virtually identical second set of these modules capture the dynamics of the populations in field 2. Parameters relevant to both healthy and diseased insects in both fields are calculated in the modules in Figure 31.3. They include:

- A calculation of the total number of adults in each fields, ALL ADULTS 1 and ALL ADULTS 2
- The ratio of the total number of adults in each region to the carrying capacity of the respective region, FRXNL CAP1 and FRXNL CAP2

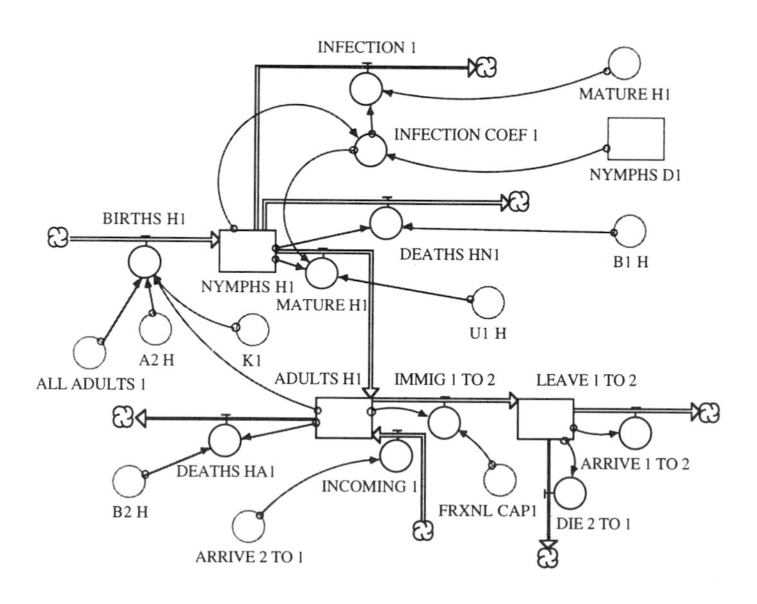

FIGURE 31.1

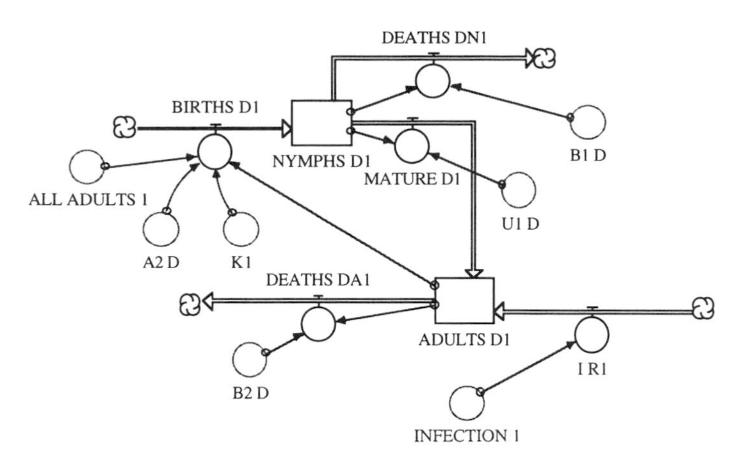

FIGURE 31.2

- Experimental maturation rates for healthy and diseades insects, F1 H, F1 D
- Model maturation rates U1 H, U1 D
- Experimental laying rates A2 H, A2 D
- Experimental daily adult survival fractions per stage, S2 H , S2 D
- Adult mortality rates B1 H, B1 D, B2 H, B2 D

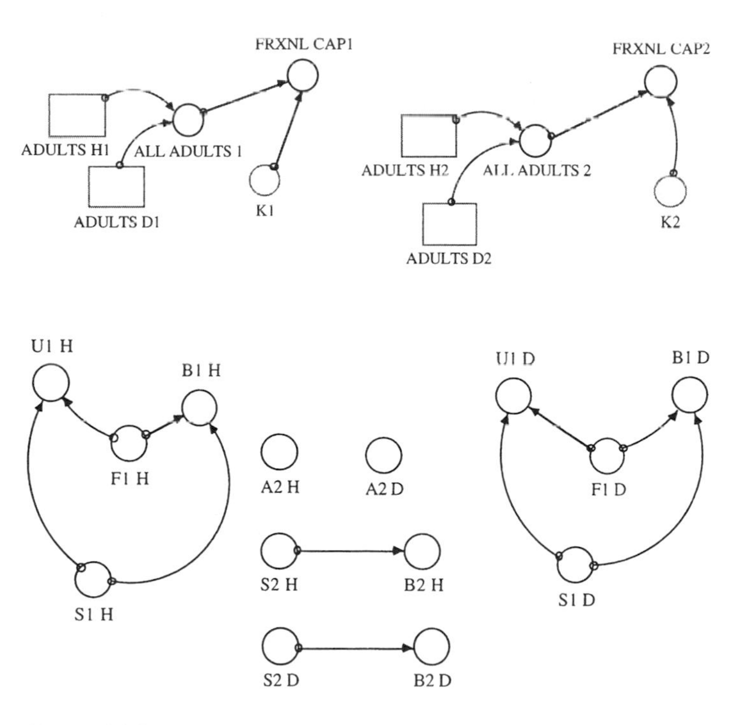

FIGURE 31.3

The latter are calculated using the exponential functions discussed in Chapter 26:

$$B1\ H = (1 - EXP(LOGN(S1\ H)*F1\ H*DT))/DT \quad (5)$$

$$B1\ D = (1 - EXP(LOGN(S1\ D)*F1\ D*DT))/DT \quad (6)$$

$$B2\ H = (1 - EXP(LOGN(S2\ H)*DT/1))/DT \quad (7)$$

$$B2\ D = (1 - EXP(LOGN(S2\ D)*DT/1))/DT \quad (8)$$

The graph in Figure 31.4 shows the combined result of the population dynamics due to natural increases and deaths as well as migration. Note how ensuing runs with the same parameters and initial conditions are significantly different. Why is this difference occurring? The diseased insects are at a "disadvantage" here. What would you change within realm of the biologically likely to favor the diseased population? Is it possible that by studying insect population dynamics from an ecological perspective we can provide a more useful means for biological control? Can you implement such a control in the model?

Over the long run, the population of each field is clearly responding to changes in the local carrying capacity. Both fields take a while to build up numbers from the low start, 0.1 adults each of healthy and diseased. Surprisingly, in the run we have graphed neither field's total population hugs the carrying capacity very well, but in other runs it sometimes did. Clearly, there are other factors at work limiting population besides carrying capacity. While we've required that 10% of healthy adults migrate, we aren't seeing the diseased population expand to fill that gap. Introduce additional factors such as seasonality into the population model.

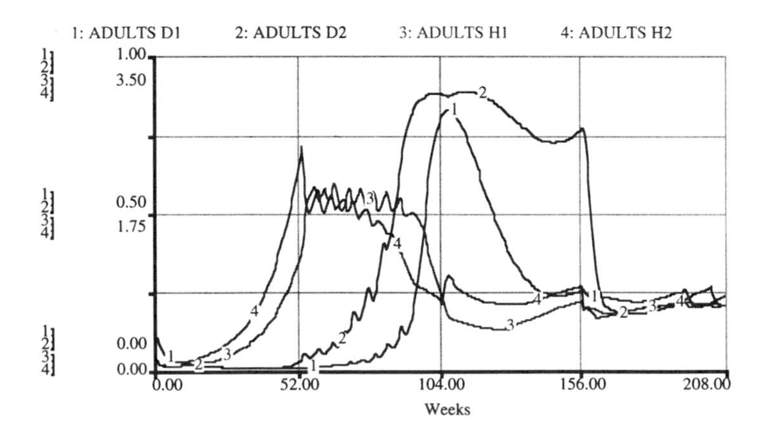

FIGURE 31.4

DISEASED AND HEALTHY IMMIGRATING INSECTS

```
ADULTS_D1(t) = ADULTS_D1(t - dt) + (MATURE_D1 + I_R1 -
DEATHS_DA1) * dt
INIT ADULTS_D1 = .1 {Initial diseased adults.}
INFLOWS:
MATURE_D1 = U1_D*NYMPHS_D1 {Individuals per Time
Period}
I_R1 = INFECTION_1 {Individuals per Time Period}
OUTFLOWS:
DEATHS_DA1 = ADULTS_D1*B2_D {Individuals per Time
Period}

ADULTS_D2(t) = ADULTS_D2(t - dt) + (MATURE_D2 + I_R2 -
DEATHS_DA2) * dt
INIT ADULTS_D2 = .1 {Initial diseased adults.}
INFLOWS:
MATURE_D2 = U1_D*NYMPHS_D2 {Individuals per Time
Period}
I_R2 = INFECTION_2 {Individuals per Time Period}
OUTFLOWS:
DEATHS_DA2 = ADULTS_D2*B2_D {Individuals per Time
Period}

ADULTS_H1(t) = ADULTS_H1(t - dt) + (MATURE_H1 +
INCOMING_1 - DEATHS_HA1 - IMMIG_1_TO_2) * dt
INIT ADULTS_H1 = .1 {Initial healthy adults}
INFLOWS:
MATURE_H1 = U1_H*NYMPHS_H1*(1-INFECTION_COEF1)
{Individuals per Time Period}
INCOMING_1 = ARRIVE_2_TO_1 {Individuals per Time
Period}
OUTFLOWS:
DEATHS_HA1 = ADULTS_H1*B2_H {Individuals per Time
Period}
IMMIG_1_TO_2 = IF ADULTS_H1 - (.1*ADULTS_H1 +
.9*FRXNL_CAP1)
> 0 THEN (.1*ADULTS_H1 + .9*FRXNL_CAP1) ELSE IF
ADULTS_H1 > 0 THEN ADULTS_H1 ELSE 0 {Individuals per
Time Period; Only healthy adults migrate. At least 10%
of the healthy adults always migrate. Under the noted
conditions the 10% healthy and an additional fraction
of the healthy adults, based empirically on the total
```

```
number of adults, also migrate. Note well the order of
the nested IF statement; the first one is checked first
and if the condition holds, the first statement is
executed and the program goes no further. Otherwise,
all the adults flee. This same statement is also true
of the adults in the other field.}

ADULTS_H2(t) = ADULTS_H2(t - dt) + (MATURE_H2 +
INCOMING_2 - DEATHS_HA2 - IMMIG_2_TO_1) * dt
INIT ADULTS_H2 = .1 {Initial healthy adults}
INFLOWS:
MATURE_H2 = U1_H*NYMPHS_H2*(1-INFECTION_COEF2)
{Individuals per Time Period}
INCOMING_2 = ARRIVE_1_TO_2 {Individuals per Time
Period}
OUTFLOWS:
DEATHS_HA2 = ADULTS_H2*B2_H {Individuals per Time
Period}
IMMIG_2_TO_1 = IF ADULTS_H2 - (.1*ADULTS_H2 +
.9*FRXNL_CAP2)
> 0 THEN (.1*ADULTS_H2 + .9*FRXNL_CAP2) ELSE IF
ADULTS_H2 > 0 THEN ADULTS_H2 ELSE 0 {Individuals per
Time Period}

CARRY_R1(t) = CARRY_R1(t - dt) + (R_COUNT1 - DUMP1) *
dt
INIT CARRY_R1 = 0
INFLOWS:
R_COUNT1 = IF MOD(time,52) = 0 THEN RANDOM(0,1)/DT ELSE
0
OUTFLOWS:
DUMP1 = IF MOD(time,52) = 0 THEN CARRY_R1/DT ELSE 0
{Ensures a new number between 0 and 1 each integer time
step.}

CARRY_R2(t) = CARRY_R2(t - dt) + (R_COUNT2 - DUMP2) *
dt
INIT CARRY_R2 = 0 {See note in Carry_R.}
INFLOWS:
R_COUNT2 = IF MOD(time,52) = 0 THEN RANDOM(0,1)/DT ELSE
0
OUTFLOWS:
DUMP2 = IF MOD(time,52) = 0 THEN CARRY_R2/DT ELSE 0
```

```
LEAVE_1_TO_2(t) = LEAVE_1_TO_2(t - dt) + (IMMIG_1_TO_2
- DIE_2_TO_1 - ARRIVE_1_TO_2) * dt
INIT LEAVE_1_TO_2 = 0
INFLOWS:
IMMIG_1_TO_2 = IF ADULTS_H1 - (.1*ADULTS_H1 +
.9*FRXNL_CAP1)
> 0 THEN (.1*ADULTS_H1 + .9*FRXNL_CAP1) ELSE IF
ADULTS_H1 > 0 THEN ADULTS_H1 ELSE 0 {Individuals per
Time Period; Only healthy adults migrate. At least 10%
of the healthy adults always migrate. Under the noted
conditions the 10% healthy and an additional fraction
of the healthy adults based empirically on the total
number of adults also migrate. Note well the order of
the nested IF statement; the first one is checked first
and if the condition holds, the first statement is
executed and the program goes no further. Otherwise,
all the adults flee. This same statement is also true
of the adults in the other field.}
OUTFLOWS:
DIE_2_TO_1 = .25*LEAVE_1_TO_2 {Individuals per Time
Period}
ARRIVE_1_TO_2 = .75*LEAVE_1_TO_2 {Individuals per Time
Period}

LEAVE_2_TO_1(t) = LEAVE_2_TO_1(t - dt) + (IMMIG_2_TO_1
- DIE_2TO1 - ARRIVE_2_TO_1) * dt
INIT LEAVE_2_TO_1 = 0
INFLOWS:
IMMIG_2_TO_1 = IF ADULTS_H2 - (.1*ADULTS_H2 +
.9*FRXNL_CAP2)
> 0 THEN (.1*ADULTS_H2 + .9*FRXNL_CAP2) ELSE IF
ADULTS_H2 > 0 THEN ADULTS_H2 ELSE 0 {Individuals per
Time Period}
OUTFLOWS:
DIE_2TO1 = .25*LEAVE_2_TO_1 {Individuals per Time
Period}
ARRIVE_2_TO_1 = .75*LEAVE_2_TO_1 {Individuals per Time
Period}

NYMPHS_D1(t) = NYMPHS_D1(t - dt) + (BIRTHS_D1 -
DEATHS_DN1 - MATURE_D1) * dt
INIT NYMPHS_D1 = 0 {Initial diseased eggs}
INFLOWS:
```

```
BIRTHS_D1 = IF (K1-ALL_ADULTS_1) > 0 THEN
A2_D*ADULTS_D1 ELSE 0 {Individuals per Time Period}
OUTFLOWS:
DEATHS_DN1 = B1_D*NYMPHS_D1 {Individuals per Time
Period}
MATURE_D1 = U1_D*NYMPHS_D1 {Individuals per Time
Period}

NYMPHS_D2(t) = NYMPHS_D2(t - dt) + (BIRTHS_D2 -
DEATHS_DN2 - MATURE_D2) * dt
INIT NYMPHS_D2 = 0 {Initial diseased eggs}
INFLOWS:
BIRTHS_D2 = IF (K2-ALL_ADULTS_2) > 0 THEN
A2_D*ADULTS_D2 ELSE 0 {Individuals per Time Period}
OUTFLOWS:
DEATHS_DN2 = B1_D*NYMPHS_D2 {Individuals per Time
Period}
MATURE_D2 = U1_D*NYMPHS_D2 {Individuals per Time
Period}

NYMPHS_H1(t) = NYMPHS_H1(t - dt) + (BIRTHS_H1 -
DEATHS_HN1 - MATURE_H1 - INFECTION_1) * dt
INIT NYMPHS_H1 = 0 {Initial Healthy eggs}
INFLOWS:
BIRTHS_H1 = IF (K1 - ALL_ADULTS_1) > 0 THEN
A2_H*ADULTS_H1 ELSE 0 {Individuals per Time Period}
OUTFLOWS:
DEATHS_HN1 = B1_H*NYMPHS_H1 {Individuals per Time Period}
MATURE_H1 = U1_H*NYMPHS_H1*(1-INFECTION_COEF1)
{Individuals per Time Period}
INFECTION_1 = INFECTION_COEF1*MATURE_H1 {Individuals
per Time Period}
NYMPHS_H2(t) = NYMPHS_H2(t - dt) + (BIRTHS_H2 -
DEATHS_HN2 - MATURE_H2 - INFECTION_2) * dt
INIT NYMPHS_H2 = 0 {Initial Healthy eggs}
INFLOWS:
BIRTHS_H2 = IF (K2-ALL_ADULTS_2) > 0 THEN
A2_H*ADULTS_H2 ELSE 0 {Individuals per Time Period}
OUTFLOWS:
DEATHS_HN2 = B1_H*NYMPHS_H2 {Individuals per Time
Period}
MATURE_H2 = U1_H*NYMPHS_H2*(1-INFECTION_COEF2)
{Individuals per Time Period}
```

```
INFECTION_2 = INFECTION_COEF2*MATURE_H2 {Individuals
per Time Period}
A2_D = .35 {Experimental laying rate. DISEASED EGGS PER
ADULT PER DAY.}
A2_H = .75 {Experimental laying rate. EGGS PER ADULT
PER DAY.}
ALL_ADULTS_1 = ADULTS_H1+ADULTS_D1
ALL_ADULTS_2 = ADULTS_H2+ADULTS_D2
B1_D = (1 - EXP(LOGN(S1_D)*F1_D*DT))/DT {Egg mortality
rate,
1/DAY. Instantaneous survival fraction + instantaneous
mortality fraction = 1.}
B1_H = (1 - EXP(LOGN(S1_H)*F1_H*DT))/DT {Egg mortality
rate,
1/DAY. Instantaneous survival fraction + instantaneous
mortality fraction = 1.}
B2_D = (1-EXP(LOGN(S2_D)*DT/1))/DT {Adult mortality
rate, 1/day. One day = T = 1 = experimental period for
which adult mortality is measured. Instantaneous sur-
vival fraction + instantaneous mortality fraction = 1.}
B2_H = (1-EXP(LOGN(S2_H)*DT/1))/DT {Adult mortality
rate, 1/day. One day = T = 1 = experimental period for
which adult mortality is measured. Instantaneous
survival fraction + instantaneous mortality fraction =
1.}
F1_D = .2 {Experimental maturation rate, 1/DAY, i.e.,
20 eggs per 100 eggs mature each day, as noted in the
experiment. In other words, a surviving egg matures on
the average in five days under the experimental
conditions.}
F1_H = .2 {Experimental maturation rate, 1/DAY, i.e.,
20 eggs per 100 eggs mature each day, as noted in the
experiment. In other words, a surviving egg matures on
the average in five days under the experimental
conditions.}
FRXNL_CAP1 = ALL_ADULTS_1/K1
DOCUMENT: This fraction is the degree to which both
healthy and diseased adults have reached their carrying
capacity.
FRXNL_CAP2 = ALL_ADULTS_2/K2
INFECTION_COEF1 = 1 - EXP(-.3*NYMPHS_H1*NYMPHS_D1)
{Constructed function giving the desired 0 to 1
probability.}
```

```
INFECTION_COEF2 = 1 - EXP(-.3*NYMPHS_H2*NYMPHS_D2)
{Constructed function giving the desired 0 to 1
probability.}
K1 = IF CARRY_R1 > .666
 THEN 2
 ELSE IF CARRY_R1 < .333
 THEN .5
 ELSE 1 {This is the carring capacity of the area the
insects area}
K2 = IF CARRY_R2 > .666
 THEN 4
 ELSE IF CARRY_R2 < .333
 THEN 1
 ELSE 2 {This is the carring capacity of the area the
insects area.}

S1_D = .5 {Experimental diseased egg survival fraction,
dimensionless, per stage. Stage = 1/F1, i.e., 30 eggs
per 100 eggs survive each 1/F1 days, as noted in the
experiment.}
S1_H = .7 {Experimental egg survival fraction,
dimensionless, per stage. Stage = 1/F1, i.e., 70 eggs
per 100 eggs survive each 1/F1 days, as noted in the
experiment.}
S2_D = .65 {Experimental daily diseased adult survival
fraction per stage, dimensionless.}
S2_H = .8 {Experimental daily adult survival fraction
per stage, dimensionless.}
U1_D = F1_D*EXP(LOGN(S1_D)*F1_D*DT) {Model maturation
rate for survivors, 1/DAY. Hatch rate = instantaneous
survival fraction*maturation rate.}
U1_H = F1_H*EXP(LOGN(S1_H)*F1_H*DT) {Model maturation
rate for survivors, 1/DAY. Hatch rate = instantaneous
survival fraction*maturation rate.}
```

32

Two-Species Colonization Model

Thou eternal fugitive, Hovering over all that live.

Ralph Waldo Emerson,
1847 *Emerson Poems*, "Ode to Beauty"
Riverside Press, Cambridge, MA.

32.1 Basic Model

In Chapter 5 we modeled spatial dynamics in a rather abstract way. Let us take up the issue of spatial dynamics in this chapter and deal specifically with the competition between two species for space. We will begin this chapter with a simple version of this model, then introduce disturbances on the physical landscape and observe the implications for population dynamics.

Assume you are the manager of forestland on which two species of trees can grow.[1] Both species are able to colonize open patches of land. The colonization coefficient C is different for each of the species. One of the species has a higher ability to colonize, but after colonization took place it is easily outcompeted by the other species. Call the species that loses in competition the INFERIOR or fugitive species, and the other one the SUPERIOR species. Both species have an extinction rate, E, which we assume—only to keep things simple—to be the same for each species. The constant E is multiplied by the number of patches occupied by a species to obtain the extinction rate of that species. Once extinction from a patch on the landscape takes place, the area that was previousy covered by a particular species is converted into an open patch.

There are three state variables of this system: One state variable for the amount of open land that can be colonized and one each for the land occupied by a Superior and Inferior species. In the model we normalize the total habitable forest area

$$\text{TOTAL} = 1 \tag{1}$$

and express the land that is open for colonization and the land that is colonized as a fraction of the total.

[1]S. Nee, and R. May, Dynamics of metapopulations: Habitat destruction and comptetitive coexistence, *Journal of Animal Ecology*, 61:37–40, 1992.

In addition to the three state variables, there are three main driving forces for the dynamics of this system that you must consider. One is the colonization of, and competition for, patches of land by the two tree species—the conversion of empty space to either INFERIOR or SUPERIOR species. The second is the removal of species through extinction. Finally, there is the conversion of INFERIOR species space through encounters with the SUPERIOR ones.

To calculate the number of patches that become occupied by inferior species, we multiply the colonization rate of the inferior species, CI, by the product of open patches and the area occupied by the inferior species. From this product we must subtract the loss of inferior species due to extinction at a rate EI,

$$\text{I COLONIZES} = \text{CI*INFERIOR*OPEN} - \text{EI*INFERIOR} \tag{2}$$

to obtain the Inferior colonization rate.

Multiplying the two state variables INFERIOR and OPEN with each other and with the colonization coefficient C to calculate the conversion of uncolonized to colonized patches is analogous to the way in which chemists calculate the product of two chemical reactions. We have made use of this idea, for example, in our epidemiology models in Chapter 20 and the host–parasite model of Chapter 30.

Again, an analogous application of the law of mass action yields the colonization rate by SUPERIOR species

$$\text{S COLONIZES} = \text{CS*SUPERIOR*OPEN} - \text{ES*SUPERIOR}, \tag{3}$$

where CS is the colonization rate of the superior species and ES is the extinction rate of the superior species. The rate at which the superior secies replaces the inferior one is

$$\text{S DISPLACES I} = \text{CS*INFERIOR*SUPERIOR}. \tag{4}$$

No resistance to this displacement is offered by the Inferior species.

The relationships between the superior and the inferior species are listed later. In general, for different extinction rates EI and ES for the inferior and superior species, the inferior species is defined by the following inequality:

$$\text{CS/ES} < \text{CI/EI}, \tag{5}$$

which can be derived by setting the derivatives in the exchange equations (2) to (4) equal to zero. The inequality means that either species must have a relatively low extinction rate or a relatively high colonization rate in order to stay in this area.

The model is shown in the diagram in Figure 32.1. Run this model for initial values of INFERIOR = 0.256 and SUPERIOR = 0.25, the colonization rates CS = 0.55 and CI = 0.7, and extinction rate ES = EI = E = 0.45.

The results of the model show that the system soon reaches a steady state in which most of the land is open patches. With ecological succession,

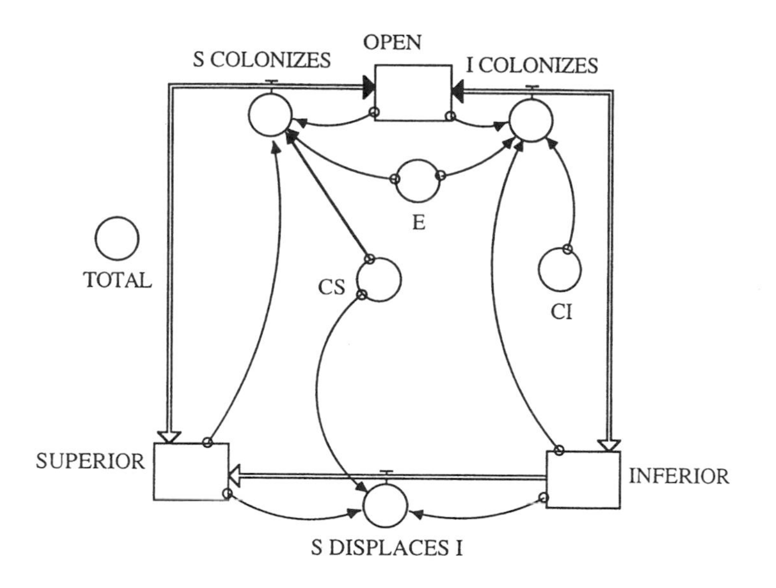

FIGURE 32.1

the inferior species is becoming increasingly replaced by superior species, leading to a dominance of the superior species in the steady state (Figs. 32.2 and 32.3).

Can you set up the model to investigate the implications of habitat loss for the colonization process? One way to set up that model is to specify a series of sensitivity runs for which the total land available for colonizations becomes smaller with each run. The following results are derived for five runs with the total habitable area declining from 1 to 0.7. Can you explain why the fraction of open patches temporarily increases in some of the runs, and then declines again?

A certain range of reduced habitable space actually favors the dominance of the inferior species. This result is due to the higher colonization rate of the inferior species, a fugitive-like species, as compared with the superior species. The inferior species is able to recover more quickly from their sudden "extinction" from certain patches and, thus, may "evade the bulldozers" more easily, whereas the chances that the superior species will get caught in a patch that is being destroyed is greater because of their slower movement. As habitable space is decreased in the example we show, the superior population declines and the inferior species declines. When the space has been preempted down to the level E/CS or below, the superior species disappears and the inferior population begins to decline from its maximum steady-state level. When

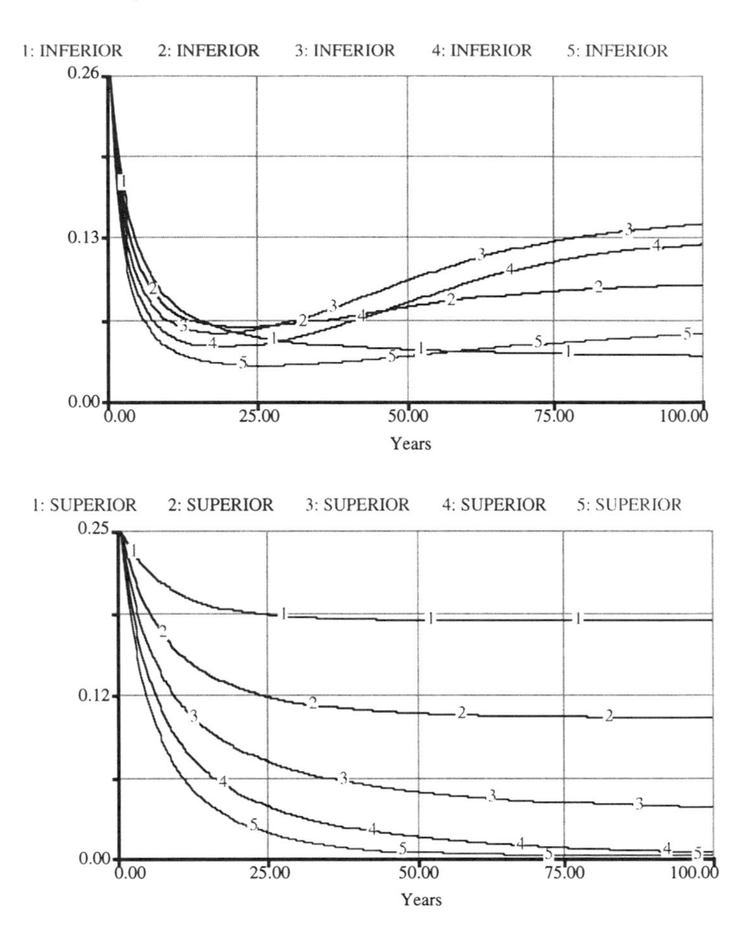

FIGURE 32.2

the habitable space is reduced to the level E/CI, both species disappear from the landscape.

Still another interesting possibility exists. Let the extinction coefficient E be the same for both species. Under a special range of choice of CS, E, and CI, only the superior species exists at the steady state, until the fractional level of habitat has been reduced below $E^*CI/(CS)^2$, in the cases where this term is less than 1. Our model indicates a subtle nuance: under those conditions where the fractional level of the habitat stands between 1 and $E^*CI/(CS)^2$, only repeated disturbance can possibly keep the inferior species in existence on this patch. Then those disturbances must not be too severe or they may speed the demise of the inferior species. The range of frequency and severity of the disturbance are critical because one can apparently only learn by experiment with numerical analysis. Try to show this "window" by selecting new values for these coefficients. The following section addresses the topic of disturbance in more detail.

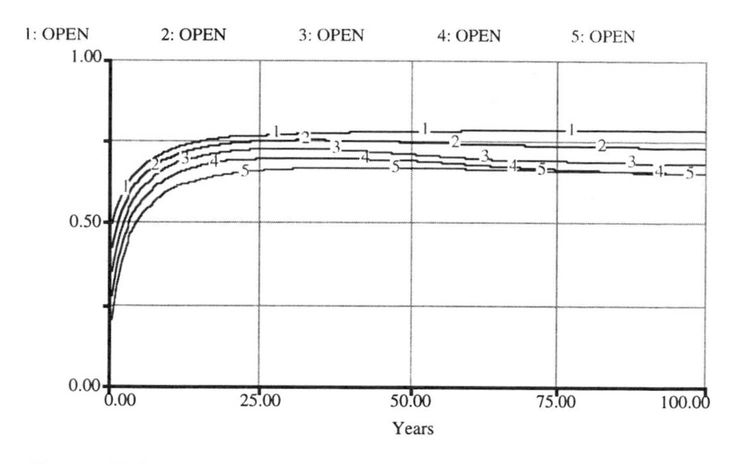

FIGURE 32.3

BASIC PATCH DYNAMICS MODEL

```
INFERIOR(t) = INFERIOR(t - dt) + (I_COLONIZES -
S_DISPLACES_I) * dt
INIT INFERIOR = .256
INFLOWS:
I_COLONIZES = CI*INFERIOR*OPEN-E*INFERIOR
OUTFLOWS:
S_DISPLACES_I = CS*INFERIOR*SUPERIOR

OPEN(t) = OPEN(t - dt) + (- S_COLONIZES - I_COLONIZES)
* dt
INIT OPEN = TOTAL - SUPERIOR - INFERIOR
OUTFLOWS:
S_COLONIZES = CS*SUPERIOR*OPEN-E*SUPERIOR
I_COLONIZES = CI*INFERIOR*OPEN-E*INFERIOR

SUPERIOR(t) = SUPERIOR(t - dt) + (S_DISPLACES_I +
S_COLONIZES) * dt
INIT SUPERIOR = .25
INFLOWS:
S_DISPLACES_I = CS*INFERIOR*SUPERIOR
S_COLONIZES = CS*SUPERIOR*OPEN-E*SUPERIOR

CI = .7
CS = .55
E = .45
TOTAL = 1
```

32.2 Two-Species Colonization Model with Fire

Real-world ecosystems are not maintained in permanent steady state. Rather, natural events, such as insect pest outbreaks or forest fires, may lead to significant changes in those systems, resetting them to a state in their early successional cycle. Let us set up the model such that forest fires occur when the forest reached a steady state and investigate the impacts of fire on the structure of the forest community (Fig. 32.4). Towards this end, we first specify a new variable that measures the change in open patches,

$$\text{D OPEN} = 5000^{*}\text{DERIVN(OPEN,1)}, \tag{1}$$

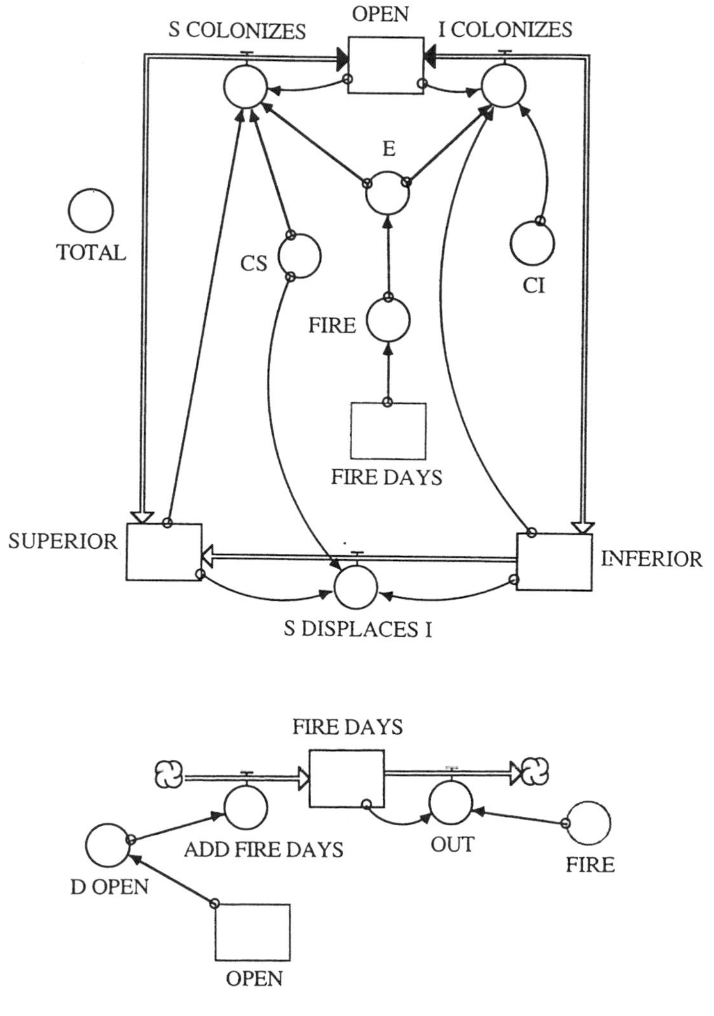

FIGURE 32.4

using the built-in function DERIVN to calculate the derivative of the state variable OPEN with respect to time. We use D OPEN to generate a random occurrence of fires,

$$\text{ADD FIRE DAYS} = \text{IF (TIME} > 1) \text{ AND (((RANDOM(0,1))}^\wedge 2 > \text{ABS(D OPEN))) THEN 1 ELSE 0} \quad (2)$$

Once 15 such fire days accumulated, we assume a fire occurs that is large enough to affect the extinction rates of species from their colonized patches. We calculate a variable FIRE as

$$\text{FIRE} = \text{IF FIRE DAYS} = 15 \text{ THEN 1 ELSE 0} \quad (3)$$

and capture its impacts on the extinction of the species from a particular forest patch by modifying the extinction rate to

$$E = .25 + \text{RANDOM}(.45,.65)^*\text{FIRE} \quad (4)$$

As the model runs, a steady state is approached. As the steady state is reached, random fires increase the proportion of empty patches. The inferior species rapidly colonize the empty habitable patches after a fire occurs. As the inferior species colonizes an increasing portion of the space, the conditions for the superior species become more favorable, leading to a replacement of inferior species. In the long run, the system returns to a near steady state, until fire breaks out again. Choose alternative parameter values to modify the severity of the fire. The case of fire leading to a temporary dominance of inferior species over superior ones is illustrated in Figure 32.5.

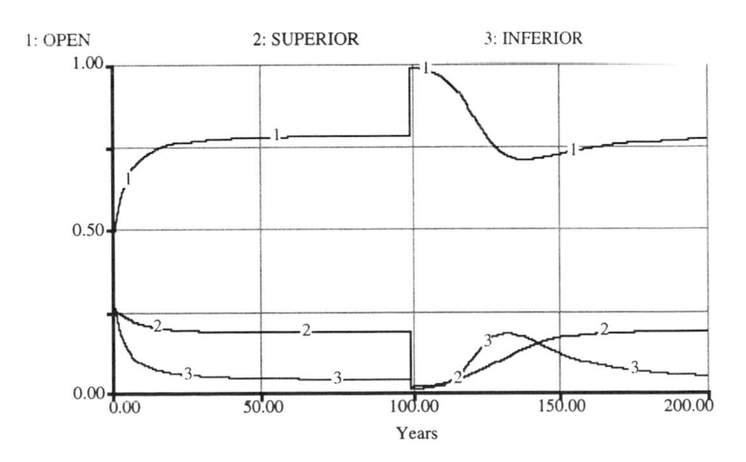

FIGURE 32.5

Investigate the combined effects of habitat loss discussed in the previous section of this chapter and forest fires through a series of sensitivity runs. What are the implications of your findings for ecosystem management?

PATCH DYNAMICS WITH FIRE

```
FIRE_DAYS(t) = FIRE_DAYS(t - dt) + (ADD_FIRE_DAYS -
OUT) * dt
INIT FIRE_DAYS = 0
INFLOWS:
ADD_FIRE_DAYS = IF (TIME > 1) AND (((RANDOM(0,1))^2 >
ABS(D_OPEN))) THEN 1 ELSE 0
OUTFLOWS:
OUT = IF FIRE=1 THEN FIRE_DAYS ELSE 0

INFERIOR(t) = INFERIOR(t - dt) + (I_COLONIZES -
S_DISPLACES_I) * dt
INIT INFERIOR = .256
INFLOWS:
I_COLONIZES = CI*INFERIOR*OPEN-E*INFERIOR
OUTFLOWS:
S_DISPLACES_I = CS*INFERIOR*SUPERIOR

OPEN(t) = OPEN(t - dt) + (- S_COLONIZES - I_COLONIZES)
* dt
INIT OPEN = TOTAL - SUPERIOR - INFERIOR
OUTFLOWS:
S_COLONIZES = CS*SUPERIOR*OPEN-E*SUPERIOR
I_COLONIZES = CI*INFERIOR*OPEN-E*INFERIOR

SUPERIOR(t) = SUPERIOR(t - dt) + (S_DISPLACES_I +
S_COLONIZES) * dt
INIT SUPERIOR = .25
INFLOWS:
S_DISPLACES_I = CS*INFERIOR*SUPERIOR
S_COLONIZES = CS*SUPERIOR*OPEN-E*SUPERIOR

CI = .7
CS = .55
E = .45 + RANDOM(.45,.65)*FIRE
FIRE = IF FIRE_DAYS=15 THEN 1 ELSE 0
TOTAL = 1
D_OPEN = 5000*DERIVN(OPEN,1)
```

32.3 Landscape and Patch Dynamics[2]

Let us expand on the model of the previous section that captured a disturbance (fire) that occurs in landscape near steady state and converted some fraction of occupied patches to empty patches. The collection of patches in the model of the previous section forms a *region*, and the modeling in this section will group the regions into an interacting set. By creating a larger landscape, made of multiple smaller regions on the landscapes, can we achieve a more steady distribution of these populations of competitiors? If we consider each smaller landscape a region of this larger landscape, how can we model the movement of species between regions, and what effect will this movement have on the equilibrium of the total landscape?

Duplicate the model of the previous chapter to generate a 3×3 grid to study the effects of adding spatial dimensions to this model. Each patch has some specific charactersistics that help to simulate a somewhat diverse landscape. One characteristic is the region's affinity for fire, which is contained as a coefficient in the D OPEN variable. This coefficient determines how close a region is to steady state, and therefore how quickly it accumulates fire days.

Assume that only the inferior species move between regions, due to their higher colonization rate. Inferior species only colognize in adjacent regions when the region is not at equilibrium (i.e., after a disturbance). We have assumed here that the colonization of inferiors in adjacent regions resembles a seeding process, that is, inferiors need not leave their own region to colonize in an adjcent region, but the success of their seeds in adjacent regions is dependent on a larger amount of open space than normal. These inferiors cannot colonize adjacent open space instantaneously either; there is a lag time in years associated with moving between regions, which we have called COL YEARS and have arbitrarily set at 10. So the idea is to have INFERIORS only as interregional colonizers, and it takes them 10 years to affect their desired boundary crossing. Between two adjacent regions, that desire is based on how many inferiors there are in the home region and how many inferiors there are in the region to be colonized.

The layout of our model is shown in Figure 32.6. The STELLA diagram (Fig. 32.7) shows, as an example, how to calculate the amount of the three

1	2	3
4	5	6
7	8	9

FIGURE 32.6

[2]We thank Brian Deal and Denny Park for their contributions to this section.

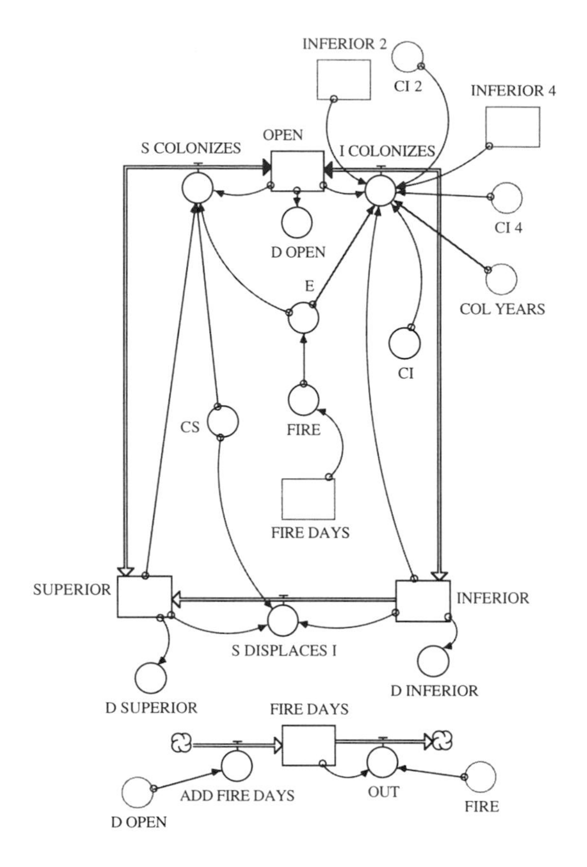

FIGURE 32.7

types of space for the upper left-hand cell (Region 1) of the 3 × 3 grid of nine interconnected models. Only the inferior species in the adjacent regions to the right and below can disperse their seeds into that portion of the habitable area. Therefore, we need to calculate I COLONIZES as a function of the presence of inferior species in those two adjacent regions, INFERIOR 2 and INFERIOR 4.

With these rules, we can begin to model the effects of patch disturbances on the overall landscape and how the landscape as a whole responds to this single-patch disturbance.[3] The results in Figures 32.8 to 32.10 are

[3]This result is part of a body of theoretical ecosystem speculation. See the following: A. Johnson, Spatiotemporal Hierarchies in Ecological Theory and Modeling, Second International Conf. on Integrating Geographic Inform. Systems and Environ. Mod., Sept 26–30, 1993, Breckenridge, CO; O. Loucks, M. Plumb-Mentjes, and D. Rodgers, Gap Processes and Large-Scale Disturbances in Sand Prairies, in: *The Ecology of Natural Disturbances and Patch Dynamics,* San Diego, CA: Academic Press,

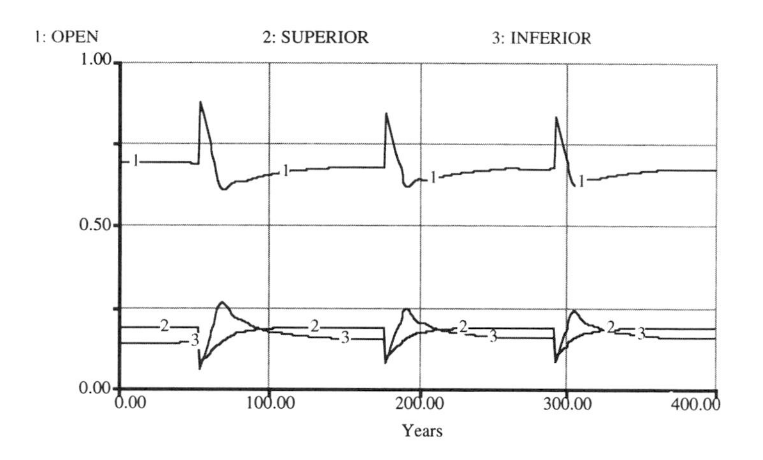

FIGURE 32.8

shown for the first two of the nine patches, for the open area—unsmoothed and smoothed with the built-in function SMTH1—and the area colonized by the inferior species over the entire landscape.

When comparing the results of this mosaic-type landscape with the single-patch model (with the same parameter values), we find that the disturbances in individual regions are damped by both the greater number of total regions, and by the colonization of adjacent inferiors. The fact that these disturbances occur out of phase helps to give the total landscape more stability than that of any individual patch. This result may be even more evident if we allow each fire to destroy a whole patch, instead of only a fraction of the patch. If this were the case, the disturbed patch would have to rely solely on the species from adjacent patches to rebuild its population. It would be interesting to see if the total landscape could maintain a somewhat steady population over time, even when whole patches are being eliminated at various times. Set up the model to investigate this case.

Modify the model of this section to allow the separate patches to have different colinization rates. How will this affect the overall dynamics of the system? Is it possible for the system as a whole to support inferior life in patches that would not be able to support this species on thier own?

1985,71–83; S. Lavorel, R. Gardner, and R. O'Neill, Analysis of Patterns in Hierarchically Structured Landscapes, *OIKOS* 67:521–528, 1993.; S. Levin, The Problem of Pattern and Scale in Ecology, *Ecology* 73-6:1943–1967, 1992; A. Hastings, Structured Models of Metapopulation Dynamics, Biology Journal of the *Linnean Society,* 42:57-71, 1991.

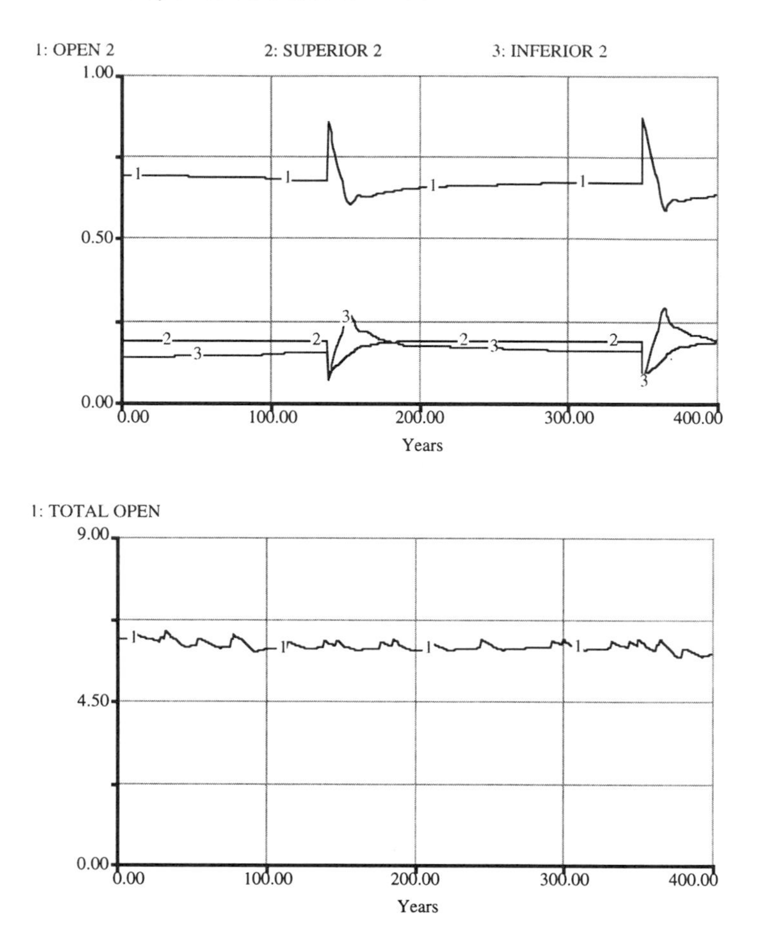

FIGURE 32.9

In this chapter we captured the competition for space. In the following four chapters we model a different type of species interaction from the one modeled here; we concentrate instead on predator–prey interactions. The first of these models deals with algae and herbivore, using hypothetical data. The second is more elaborate; it is built on real data for grass carp populations. The third predator–prey model concentrates on population management methods that are built on predator–prey interactions. Finally, in Chapter 36 we will return to the issue of spatial dynamics already discussed here in the context of spatial competition.

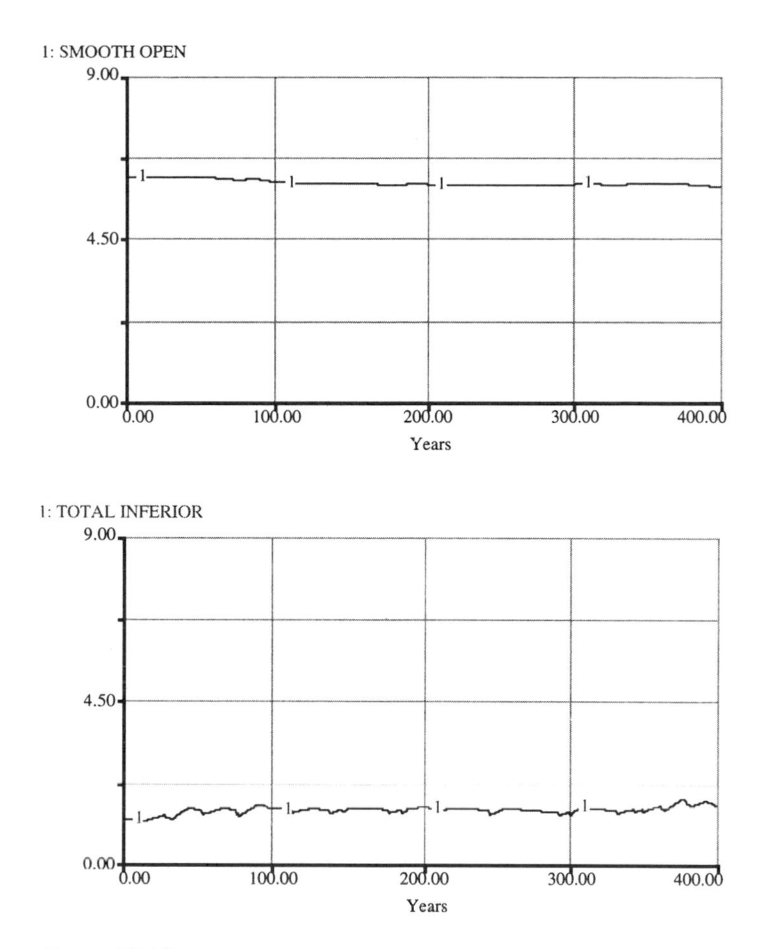

FIGURE 32.10

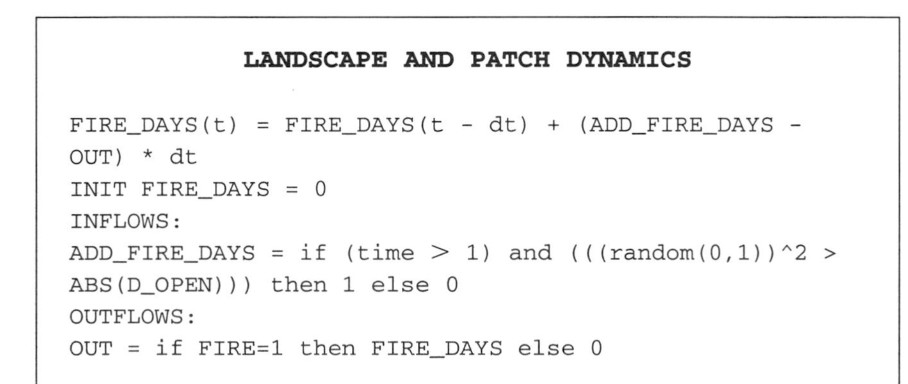

LANDSCAPE AND PATCH DYNAMICS

```
FIRE_DAYS(t) = FIRE_DAYS(t - dt) + (ADD_FIRE_DAYS -
OUT) * dt
INIT FIRE_DAYS = 0
INFLOWS:
ADD_FIRE_DAYS = if (time > 1) and (((random(0,1))^2 >
ABS(D_OPEN))) then 1 else 0
OUTFLOWS:
OUT = if FIRE=1 then FIRE_DAYS else 0
```

```
FIRE_DAYS_2(t) = FIRE_DAYS_2(t - dt) + (ADD_FIRE_DAYS_2
- OUT_2) * dt
INIT FIRE_DAYS_2 = 0
INFLOWS:
ADD_FIRE_DAYS_2 = if (time > 1) and (((random(0,1))^2
> ABS(D_OPEN_2))) then 1 else 0
OUTFLOWS:
OUT_2 = if FIRE_2=1 then FIRE_DAYS_2 else 0

FIRE_DAYS_3(t) = FIRE_DAYS_3(t - dt) + (ADD_FIRE_DAYS_3
- OUT_3) * dt
INIT FIRE_DAYS_3 = 0
INFLOWS:
ADD_FIRE_DAYS_3 = if (time > 1) and (((random(0,1))^2
> ABS(D_OPEN_3))) then 1 else 0
OUTFLOWS:
OUT_3 = if FIRE_3=1 then FIRE_DAYS_3 else 0

FIRE_DAYS_4(t) = FIRE_DAYS_4(t - dt) + (ADD_FIRE_DAYS_4
- OUT_4_4) * dt
INIT FIRE_DAYS_4 = 0
INFLOWS:
ADD_FIRE_DAYS_4 = if (time > 1) and (((random(0,1))^2
> ABS(D_OPEN_4))) then 1 else 0
OUTFLOWS:
OUT_4_4 = if FIRE_4=1 then FIRE_DAYS_4 else 0

FIRE_DAYS_5(t) = FIRE_DAYS_5(t - dt) + (ADD_FIRE_DAYS_5
- OUT_5) * dt
INIT FIRE_DAYS_5 = 0
INFLOWS:
ADD_FIRE_DAYS_5 = if (time > 1) and (((random(0,1))^2 >
ABS(D_OPEN_5))) then 1 else 0
OUTFLOWS:
OUT_5 = if FIRE_5=1 then FIRE_DAYS_5 else 0

FIRE_DAYS_6(t) = FIRE_DAYS_6(t - dt) + (ADD_FIRE_DAYS_6
- OUT_6) * dt
INIT FIRE_DAYS_6 = 0
INFLOWS:
ADD_FIRE_DAYS_6 = if (time > 1) and (((random(0,1))^2
> ABS(D_OPEN_6))) then 1 else 0
OUTFLOWS:
```

```
OUT_6 = if FIRE_6=1 then FIRE_DAYS_6 else 0

FIRE_DAYS_7(t) = FIRE_DAYS_7(t - dt) + (ADD_FIRE_DAYS_7
- OUT_7) * dt
INIT FIRE_DAYS_7 = 0
INFLOWS:
ADD_FIRE_DAYS_7 = if (time > 1) and (((random(0,1))^2
> ABS(D_OPEN_7))) then 1 else 0
OUTFLOWS:
OUT_7 = if FIRE_7=1 then FIRE_DAYS_7 else 0

FIRE_DAYS_8(t) = FIRE_DAYS_8(t - dt) + (ADD_FIRE_DAYS_8
- OUT_8) * dt
INIT FIRE_DAYS_8 = 0
INFLOWS:
ADD_FIRE_DAYS_8 = if (time > 1) and (((random(0,1))^2
> ABS(D_OPEN_8))) then 1 else 0
OUTFLOWS:
OUT_8 = if FIRE_8=1 then FIRE_DAYS_8 else 0

FIRE_DAYS_9(t) = FIRE_DAYS_9(t - dt) + (ADD_FIRE_DAYS_9
- OUT_9) * dt
INIT FIRE_DAYS_9 = 0
INFLOWS:
ADD_FIRE_DAYS_9 = if (time > 1) and (((random(0,1))^2
> ABS(D_OPEN_9))) then 1 else 0
OUTFLOWS:
OUT_9 = if FIRE_9=1 then FIRE_DAYS_9 else 0

INFERIOR(t) = INFERIOR(t - dt) + (I_COLONIZES -
S_DISPLACES_I) * dt
INIT INFERIOR = .1304
INFLOWS:
I_COLONIZES = (CI*INFERIOR*OPEN-E*INFERIOR) +
(DELAY(ABS((.69-
OPEN))/2*CI_2*INFERIOR_2*OPEN,COL_YEARS))+
(DELAY(ABS((.69-
OPEN))/2*CI_4*INFERIOR_4*OPEN,COL_YEARS))
OUTFLOWS:
S_DISPLACES_I = CS*INFERIOR*SUPERIOR
INFERIOR_2(t) = INFERIOR_2(t - dt) + (I_COLONIZES_2 -
S_DISPLACES_I_2) * dt
INIT INFERIOR_2 = .1304
```

```
INFLOWS:
I_COLONIZES_2 = (CI_2*INFERIOR_2*OPEN_2-E_2*INFERIOR_2)
+ (DELAY(ABS((.69-OPEN_2))/2.5*CI*INFERIOR*OPEN_2
,COL_YEARS)) + (DELAY(ABS((.69-
OPEN_2))/2.5*CI_3*INFERIOR_3*OPEN_2 ,COL_YEARS))+
(DELAY(ABS((.69-
OPEN_2))/2.5*CI_5*INFERIOR_5*OPEN_2,COL_YEARS))
OUTFLOWS:
S_DISPLACES_I_2 = CS_2*INFERIOR_2*SUPERIOR_2

INFERIOR_3(t) = INFERIOR_3(t - dt) + (I_COLONIZES_3 -
S_DISPLACES_I_3) * dt
INIT INFERIOR_3 = .1304
INFLOWS:
I_COLONIZES_3 = (CI_3*INFERIOR_3*OPEN_3-
E_3*INFERIOR_3)+
(DELAY(ABS((.69-
OPEN_3))/2*CI_2*INFERIOR_2*OPEN_3,COL_YEARS))+
(DELAY(ABS((.69-
OPEN_3))/2*CI_6*INFERIOR_6*OPEN_3,COL_YEARS))
OUTFLOWS:
S_DISPLACES_I_3 = CS_3*INFERIOR_3*SUPERIOR_3

INFERIOR_4(t) = INFERIOR_4(t - dt) + (I_COLONIZES_4 -
S_DISPLACES_I_4) * dt
INIT INFERIOR_4 = .1304
INFLOWS:
I_COLONIZES_4 = (CI_4*INFERIOR_4*OPEN_4-
E_4*INFERIOR_4)+
(DELAY(ABS((.69-
OPEN_4))/2.5*CI*INFERIOR*OPEN_4,COL_YEARS))+
(DELAY(ABS((.69-
OPEN_4))/2.5*CI_5*INFERIOR_5*OPEN_4,COL_YEARS))+
(DELAY(ABS((.69-
OPEN_4))/2.5*CI_7*INFERIOR_7*OPEN_4,COL_YEARS))
OUTFLOWS:
S_DISPLACES_I_4 = CS_4*INFERIOR_4*SUPERIOR_4

INFERIOR_5(t) = INFERIOR_5(t - dt) + (I_COLONIZES_5 -
S_DISPLACES_I_5) * dt
INIT INFERIOR_5 = .1304
INFLOWS:
```

```
I_COLONIZES_5 = (CI_5*INFERIOR_5*OPEN_5-E_5*INFERIOR_5)
+ (DELAY(ABS((.69-
OPEN_5))/3*CI_2*INFERIOR_2*OPEN_5,COL_YEARS))+
(DELAY(ABS((.69-
OPEN_5))/3*CI_4*INFERIOR_4*OPEN_5,COL_YEARS))+
(DELAY(ABS((.69-
OPEN_5))/3*CI_6*INFERIOR_6*OPEN_5,COL_YEARS))+
(DELAY(ABS((.69-
OPEN_5))/3*CI_8*INFERIOR_8*OPEN_5,COL_YEARS))
OUTFLOWS:
S_DISPLACES_I_5 = CS_5*INFERIOR_5*SUPERIOR_5

INFERIOR_6(t) = INFERIOR_6(t - dt) + (I_COLONIZES_6 -
S_DISPLACES_I_6) * dt
INIT INFERIOR_6 = .1304
INFLOWS:
I_COLONIZES_6 = (CI_6*INFERIOR_6*OPEN_6-E_6*INFERIOR_6)
+ (DELAY(ABS((.69-
OPEN_6))/2.5*CI_5*INFERIOR_5*OPEN_6,COL_YEARS))+
(DELAY(ABS((.69-
OPEN_6))/2.5*CI_3*INFERIOR_3*OPEN_6,COL_YEARS))+
(DELAY(ABS((.69-
OPEN_6))/2.5*CI_9*INFERIOR_9*OPEN_6,COL_YEARS))
OUTFLOWS:
S_DISPLACES_I_6 = CS_6*INFERIOR_6*SUPERIOR_6

INFERIOR_7(t) = INFERIOR_7(t - dt) + (I_COLONIZES_7 -
S_DISPLACES_I_7) * dt
INIT INFERIOR_7 = .1304
INFLOWS:
I_COLONIZES_7 = (CI_7*INFERIOR_7*OPEN_7-E_7*INFERIOR_7)
+ (DELAY(ABS((.69-
OPEN_7))/2*CI_4*INFERIOR_4*OPEN_7,COL_YEARS))+
(DELAY(ABS((.69-
OPEN_7))/2*CI_8*INFERIOR_8*OPEN_7,COL_YEARS))
OUTFLOWS:
S_DISPLACES_I_7 = CS_7*INFERIOR_7*SUPERIOR_7

INFERIOR_8(t) = INFERIOR_8(t - dt) + (I_COLONIZES_8 -
S_DISPLACES_I_8) * dt
INIT INFERIOR_8 = .1304
INFLOWS:
```

```
I_COLONIZES_8 = (CI_8*INFERIOR_8*OPEN_8-E_8*INFERIOR_8)
+ (DELAY(ABS((.69-
OPEN_8))/2.5*CI_5*INFERIOR_5*OPEN_8,COL_YEARS))+
(DELAY(ABS((.69-
OPEN_8))/2.5*CI_7*INFERIOR_7*OPEN_8,COL_YEARS))+
(DELAY(ABS((.69-
OPEN_8))/2.5*CI_9*INFERIOR_9*OPEN_8,COL_YEARS))
OUTFLOWS:
S_DISPLACES_I_8 = CS_8*INFERIOR_8*SUPERIOR8

INFERIOR_9(t) = INFERIOR_9(t - dt) + (I_COLONIZES_9 -
S_DISPLACES_I_9) * dt
INIT INFERIOR_9 = .1304
INFLOWS:
I_COLONIZES_9 = (CI_9*INFERIOR_9*OPEN_9-E_9*INFERIOR_9)
+ (DELAY(ABS((.69-
OPEN_9))/2*CI_6*INFERIOR_6*OPEN_9,COL_YEARS))+
(DELAY(ABS((.69-
OPEN_9))/2*CI_8*INFERIOR_8*OPEN_9,COL_YEARS))
OUTFLOWS:
S_DISPLACES_I_9 = CS_9*INFERIOR_9*SUPEIOR_9

OPEN(t) = OPEN(t - dt) + (- S_COLONIZES - I_COLONIZES)
* dt
INIT OPEN = 1 - SUPERIOR - INFERIOR
OUTFLOWS:
S_COLONIZES = (CS*SUPERIOR*OPEN - E*SUPERIOR)
I_COLONIZES = (CI*INFERIOR*OPEN-E*INFERIOR) +
(DELAY(ABS((.69-
OPEN))/2*CI_2*INFERIOR_2*OPEN,COL_YEARS))+
(DELAY(ABS((.69-
OPEN))/2*CI_4*INFERIOR_4*OPEN,COL_YEARS))

OPEN_2(t) = OPEN_2(t - dt) + (- S_COLONIZES_2 -
I_COLONIZES_2) * dt
INIT OPEN_2 = 1 - SUPERIOR_2 - INFERIOR_2
OUTFLOWS:
S_COLONIZES_2 = CS_2*SUPERIOR_2*OPEN_2 - E_2*SUPERIOR_2
I_COLONIZES_2 = (CI_2*INFERIOR_2*OPEN_2-E_2*INFERIOR_2)
+ (DELAY(ABS((.69-OPEN_2))/2.5*CI*INFERIOR*OPEN_2
,COL_YEARS)) + (DELAY(ABS((.69-
OPEN_2))/2.5*CI_3*INFERIOR_3*OPEN_2 ,COL_YEARS))+
```

```
(DELAY(ABS((.69-
OPEN_2))/2.5*CI_5*INFERIOR_5*OPEN_2,COL_YEARS))

OPEN_3(t) = OPEN_3(t - dt) + (- S_COLONIZES_3 -
I_COLONIZES_3) * dt
INIT OPEN_3 = 1 - SUPERIOR_3 - INFERIOR_3
OUTFLOWS:
S_COLONIZES_3 = CS_3*SUPERIOR_3*OPEN_3 - E_3*SUPERIOR_3
I_COLONIZES_3 = (CI_3*INFERIOR_3*OPEN_3-
E_3*INFERIOR_3)+
(DELAY(ABS((.69-
OPEN_3))/2*CI_2*INFERIOR_2*OPEN_3,COL_YEARS))+
(DELAY(ABS((.69-
OPEN_3))/2*CI_6*INFERIOR_6*OPEN_3,COL_YEARS))

OPEN_4(t) = OPEN_4(t - dt) + (- S_COLONIZES_4 -
I_COLONIZES_4) * dt
INIT OPEN_4 = 1 - SUPERIOR_4 - INFERIOR_4
OUTFLOWS:
S_COLONIZES_4 = CS_4*SUPERIOR_4*OPEN_4 - E_4*SUPERIOR_4
I_COLONIZES_4 = (CI_4*INFERIOR_4*OPEN_4-
E_4*INFERIOR_4)+
(DELAY(ABS((.69-
OPEN_4))/2.5*CI*INFERIOR*OPEN_4,COL_YEARS))+
(DELAY(ABS((.69-
OPEN_4))/2.5*CI_5*INFERIOR_5*OPEN_4,COL_YEARS))+
(DELAY(ABS((.69-
OPEN_4))/2.5*CI_7*INFERIOR_7*OPEN_4,COL_YEARS))

OPEN_5(t) = OPEN_5(t - dt) + (- S_COLONIZES_5 -
I_COLONIZES_5) * dt
INIT OPEN_5 = 1 - SUPERIOR_5 - INFERIOR_5
OUTFLOWS:
S_COLONIZES_5 = CS_5*SUPERIOR_5*OPEN_5 - E_5*SUPERIOR_5
I_COLONIZES_5 = (CI_5*INFERIOR_5*OPEN_5-E_5*INFERIOR_5)
+ (DELAY(ABS((.69-
OPEN_5))/3*CI_2*INFERIOR_2*OPEN_5,COL_YEARS))+
(DELAY(ABS((.69-
OPEN_5))/3*CI_4*INFERIOR_4*OPEN_5,COL_YEARS))+
(DELAY(ABS((.69-
OPEN_5))/3*CI_6*INFERIOR_6*OPEN_5,COL_YEARS))+
(DELAY(ABS((.69-
OPEN_5))/3*CI_8*INFERIOR_8*OPEN_5,COL_YEARS))
```

```
OPEN_6(t) = OPEN_6(t - dt) + (- S_COLONIZES_6 -
I_COLONIZES_6) * dt
INIT OPEN_6 = 1 - SUPERIOR_6 - INFERIOR_6
OUTFLOWS:
S_COLONIZES_6 = CS_6*SUPERIOR_6*OPEN_6 - E_6*SUPERIOR_6
I_COLONIZES_6 = (CI_6*INFERIOR_6*OPEN_6-E_6*INFERIOR_6)
+ (DELAY(ABS((.69-
OPEN_6))/2.5*CI_5*INFERIOR_5*OPEN_6,COL_YEARS))+
(DELAY(ABS((.69-
OPEN_6))/2.5*CI_3*INFERIOR_3*OPEN_6,COL_YEARS))+
(DELAY(ABS((.69-
OPEN_6))/2.5*CI_9*INFERIOR_9*OPEN_6,COL_YEARS))

OPEN_7(t) = OPEN_7(t - dt) + (- S_COLONIZES_7 -
I_COLONIZES_7) * dt
INIT OPEN_7 = 1 - SUPERIOR_7 - INFERIOR_7
OUTFLOWS:
S_COLONIZES_7 = CS_7*SUPERIOR_7*OPEN_7 - E_7*SUPERIOR_7
I_COLONIZES_7 = (CI_7*INFERIOR_7*OPEN_7-E_7*INFERIOR_7)
+ (DELAY(ABS((.69-
OPEN_7))/2*CI_4*INFERIOR_4*OPEN_7,COL_YEARS))+
(DELAY(ABS((.69-
OPEN_7))/2*CI_8*INFERIOR_8*OPEN_7,COL_YEARS))

OPEN_8(t) = OPEN_8(t - dt) + (- S_COLONIZES_8 -
I_COLONIZES_8) * dt
INIT OPEN_8 = 1 - SUPERIOR8 - INFERIOR_8
OUTFLOWS:
S_COLONIZES_8 = CS_8*SUPERIOR8*OPEN_8 - E_8*SUPERIOR8
I_COLONIZES_8 = (CI_8*INFERIOR_8*OPEN_8-E_8*INFERIOR_8)
+ (DELAY(ABS((.69-
OPEN_8))/2.5*CI_5*INFERIOR_5*OPEN_8,COL_YEARS))+
(DELAY(ABS((.69-
OPEN_8))/2.5*CI_7*INFERIOR_7*OPEN_8,COL_YEARS))+
(DELAY(ABS((.69-
OPEN_8))/2.5*CI_9*INFERIOR_9*OPEN_8,COL_YEARS))

OPEN_9(t) = OPEN_9(t - dt) + (- S_COLONIZES_9 -
I_COLONIZES_9) * dt
INIT OPEN_9 = 1 - SUPEIOR_9 - INFERIOR_9
OUTFLOWS:
S_COLONIZES_9 = CS_9*SUPEIOR_9*OPEN_9 - E_9*SUPEIOR_9
```

```
I_COLONIZES_9 = (CI_9*INFERIOR_9*OPEN_9-E_9*INFERIOR_9)
+ (DELAY(ABS((.69-
OPEN_9))/2*CI_6*INFERIOR_6*OPEN_9,COL_YEARS))+
(DELAY(ABS((.69-
OPEN_9))/2*CI_8*INFERIOR_8*OPEN_9,COL_YEARS))

SUPEIOR_9(t) = SUPEIOR_9(t - dt) + (S_DISPLACES_I_9 +
S_COLONIZES_9) * dt
INIT SUPEIOR_9 = .1818
INFLOWS:
S_DISPLACES_I_9 = CS_9*INFERIOR_9*SUPEIOR_9
S_COLONIZES_9 = CS_9*SUPEIOR_9*OPEN_9 - E_9*SUPEIOR_9

SUPERIOR(t) = SUPERIOR(t - dt) + (S_DISPLACES_I +
S_COLONIZES) * dt
INIT SUPERIOR = .1818
INFLOWS:
S_DISPLACES_I = CS*INFERIOR*SUPERIOR
S_COLONIZES = (CS*SUPERIOR*OPEN - E*SUPERIOR)

SUPERIOR8(t) = SUPERIOR8(t - dt) + (S_DISPLACES_I_8 +
S_COLONIZES_8) * dt
INIT SUPERIOR8 = .1818
INFLOWS:
S_DISPLACES_I_8 = CS_8*INFERIOR_8*SUPERIOR8
S_COLONIZES_8 = CS_8*SUPERIOR8*OPEN_8 - E_8*SUPERIOR8

SUPERIOR_2(t) = SUPERIOR_2(t - dt) + (S_DISPLACES_I_2 +
S_COLONIZES_2) * dt
INIT SUPERIOR_2 - .1818
INFLOWS:
S_DISPLACES_I_2 = CS_2*INFERIOR_2*SUPERIOR_2
S_COLONIZES_2 = CS_2*SUPERIOR_2*OPEN_2 - E_2*SUPERIOR_2

SUPERIOR_3(t) = SUPERIOR_3(t - dt) + (S_DISPLACES_I_3 +
S_COLONIZES_3) * dt
INIT SUPERIOR_3 = .1818
INFLOWS:
S_DISPLACES_I_3 = CS_3*INFERIOR_3*SUPERIOR_3
S_COLONIZES_3 = CS_3*SUPERIOR_3*OPEN_3 - E_3*SUPERIOR_3

SUPERIOR_4(t) = SUPERIOR_4(t - dt) + (S_DISPLACES_I_4 +
S_COLONIZES_4) * dt
```

```
INIT SUPERIOR_4 = .1818
INFLOWS:
S_DISPLACES_I_4 = CS_4*INFERIOR_4*SUPERIOR_4
S_COLONIZES_4 = CS_4*SUPERIOR_4*OPEN_4 - E_4*SUPERIOR_4

SUPERIOR_5(t) = SUPERIOR_5(t - dt) + (S_DISPLACES_I_5 +
S_COLONIZES_5) * dt
INIT SUPERIOR_5 = .1818
INFLOWS:
S_DISPLACES_I_5 = CS_5*INFERIOR_5*SUPERIOR_5
S_COLONIZES_5 = CS_5*SUPERIOR_5*OPEN_5 - E_5*SUPERIOR_5

SUPERIOR_6(t) = SUPERIOR_6(t - dt) + (S_DISPLACES_I_6 +
S_COLONIZES_6) * dt
INIT SUPERIOR_6 = .1818
INFLOWS:
S_DISPLACES_I_6 = CS_6*INFERIOR_6*SUPERIOR_6
S_COLONIZES_6 = CS_6*SUPERIOR_6*OPEN_6 - E_6*SUPERIOR_6

SUPERIOR_7(t) = SUPERIOR_7(t - dt) + (S_DISPLACES_I_7 +
S_COLONIZES_7) * dt
INIT SUPERIOR_7 = .1818
INFLOWS:
S_DISPLACES_I_7 = CS_7*INFERIOR_7*SUPERIOR_7
S_COLONIZES_7 = CS_7*SUPERIOR_7*OPEN_7 - E_7*SUPERIOR_7

CI = .8
CI_2 = .8
CI_3 = .8
CI_4 = .8
CI_5 = .8
CI_6 = .8
CI_7 = .8
CI_8 = .8
CI_9 = .8
COL_YEARS = 10
CS = .55
CS_2 = .55
CS_3 = .55
CS_4 = .55
CS_5 = .55
CS_6 = .55
CS_7 = .55
```

```
CS_8 = .55
CS_9 = .55
D_INFERIOR = 100*DERIVN(INFERIOR,1)
D_INFERIOR_2 = 100*DERIVN(INFERIOR_2,1)
D_INFERIOR_3 = 100*DERIVN(INFERIOR_3,1)
D_INFERIOR_4 = 100*DERIVN(INFERIOR_4,1)
D_INFERIOR_5 = 100*DERIVN(INFERIOR_5,1)
D_INFERIOR_6 = 100*DERIVN(INFERIOR_6,1)
D_INFERIOR_7 = 100*DERIVN(INFERIOR_7,1)
D_INFERIOR_8 = 100*DERIVN(INFERIOR_8,1)
D_INFERIOR_9 = 100*DERIVN(INFERIOR_9,1)
D_OPEN = 5000*DERIVN(OPEN,1)
D_OPEN_2 = 10000*DERIVN(OPEN_2,1)
D_OPEN_3 = 15000*DERIVN(OPEN_3,1)
D_OPEN_4 = 20000*DERIVN(OPEN_4,1)
D_OPEN_5 = 5000*DERIVN(OPEN_5,1)
D_OPEN_6 = 10000*DERIVN(OPEN_6,1)
D_OPEN_7 = 15000*DERIVN(OPEN_7,1)
D_OPEN_8 = 20000*DERIVN(OPEN_8,1)
D_OPEN_9 = 5000*DERIVN(OPEN_9,1)
D_SUPERIOR = 100*DERIVN(SUPERIOR,1)
D_SUPERIOR_2 = 100*DERIVN(SUPERIOR_2,1)
D_SUPERIOR_3 = 100*DERIVN(SUPERIOR_3,1)
D_SUPERIOR_4 = 100*DERIVN(SUPERIOR_4,1)
D_SUPERIOR_5 = 100*DERIVN(SUPERIOR_5,1)
D_SUPERIOR_6 = 100*DERIVN(SUPERIOR_6,1)
D_SUPERIOR_7 = 100*DERIVN(SUPERIOR_7,1)
D_SUPERIOR_8 = 100*DERIVN(SUPERIOR8,1)
D_SUPERIOR_9 = 100*DERIVN(SUPEIOR_9,1)
E = .45 + random(0.45,0.65)*FIRE
E_2 = .45 + random(0.45,0.65)*FIRE_2
E_3 = .45 + random(0.45,0.65)*FIRE_3
E_4 = .45 + random(0.45,0.65)*FIRE_4
E_5 = .45 + random(0.45,0.65)*FIRE_5
E_6 = .45 + random(0.45,0.65)*FIRE_6
E_7 = .45 + random(0.45,0.65)*FIRE_7
E_8 = .45 + random(0.45,0.65)*FIRE_8
E_9 = .45 + random(0.45,0.65)*FIRE_9
FIRE = If FIRE_DAYS=15 then 1 else 0
FIRE_2 = If FIRE_DAYS_2=15 then 1 else 0
FIRE_3 = If FIRE_DAYS_3=15 then 1 else 0
FIRE_4 = IF TIME = 10 THEN 1 ELSE If FIRE_DAYS_4=15
then 1 else 0
```

```
FIRE_5 = If FIRE_DAYS_5=15 then 1 else 0
FIRE_6 = If FIRE_DAYS_6=15 then 1 else 0
FIRE_7 = If FIRE_DAYS_7=15 then 1 else 0
FIRE_8 = IF TIME = 25 THEN 1 ELSE IF FIRE_DAYS_8=15
THEN 1 ELSE 0
FIRE_9 = If FIRE_DAYS_9=15 then 1 else 0
SMOOTH_OPEN = SMTH1(TOTAL_OPEN,40)
TOTAL_INFERIOR =
INFERIOR+INFERIOR_2+INFERIOR_3+INFERIOR_4+INFERIOR_5+IN
FERIOR_6+INFERIOR_7+INFERIOR_8+INFERIOR_9
TOTAL_OPEN =
OPEN+OPEN_2+OPEN_3+OPEN_4+OPEN_5+OPEN_6+OPEN_7+OPEN_8+
OPEN_9
```

33

Herbivore–Algae Predator–Prey Model

That the carnivore may live herbivores must die.

H. Spencer,
Data of Ethics, 1879

Let us develop a simple predator–prey model to see that even without migration, the system can exhibit a wide range of responses, not just a simple population crash. Assume that the prey are algae in a pond on which an herbivore grazes. The data for this problem have been invented. Its input data, parameters, and initial conditions would normally be determined by experiment.

The model consists of two main parts, one for the change in the algae population and one for the herbivore. The algae-growth portion of the model we have seen before in various forms. The growth rate is a function of the algal density, ALGAE. This function is monotonic and declining (Fig. 33.1). Algal growth is calculated as the product of the density and the growth rate.

The algae density is reduced through consumption by the herbivore. The consumption per head is a nonlinear function of the algal density: The greater the density, the higher the consumption per head (Fig. 33.2). The

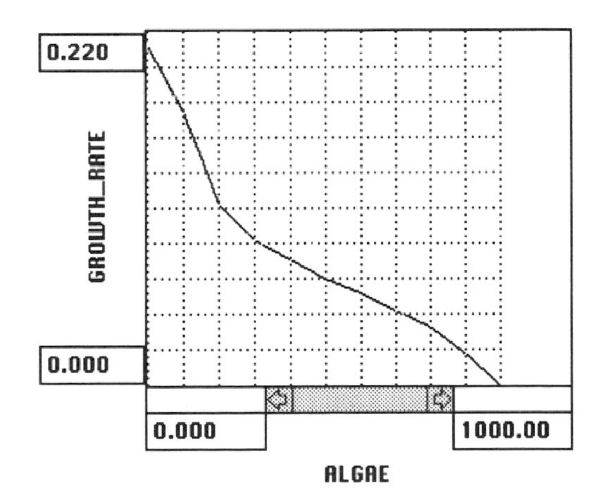

FIGURE 33.1

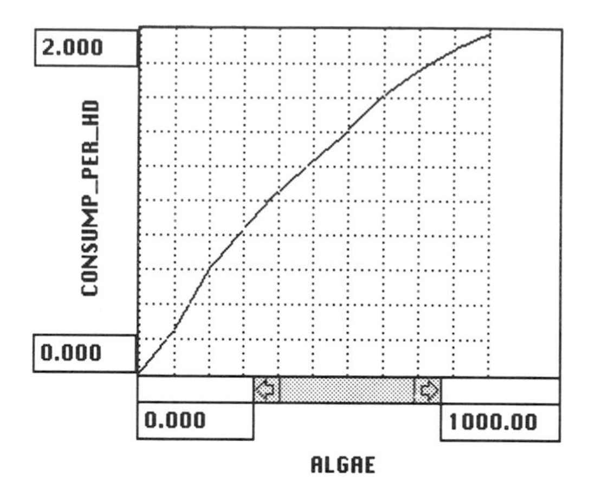

FIGURE 33.2

consumption rate is simply the product of the number of herbivores and the consumption per head.

The herbivore death rate is determined by their average life span, which is a nonlinear function of the consumption per head: The higher the consumption per head, the longer the life span, within limits (Fig. 33.3). Indirectly, the denser the algae, the lower the herbivore death rate.

The herbivore growth rate is a product of the herbivore stock and the fractional herbivore growth rate, FCN HERB GROW. To increase realism of

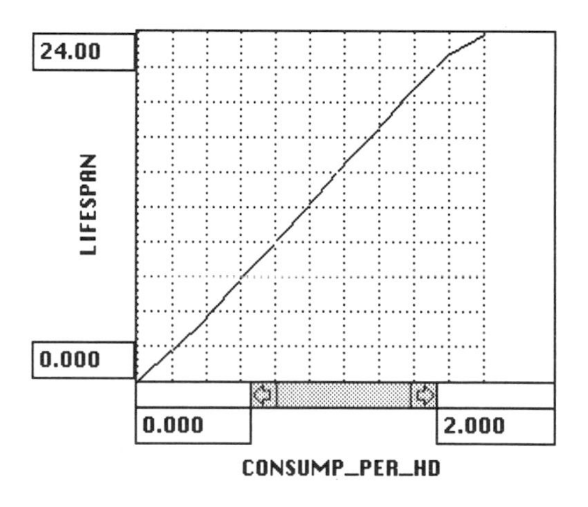

FIGURE 33.3

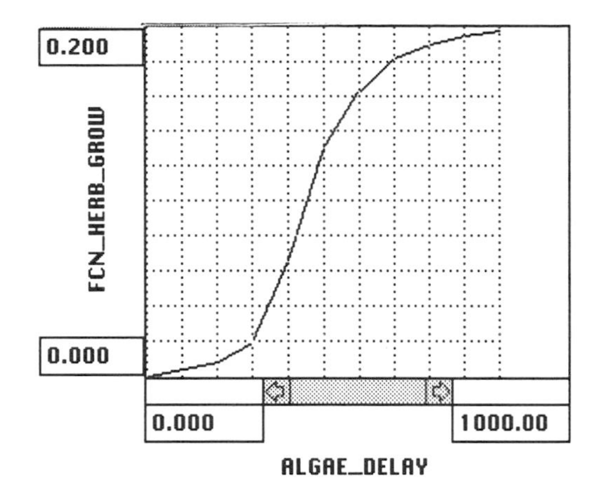

FIGURE 33.4

the model, we make FCN HERB GROW a function of the algae density in the previous time period (Fig. 33.4). This is done by producing an additional stock called ALGAE DELAY (Fig. 33.5). In general, it makes sense to represent herbivore behavior in this way. Herbivore gestation time reflects the origin of this lagged behavior.

The first graph in Figure 33.6 shows the wide swings in algal density and herbivore population over time. The second graph in Figure 33.6, a plot of algal density against the herbivore population, shows the limit cycle resulting from this particular choice of variables.

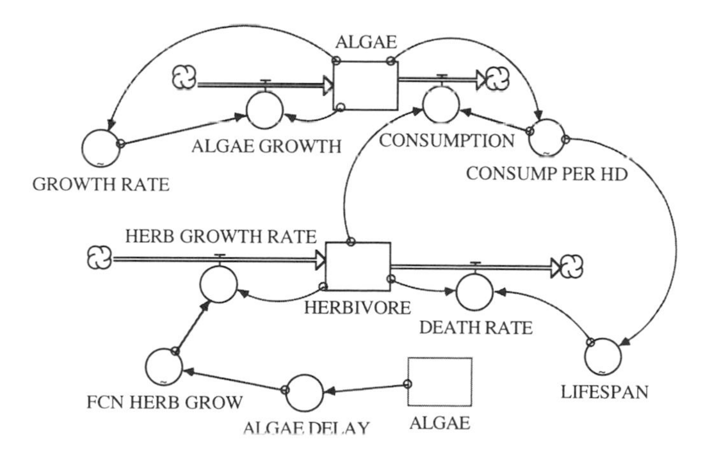

FIGURE 33.5

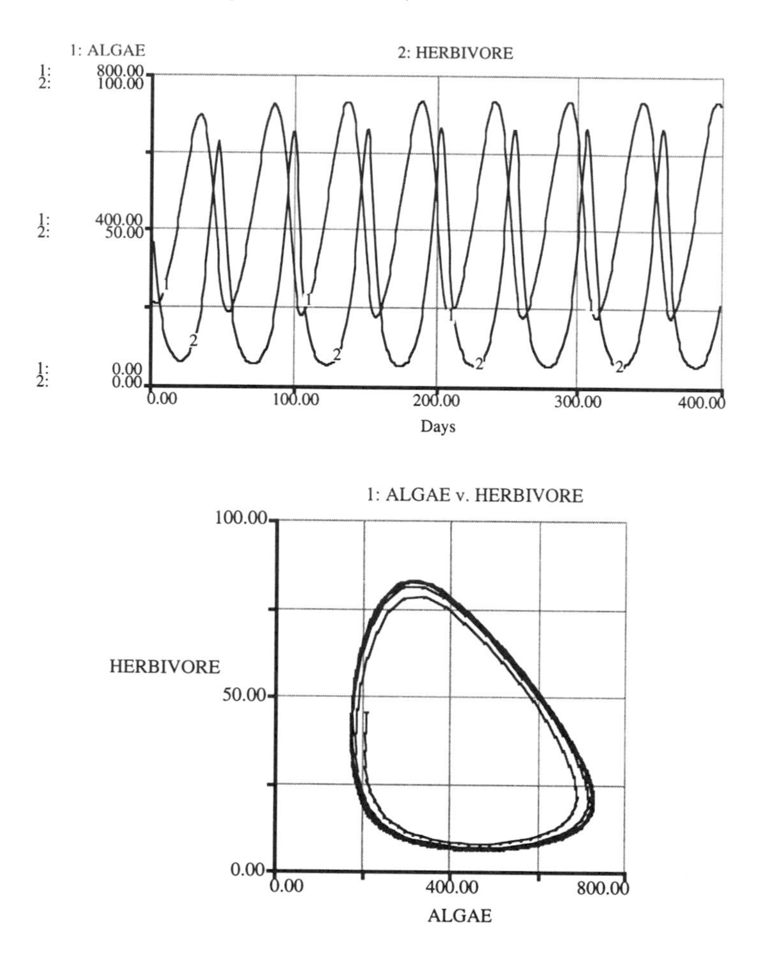

FIGURE 33.6

Now it is your turn to try changing things. Can you make the herbivore crash and not re-emerge? Try to maximize the herbivore population. Can you do this by adjusting only the variable FCN HERB GROW, without changing the maximum and minimum rates?

HERBIVORE–ALGAE MODEL

```
ALGAE(t) = ALGAE(t - dt) + (ALGAE_GROWTH - CONSUMPTION)
* dt
INIT ALGAE = 210 {Algae per Area}
INFLOWS:
```

```
ALGAE_GROWTH = ALGAE*GROWTH_RATE {Algae per Area per
Time Period}
OUTFLOWS:
CONSUMPTION = HERBIVORE*CONSUMP_PER_HD {Algae per Area
per Time Period}

HERBIVORE(t) = HERBIVORE(t - dt) + (HERB_GROWTH_RATE -
DEATH_RATE) * dt
INIT HERBIVORE = 45 {Individuals}
INFLOWS:
HERB_GROWTH_RATE = HERBIVORE*FCN_HERB_GROW {Individuals
per Time Period}
OUTFLOWS:
DEATH_RATE = HERBIVORE/LIFESPAN {Individuals per Time
Period}

ALGAE_DELAY = DELAY(ALGAE,2) {Individuals}
CONSUMP_PER_HD = GRAPH(ALGAE)
(0.00, 0.00), (100, 0.25), (200, 0.6), (300, 0.83),
(400, 1.06), (500, 1.24), (600, 1.41), (700, 1.61),
(800, 1.77), (900, 1.89), (1000, 1.98)
FCN_HERB_GROW = GRAPH(ALGAE_DELAY)
(0.00, 0.00), (100, 0.0035), (200, 0.0075), (300,
0.019), (400, 0.065), (500, 0.13), (600, 0.163), (700,
0.181), (800, 0.19), (900, 0.195), (1000, 0.198)
GROWTH_RATE = GRAPH(ALGAE)
(0.00, 0.21), (100, 0.168), (200, 0.112), (300,
0.0902), (400, 0.0781), (500, 0.066), (600, 0.0572),
(700, 0.0462), (800, 0.0363), (900, 0.0198), (1000,
0.00)
LIFESPAN = GRAPH(CONSUMP_PER_HD)
(0.00, 0.00), (0.2, 2.16), (0.4, 4.32), (0.6, 6.96),
(0.8, 9.48), (1.00, 12.1), (1.20, 14.9), (1.40, 17.3),
(1.60, 20.2), (1.80, 22.6), (2.00, 23.8)
```

34

The Grass Carp

The Carp is the Queen of Rivers: a stately, a good, and a very subtle fish.
Isaac Walton, 1653
The Compleat Angler,
E.P. Dutton & Co., NY, i. ix

The grass carp model is a large one. It combines insight from the herbivore–algae model discussed in the previous chapter with the need for human management of the predator–prey releationships. A management practice known as biomanipulation has sprung from the idea of manipulating predator–prey relationships and is gaining popularity among lake management organizations. The model was motivated by the need of controlling the growth of grass in ponds and lakes. Nutrient-rich waters flow into these bodies, producing prodigious growth rates of a variety of plants. Plant growth is so luxuriant that sport fish cannot find food. To control the grass, carp are introduced to eat the plants. To prevent the waterway from being overrun by the carp, they are bred to be sterile. Such sterile carp are called *triploid* in this model. The carp can overdo it as well; if they eat all of the plants, they will starve and the young sport fish become easy prey for the large fish. When the carp reduce the grass biomass to about 35% to 45% of the unregulated biomass, on average, the optimum level of control is reached. The problem becomes one of finding the appropriate number, size, and time of introduction of the carp into the waterway. If such a method can be found, then biological control can successfully displace chemical vegetation control.

Nursery–raised carp are commonly sold at 200 grams, the minimum size for safe transfer to a new waterway. The time horizon for most waterways is assumed to be 5 years, that is, for convenience the stocking should occur about every 5 years. Thus, to predict the full effect of the fish, a 10 year period is set as the modeling time: Fish are introduced in the spring of the first and the fifth year. Spring is chosen to maximize the survival rate of the young fish. An area of 1000 square meters is chosen as the basic unit of waterway area. The targeted water area is that portion of the waterway up to 6 meters in depth[1].

---

[1]The data and results of the modeling process are reported in M.J. Wiley, P.P. Tazik, and S.T. Sobaski, *Controlling Aquatic Vegetation with Triploid Grass Carp*, Champaign, IL: Illinois Natural History Survey, Circular 57, 1987.

There are three basic parts to this model: the number of fish, the average size of the fish, and the plant biomass. Let us first turn to the model component that deals with the number of fish.

In Figure 34.1, the input to the number of carp, the STOCK RATE, is the specification of when and how many 200 gram fish are added to the waterway. The output flows that diminish the number of fish are grouped under the control MORTALITY. They are

- PRED MORT, the consumption of the carp by the larger sport fish
- WINTER MORT, the death of fish due to the harshness of the winter months
- STARVE MORT, the death of the carp due to lack of food

Predator mortality is a function of the winter period, WINTER; a mortality coefficient MORT COEF; and the number of fish. The mortality coefficient, in turn, is a function of the average age and size of the fish and a variable that causes the mortality rate to increase as the fish ages, MORT COEF AGE. Winter mortality is a function of the number of fish; the winter period,

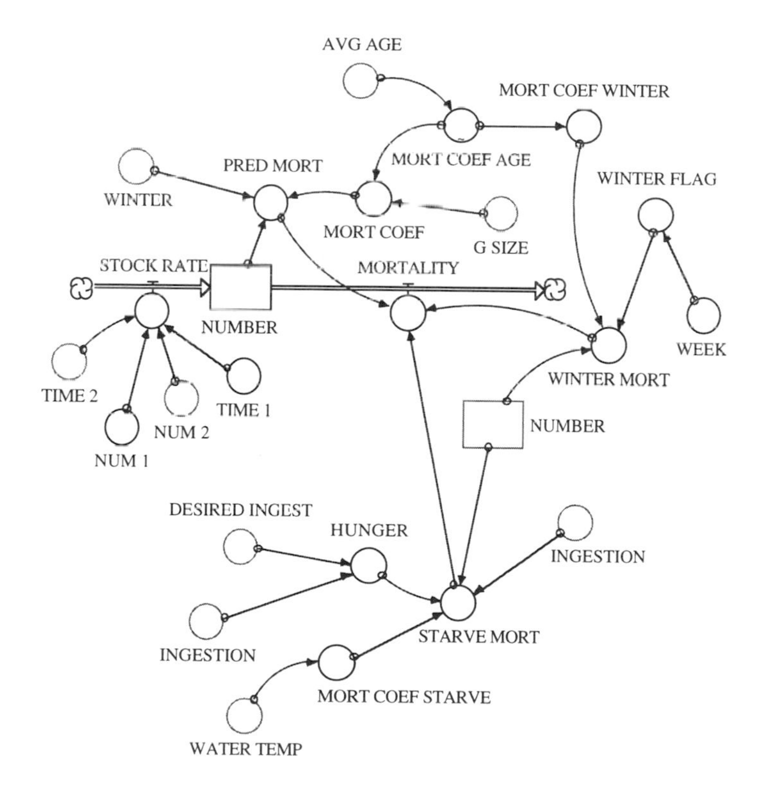

Figure 34.1

WINTER FLAG; and the above-mentioned aging coefficient. The WINTER FLAG indicates when winter occurs. It is defined as

$$\text{WINTER FLAG} = \text{IF } (\text{WEEK} \leq 12) \text{ OR } (\text{WEEK} \geq 52) \text{ THEN } 1 \text{ ELSE } 0. \quad (1)$$

The starvation mortality is a function of the number of fish, a mortality coefficient based on the water temperature and a determination of the food scarcity, HUNGER. The hunger variable is determined by comparing the desired and actual ingestion rates, as discussed later.

Figure 34.2 shows how the temperature is derived. Average weekly air temperatures (10 year sample) for Region 11 Illinois were fit with a fourth-order polynomial. The resulting equation is

$$\begin{aligned}\text{AIR TEMP} = & -2.8474 - 1.025*\text{WEEK} + .2114*(\text{WEEK})\wedge 2 \\ & -.0066*(\text{WEEK})\wedge 3 + 5.548\text{E-}5*(\text{WEEK})\wedge 4 \\ & + \text{NORMAL}(1,3). \quad (2)\end{aligned}$$

Winter is defined as that period when the average weekly temperature dropped below 8°C.

The next module displays the procedure for calculating the average size of the fish in the waterway in calories and grams. The fish ingest vegetation measured in calories. They respire and excrete substances measured in calories as well. If the ingested calories exceed the calories excreted and respired, the fish will gain weight. If ingestion is smaller, the average size declines. The excess calories are converted to grams of flesh at different rates, and the conversion efficiency declines as the fish grows larger. The conversion efficiencies for small and large fish, G SIZES and G SIZEL, are shown in the graphs in Figure 34.3.

The fish change their diet to one of declining caloric density as they grow older. These changes are captured by FCC1, FCC2, and FCC3 (Fig. 34.4):

$$\text{FCC1} = \text{IF (G SIZE} < 30) \text{ AND (G SIZE} \geq 0) \text{ THEN } 500 \text{ ELSE FCC2} \quad (3)$$

$$\text{FCC2} = \text{IF (G SIZE} \geq 30) \text{ AND (G SIZE} < 100) \text{ THEN } 1000 \text{ ELSE FCC3} \quad (4)$$

$$\text{FCC3} = \text{IF G SIZE} \geq 100 \text{ THEN } 430 \text{ ELSE } 0 \quad (5)$$

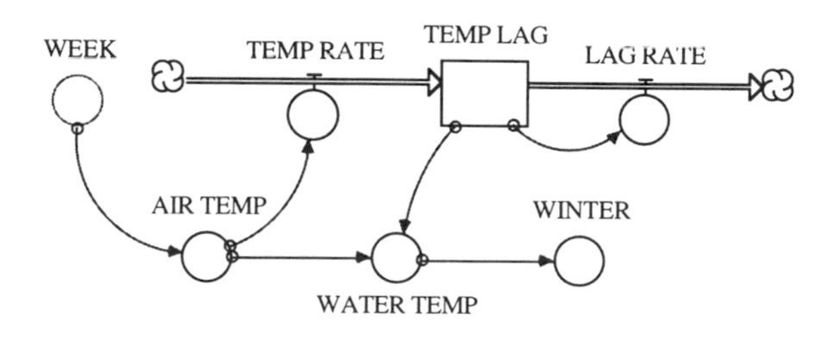

FIGURE 34.2

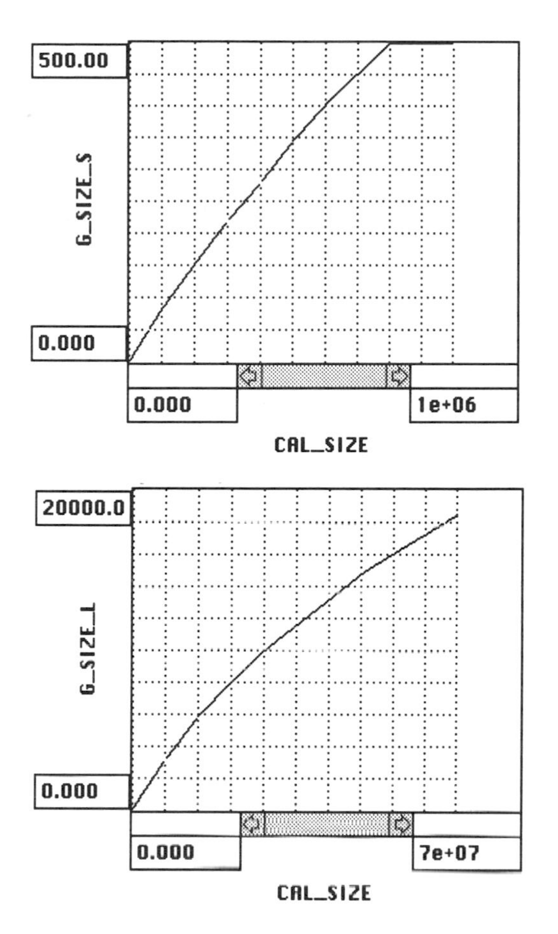

FIGURE 34.3

Actual ingestion is determined in the plant module (shown later), but the desired ingestion, DESIRED INGEST, is a function of the caloric density; the average eating rate, BWPD; a factor for comparing triploid eating rates to the natural carp eating rate, PLOIDY; the grams size, G SIZE; and voraciousness factors, TC and TC1, which depend on the water temperature:

$$\text{DESIRED INGEST} = 7*\text{BWPD}*\text{G SIZE}*\text{TC}*\text{PLOIDY}*\text{FCC1} \tag{6}$$

$$\text{BWPD} = \text{IF G SIZE} < 15000 \text{ THEN } 0.52 \text{ ELSE } .21 \tag{7}$$

$$\text{PLOIDY} = 0.91 \tag{8}$$

$$\text{TC} = \text{IF (WATER TEMP} \geq 11) \text{ AND (WATER TEMP} \leq 25) \text{ THEN } -2.8591+1.19889*\text{LOGN(WATER TEMP) ELSE TC1} \tag{9}$$

$$\text{TC1} = \text{IF WATER TEMP} > 25 \text{ THEN 1 ELSE 0} \tag{10}$$

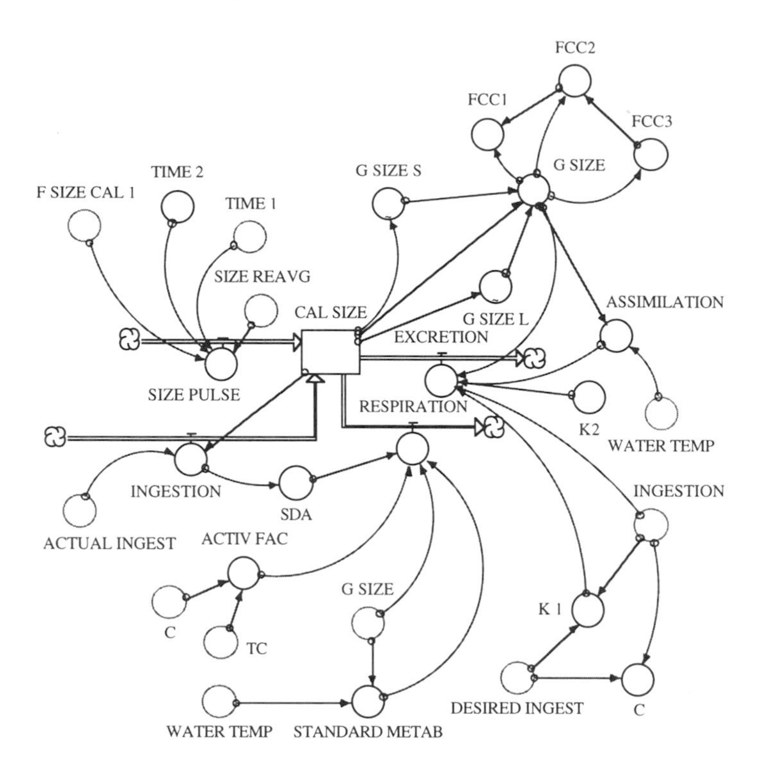

FIGURE 34.4

The respiration energy rate in the module in Figure 34.4 depends on the digestion effort, SDA; the activity level ACTIV FAC; and the standard metabolism, STANDARD METAB, the latter being dependent on the gram size and water temperature.

RESPIRATION = IF G_SIZE > 0 THEN
7*(STANDARD METAB*ACTIV FAC) + SDA ELSE 0 (11)

SDA = .06*INGESTION (12)

ACTIV FAC = TC/(1.06*C)+1 (13)

STANDARD METAB = IF (WATER TEMP ≥ 0) AND (G SIZE > 1)
THEN 82.2*.026*(G SIZE^(.645))*WATER
TEMP^1.07 ELSE 0, (14)

where conversions are made from milligrams of oxygen per fish-hour to standard calories per fish-day.

The excretion rate is a function of the amount of energy not assimilated and the fish size. The assimilation level depends in a complex way on the ratio of the desired to actual ingestion (Fig. 34.5).

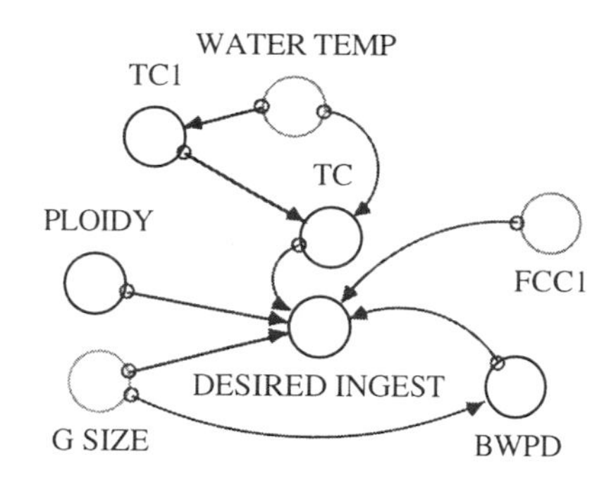

FIGURE 34.5

$$\text{EXCRETION RATE} = \text{IF G SIZE} > 0 \text{ THEN } (1 - \text{ASSIMILATION*K 1}) \text{*K2*INGESTION ELSE } 0 \quad (15)$$

In Figure 34.6, the plant model is shown. Growth of the plant is a simple logistic form with a specified upper limit on the maximum plant density, PCC = 500,000, and a specified growth rate G = 0.125. The growth rate is controlled by the temperature, TEMP SWITCH G:

$$\text{TEMP SWITCH G} = \text{IF WATER TEMP} > 15 \text{ THEN 1 ELSE 0.} \quad (16)$$

The temperature must be above a specified level for growth to occur. Plant mortality is controlled by the number of cumulative degree-days which produces a mortality rate for the plant, INSTANT MORT,

$$\text{INSTANT MORT} = \text{A*CUM DEG DAYS}^{\wedge}\text{B.} \quad (17)$$

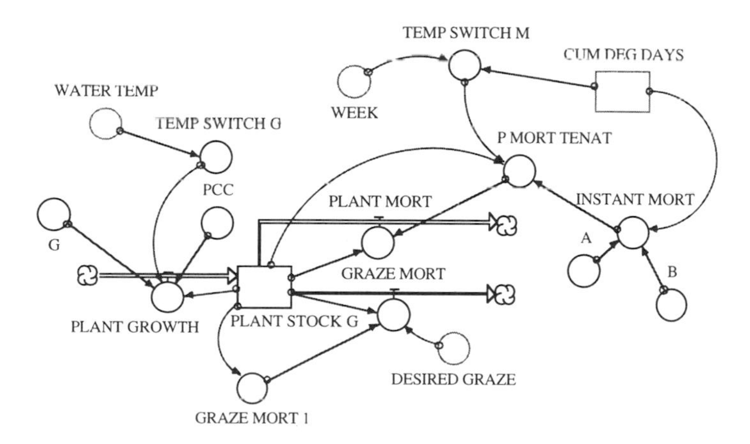

FIGURE 34.6

A minimum level of vegetation (5000 g) is preserved for regeneration, and care must be taken to ensure that plant mortality does not exceed this limit, P MORT TENAT:

$$\text{P MORT TENAT} = 7*\text{INSTANT MORT}*\text{PLANT STOCK G} *\text{TEMP SWITCH M}. \quad (18)$$

The most complex part of the model is the removal of vegetation by fish grazing, GRAZE MORT (Figs. 34.6 and 34.7). Grazing mortality must not exceed the minimum level GRAZE MORT1, and the desired ingestion is consequently controlled. The desired grazing rate is the desired ingestion converted to grams of wet plant material from a dry caloric base. Actual ingestion is the allowed grazing rate converted back to dry calories:

$$\text{GRAZE MORT} = \text{IF P MORT TENAT} \geq \text{PLANT STOCK G} - 5000 \text{ THEN (PLANT STOCK G} - 5000) \text{ ELSE P MORT TENAT} \quad (19)$$

$$\text{GRAZE MORT1} = \text{IF PLANT STOCK G} - 5000 > 0 \text{ THEN PLANT STOCK G} - 5000 \text{ ELSE } 0 \quad (20)$$

Figure 34.8 gives the two ways to find the average percentage plant biomass consumed by the fish for the 10 year period. The first module records the collective actual peak biomass levels for the 10 periods and divides by 10 and by the annual undisturbed peak in biomass. The second module simply integrates the area under the plant biomass-time curve and divides the sum by 10 times the area under the undisturbed annual biomass curve. These are the two similar measures of the success of the stocking program being tested.

Figure 34.9 shows how the average yearly age AVG AGE of the current stock of fish at the second stocking time T2 is calculated. Before and after this time, the average age of the fish is proportional to the TIME variable. The average age is used to change the winter mortality rate (see earlier).

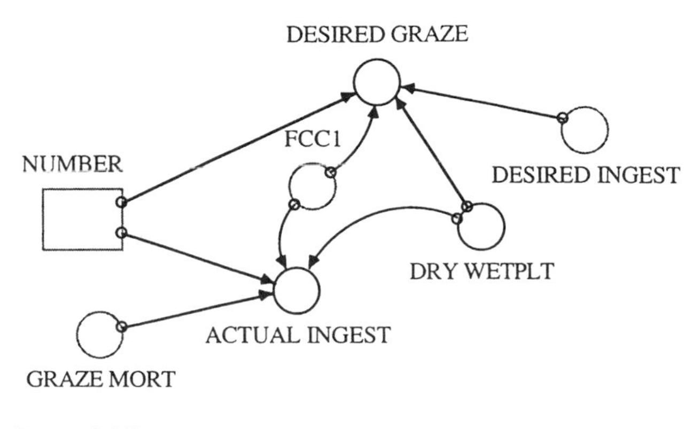

FIGURE 34.7

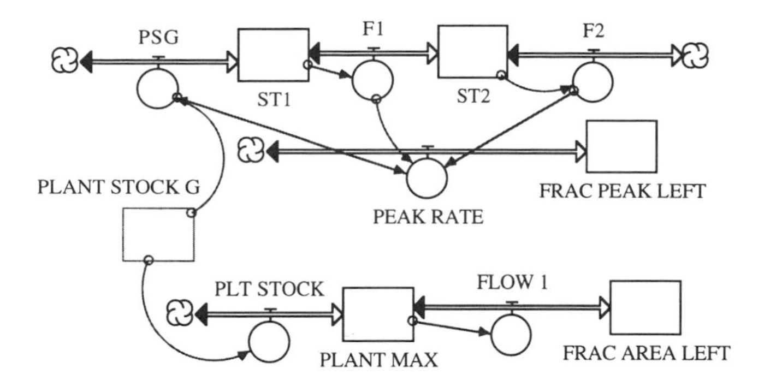

FIGURE 34.8

Figure 34.10 gives the size averaging module. In this module, the average size of the fish is thought to be sufficiently accurate. The alternative is to model each of the stockings independently.

The last module is set up for the degree–day calculation (Fig. 34.11). The base here is 0°C.

Now all the necessary components of the model are laid out. Run the model as suggested earlier. It will yield the results illustrated in Figure 34.12.

Now, open the model and experiment with different stocking numbers. Try using a smaller number in the second stocking period to smooth out the vegetative peak variation. You will find that the weakest part of the model is the part where known air temperatures are converted into corresponding water temperatures. You will find that we have experimented with this connection, using random variations on the average weekly temperatures and a variety of lag times between the air and water temperatures.

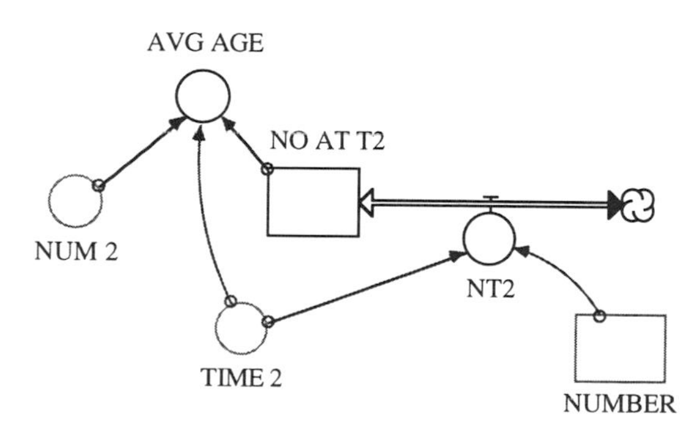

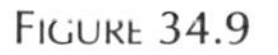
FIGURE 34.9

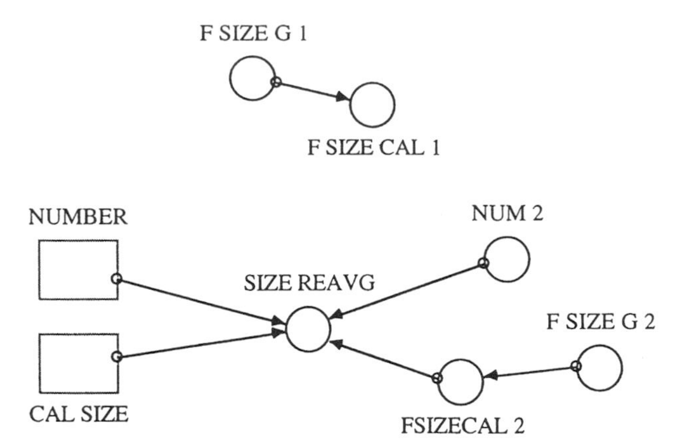

FIGURE 34.10

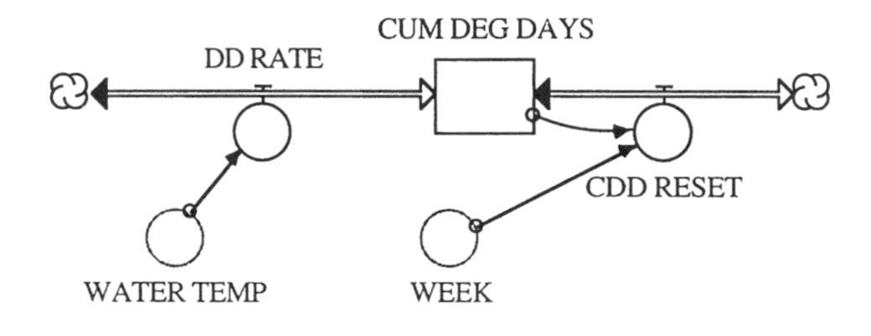

FIGURE 34.11

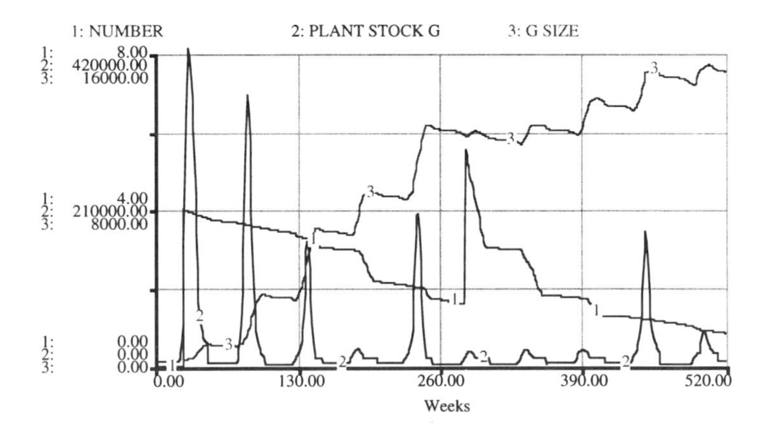

FIGURE 34.12

GRASS CARP MODEL

```
CAL_SIZE(t) = CAL_SIZE(t - dt) + (INGESTION +
SIZE_PULSE - EXCRETION - RESPIRATION) * dt
INIT CAL_SIZE = 0
INFLOWS:
INGESTION = IF CAL_SIZE > 0 THEN ACTUAL_INGEST ELSE 0
{Calories per Fish-Week}
SIZE_PULSE = PULSE(F_SIZE_CAL_1, TIME_1, 1000)
+ PULSE(SIZE_REAVG, TIME_2, 1000)
{This is the pulsing of the two control sizes in std.
cals, into the average calorie size state variable
box.}
OUTFLOWS:
EXCRETION = IF G_SIZE > 0 THEN (1-
ASSIMILATION*K_1)*K2*INGESTION ELSE 0
{Calories per Fish-Week}
RESPIRATION = IF G_SIZE > 0 THEN
7*(STANDARD_METAB*ACTIV_FAC)+SDA ELSE 0
{Standard Calories per Fish-Week}

CUM_DEG_DAYS(t) = CUM_DEG_DAYS(t - dt) + (DD_RATE -
CDD_RESET) * dt
INIT CUM_DEG_DAYS = 0
INFLOWS:
DD_RATE = IF WATER_TEMP ≥ 0 THEN (WATER_TEMP)*7 ELSE 0
OUTFLOWS:
CDD_RESET = IF WEEK = 52 THEN CUM_DEG_DAYS ELSE 0
{This control dumps the Cumulative Degree Days on Jan.
1st so that another accumulation can begin. }

FRAC_AREA_LEFT(t) = FRAC_AREA_LEFT(t - dt) + (FLOW_1) *
dt
INIT FRAC_AREA_LEFT = 0
INFLOWS:
FLOW_1 = IF TIME = 519 THEN PLANT_MAX/45300000 ELSE 0
{Divides the cumulative area under the plant stock
(Grams) vs. time curve by the area under the standard
curve (10 years), no variation in the avg. temp. curve}

FRAC_PEAK_LEFT(t) = FRAC_PEAK_LEFT(t - dt) +
(PEAK_RATE) * dt
INIT FRAC_PEAK_LEFT = 0
```

```
INFLOWS:
PEAK_RATE = IF (F1 > PSG) AND (F1 > F2) AND (F1 >
50000) THEN
F1/4250000 ELSE 0

NO_AT_T2(t) = NO_AT_T2(t - dt) + (NT2) * dt
INIT NO_AT_T2 = 0 {This section calculates the average
age of the fish in years. It averages the fish of the
second pulse with the age of the remaining fish from
the first pulse.}
INFLOWS:
NT2 = IF TIME = TIME_2 THEN NUMBER ELSE 0

NUMBER(t) = NUMBER(t - dt) + (STOCK_RATE - MORTALITY)*dt
INIT NUMBER = 0
INFLOWS:
STOCK_RATE = PULSE(NUM_1,TIME_1,1000) +
PULSE(NUM_2,TIME_2,1000) {These are the Pulse
functions. They only work with Euler integration and dt
=1.00. See Specs Menu.}
OUTFLOWS:
MORTALITY = PRED_MORT + WINTER_MORT + STARVE_MORT

PLANT_MAX(t) = PLANT_MAX(t - dt) + (PLT_STOCK - FLOW_1)
* dt
INIT PLANT_MAX = 0
INFLOWS:
PLT_STOCK = PLANT_STOCK_G {This section computes the
total area under the plant curve for the 10 year test
run and then divides it by the total area under the
undisturbed
(by grazing) curve of plant growth.}
OUTFLOWS:
FLOW_1 = IF TIME = 519 THEN PLANT_MAX/45300000 ELSE 0
{Divides the cumulative area under the plant stock
(Grams) vs. time curve by the area under the standard
curve (10 years), no variation in the avg. temp. curve}

PLANT_STOCK_G(t) = PLANT_STOCK_G(t - dt) +
(PLANT_GROWTH - PLANT_MORT - GRAZE_MORT) * dt
INIT PLANT_STOCK_G = 5000 {Grams Dry Weight per 1000
Square Meters. To change veg. type, change variables:
A, B, FCC3, G, PC, PCC, TEMP SWITCHG and DRY WETPLT}
```

```
INFLOWS:
PLANT_GROWTH = IF PLANT_STOCK_G > 0 THEN
7*TEMP_SWITCH_G*G*PLANT_STOCK_G*(1-PLANT_STOCK_G/PCC)
ELSE 0 {Grams dry weight per 1000 Square Meters-Week}
OUTFLOWS:
PLANT_MORT = IF P_MORT_TENAT ≥ PLANT_STOCK_G - 5000
THEN (PLANT_STOCK_G -5000) ELSE P_MORT_TENAT

GRAZE_MORT = IF DESIRED_GRAZE ≤ PLANT_STOCK_G - 5000
THEN DESIRED_GRAZE ELSE GRAZE_MORT_1 {Dry Weight Grams
per Week. A 5000 Gram/Square Meter-Week reserve is
maintained.}
ST1(t) = ST1(t - dt) + (PSG - F1) * dt
INIT ST1 = 0
INFLOWS:
PSG = PLANT_STOCK_G {This section computes the max
plant peak in each year, sums those peaks over the 10
year period of the run and then divides this sum by the
sum of 10 years of plant peaks that are undisturbed by
grazing.}
OUTFLOWS:
F1 = ST1
ST2(t) = ST2(t - dt) + (F1 - F2) * dt
INIT ST2 = 0
INFLOWS:
F1 = ST1
OUTFLOWS:
F2 = ST2
TEMP_LAG(t) = TEMP_LAG(t - dt) + (TEMP_RATE - LAG_RATE)
* dt
INIT TEMP_LAG = 0 {Degrees C}
INFLOWS:
TEMP_RATE = AIR_TEMP {Degrees C per Time Period}
OUTFLOWS:
LAG_RATE = TEMP_LAG/3 {Degrees C per Time Period}

A = .11E-12
ACTIV_FAC = TC/(1.06*C)+1
ACTUAL_INGEST = IF NUMBER > 0 THEN
GRAZE_MORT*FCC1/DRY_WETPLT/NUMBER ELSE 0 {Converting
the actually allowed ingestion back to Wet Standard
Calorics per Fish-Week from Dry Vegetation Grams per
Week.}
```

```
AIR_TEMP = -2.8474-1.025*WEEK+.2114*(WEEK)^2-
.0066*(WEEK)^3+5.548E-5*(WEEK)^4 + NORMAL(1,3) {Degrees
C}
ASSIMILATION = IF (WATER_TEMP > 1) AND (G_SIZE > 1)
THEN
 -.026-.058*LOGN(G_SIZE) + .213*LOGN(WATER_TEMP) ELSE 0
AVG_AGE = IF TIME ≥ TIME_2 THEN
(NO_AT_T2*TIME_2/52)/(NO_AT_T2 + NUM_2) + (TIME -
TIME_2)/52 ELSE TIME/52
B = 3.45
BWPD = IF G_SIZE <15000 THEN 0.52 ELSE .21
{Average consumption rate for Elodea; Grams Wet
Vegetation per Gram Fresh Fish-Day}
C = IF DESIRED_INGEST > INGESTION THEN 2 ELSE 1
{Reduces assimilation rate during starvation}

DESIRED_GRAZE = NUMBER*DESIRED_INGEST*DRY_WETPLT/FCC1
{Dry Vegetation Grams per Week. Conversion: from
Standard Calories by /FCC1; from Wet Grams to Dry Grams
by *DRY WETPLT}
DESIRED_INGEST = 7*BWPD*G_SIZE*TC*PLOIDY*FCC1 {Regular
Calories per Fish-Week}
DRY_WETPLT = .24
{This is the dry to wet weight ratio for Elodea.}
FCC1 = IF (G_SIZE < 30) AND (G_SIZE≥0) THEN 500 ELSE
FCC2 {Calories per gram of wet weight of Elodea c.}
FCC2 = IF (G_SIZE ≥ 30) AND (G_SIZE < 100) THEN 1000
ELSE FCC3 {Calories per gram of wet weight of Elodea
c.}
FCC3 = IF G_SIZE ≥ 100 THEN 430 ELSE 0 {430 Calories
per gram of wet weight of Elodea c.}
FSIZECAL_2 = 827.18*F_SIZE_G_2^1.0968
+ .0115*F_SIZE_G_2^2.1936
F_SIZE_CAL_1 = 827.18*F_SIZE_G_1^1.0968
+ .0115*F_SIZE_G_1^2.1936
{See Ref. under FSizeCal_2; eqn is multiplied by
FSize_g_1 to get total Standard Calories per Fish}
F_SIZE_G_1 = 200
{Grams fresh weight, per fish. This variable is set by
the user.}
F_SIZE_G_2 = 200
{The size of the average fish in the second pulse, in g
fresh weight. This variable is set by the user.}
```

```
G = .125 {Instantaneous growth rate of Elodea c., Grams
per Gram-Day}
GRAZE_MORT_1 = IF PLANT_STOCK_G - 5000 > 0 THEN
PLANT_STOCK_G - 5000 ELSE 0
G_SIZE = IF CAL_SIZE < 700000 THEN G_SIZE_S ELSE
G_SIZE_L
{I broke the Wiley/Wike function into 2 parts for
better accuracy. This relation controls the conversion
of net cal. to fresh g of fish.}
HUNGER = 1- INGESTION/(DESIRED_INGEST+1.0)

{The 1.0 keeps the ratio from becoming indefinite.}
INSTANT_MORT = A*CUM_DEG_DAYS^B
K2 = .97 {Calibration coefficient}
K_1 = 1 + (.2-.2*(INGESTION/(DESIRED_INGEST+1.0)))
{The 1.0 keeps 0/0 from being an indefinite number.}
MORT_COEF = IF (G_SIZE > 1.0) AND (G_SIZE < 100) THEN
.04645-.00705*LOGN(G_SIZE) ELSE MORT_COEF_AGE
MORT_COEF_STARVE = IF WATER_TEMP < 20 THEN 0 ELSE
.005479
MORT_COEF_WINTER = MORT_COEF_AGE
NUM_1 = 4
{Number of fish per 1000 sq. m. in the first pulse.
This variable is set by the user.}
NUM_2 = 4
{Number of fish per 1000 sq. m. in the second pulse.
This variable is set by the user.}
PCC = 500000 {Carrying capacity for Elodea; Dry Grams
per 1000 Square Meters.}
PLOIDY = .91 {This is the factor for comparing triploid
eating rates to the natural carp eating rate.}
PRED_MORT = IF NUMBER > 0 THEN 7*MORT_COEF*NUMBER*(1-
WINTER) ELSE 0
{Number per week}
P_MORT_TENAT =
7*INSTANT_MORT*PLANT_STOCK_G*TEMP_SWITCH_M
{Grams per 1000 Square Meters}
SDA = .06*INGESTION {Grams Dry Vegetation Equivalent
per Fish-Week}
SIZE_REAVG = (NUMBER*CAL_SIZE + NUM_2*FSIZECAL_2)/
(NUM_2 + NUMBER) - CAL_SIZE
{Re-averages the caloric size when the second pulse
occurs.}
```

```
STANDARD_METAB = IF (WATER_TEMP ≥ 0) AND (G_SIZE > 1)
THEN 82.2*.026*(G_SIZE^(.645))*WATER_TEMP^1.07 ELSE 0
{82.2 converts from Milligram Oxygen per Fish-Hour to
Standard Calories per Fish-Day}
STARVE_MORT = IF (NUMBER > 0) AND (INGESTION > 0) THEN
7*HUNGER*MORT_COEF_STARVE*NUMBER ELSE 0
{Number per week}
TC = IF (WATER_TEMP ≥ 11) AND (WATER_TEMP ≤ 25) THEN -
2.8591+1.19889*LOGN(WATER_TEMP) ELSE TC1
TC1 = IF WATER_TEMP > 25 THEN 1 ELSE 0
TEMP_SWITCH_G = IF WATER_TEMP > 15 THEN 1 ELSE 0
{Temp. growth threshold for Elodea C., Degrees C}
TEMP_SWITCH_M = IF (CUM_DEG_DAYS ≥ 800) AND
(CUM_DEG_DAYS ≤ 10000) AND (WEEK > 1) AND (WEEK < 51)
THEN 1 ELSE 0
TIME_1 = 21
{Time in weeks to the first pulse. Usually the best
time is April or 17 weeks into the year. This variable
set by user.}
TIME_2 = 281
{Time of the second pulse, in weeks. This variable set
by user.}
WATER_TEMP = 3.06 + 0.32*TEMP_LAG + 0*AIR_TEMP {Degrees
C}
WEEK = MOD(TIME,52) + 1 {Determines the number (1 to
52) of the week of the year}
WINTER = IF WATER_TEMP < 8 THEN 1 ELSE 0
WINTER_FLAG = IF (WEEK≤12) OR (WEEK≥52) THEN 1 ELSE 0
{Indicates when winter occurs: December 21 thru March
21}
WINTER_MORT = IF NUMBER > 0 THEN
7*MORT_COEF_WINTER*WINTER_FLAG*NUMBER ELSE 0
{Number/Week}
G_SIZE_L = GRAPH(CAL_SIZE)
(0.00, 5.67e-317), (7e+06, 3100), (1.4e+07, 5800),
(2.1e+07, 8000), (2.8e+07, 10000), (3.5e+07, 11600),
(4.2e+07, 13200), (4.9e+07, 14800), (5.6e+07, 16100),
(6.3e+07, 17300), (7e+07, 18500)
G_SIZE_S = GRAPH(CAL_SIZE)
(0.00, 6.05e-317), (100000, 78.0), (200000, 148),
(300000, 218), (400000, 277), (500000, 342), (600000,
400), (700000, 452), (800000, 500), (900000, 500),
(1e+06, 500)
```

```
MORT_COEF_AGE = GRAPH(AVG_AGE)
(0.00, 0.000495), (1.20, 0.0002), (2.40, 0.0002),
(3.60, 0.00015), (4.80, 0.000395), (6.00, 0.001),
(7.20, 0.001), (8.40, 0.001), (9.60, 0.001), (10.8,
0.001), (12.0, 0.001)
```

35

Recruitment and Trophic Dynamics of Gizzard Shad

> The exquisite manipulation of the master gives to each atom of the multitude its own character and expression.
>
> Ruskin Modern Painters,
> *Modern Painters,* 1843,
> J.M. Dent, London (1995) ii. iii. ii. Sect. 10.

The previous chapter provides one example of ways in which ecosystems can be managed via the deliberate manipulation of food webs. In this chapter, we develop a model of the addition of piscivorous (fish-eating) predators to a system in order to enhance water quality by reducing algal biomass. These effects are obtained when predators diminish, for example, planktivore biomass, which in turn increase zooplankton production. Increased numbers of zooplankton then result in lower numbers of algae, which increases water quality for human uses.

Unfortunately, numerous exceptions have been found within this simple cascading mechanism. For instance, due to rapid turnover of primary production, highly eutrophic systems are not easily limited by top-down regulation. Systems with many littoral plants also resist biomanipulation, because these plants serve as a reservoir of production and nutrients apart from the limnetic community.

The model developed in this chapter examines the recruitment and trophic dynamics of a freshwater clupeid, gizzard shad, in a flood control reservoir. Due to rapid growth, omnivorous food habits, and a high fecundity, populations of this fish species may escape both regulation by predators and competition for food resources. As a result they may often impact predator and zooplankton populations more than they are impacted themselves. Such "middle-out" effects in the trophic cascade may thwart any attempt to improve water quality by adding predators to a system containing a substantial population of gizzard shad. Thus the question arises, Under what conditions could a lake manager hope to control gizzard shad populations and improve water quality through biomanipulation? In this chapter we examine the effect of various biomanipulation regimes, the effect of predator death rates, and gizzard shad egg survival.

Let us model a community of primary producers, zooplankton, planktivores (gizzard shad), and a predator species that forages optimal size classes of gizzard shad. The model is set up to run for approximately 10 years of simulated time. Gizzard shad growth is dependent on zooplankton density for the first two size classes (larval 5 to 20 mm, and early juvenile 20 to 40 mm). The larval and juvenile growth rates are graphically specified in Figure 35.1.

Predation rates remain constant for a given abundance of predators. Thus survival of gizzard shad depends on the ability of the fish to outgrow predation. Growth is especially important during the first three life stages (larval, early juvenile, and late juvenile), but fish will transfer at set time intervals to higher size classes from the late juvenile stage onward. This maturation pattern reflects the foraging shift from zooplankton to detritus during the first year of life, and the dimishing importance of density-dependent events for adult detritivores. Egg survival determined under the control variable SURVIVAL and WINTERKILL is controlled by SEVERITY. SURVIVAL is set to 0.008 and reflects year-to-year variation in egg survival

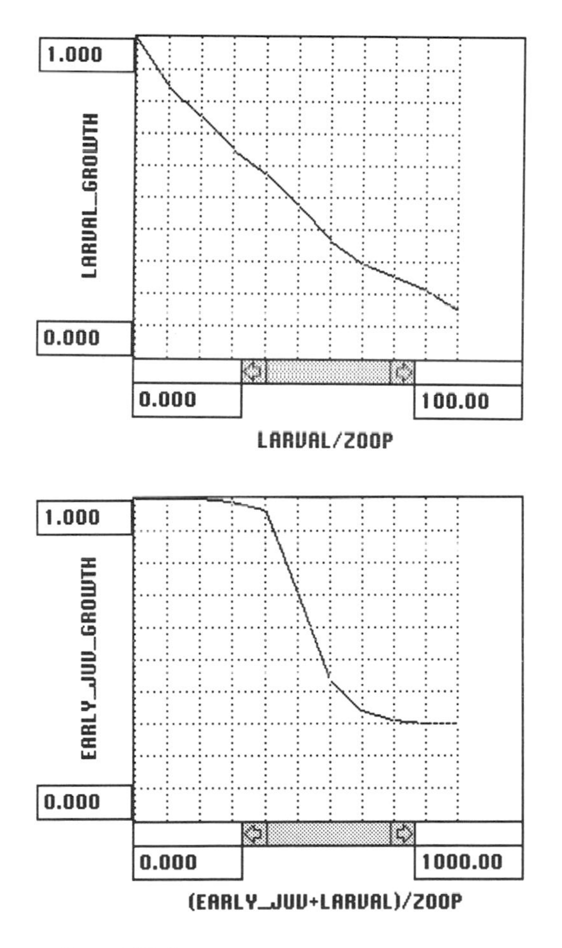

FIGURE 35.1

due to flooding. SEVERITY is set equal to 0.1 and reflects year-to-year variation in winter severity and subsequent winter kill:

$$\text{WINTERKILL} = \text{PULSE(LARVAL*SEVERITY,50,52)} \tag{1}$$

Diminishing predation on late juvenile fish as the summer progresses is indicated by the controller variable SEASONAL SHIFT, which is a pulse that diminishes over time:

$$\text{SEASONAL SHIFT} = 1 - \text{PULSE(1,8,52)} \tag{2}$$

Predation on larvae and small fish, that is, early and late juveniles, are specified as shown in the graphs in Figure 35.2.

Predators are divided into three categories—juveniles (young of year or YOY), small predators, and adults (Fig. 35.3 and 35.4).

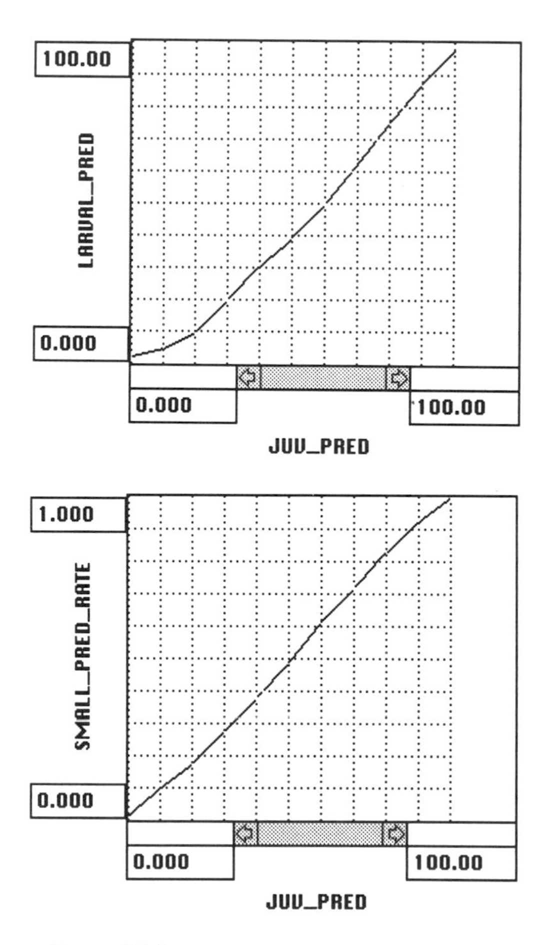

FIGURE 35.2

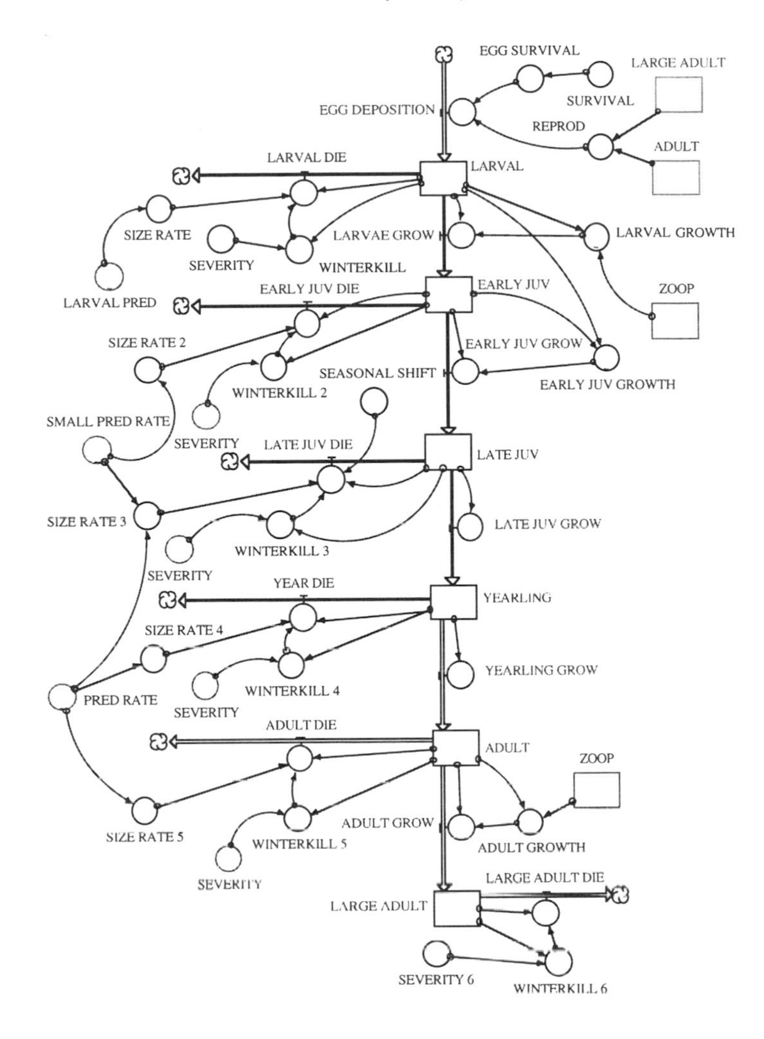

FIGURE 35.3

Death of gizzard shad is determined by the density of predators of various size classes. The predation rate is given graphically in Figure 35.5.

Large adult gizzard shad are not removed by predation at all. Regulation strength of predators is modified at the variable REGULATION STRENGTH. This control variable affects death rates of the predator and its ability to control gizzard shad. It may be said that the control is directly related to the spatial seperation of the two species (gizzard shad occur in both the limnetic and littoral zones, and largemouth bass, the main predator of gizzard shad, occur primarily in the littoral zone).

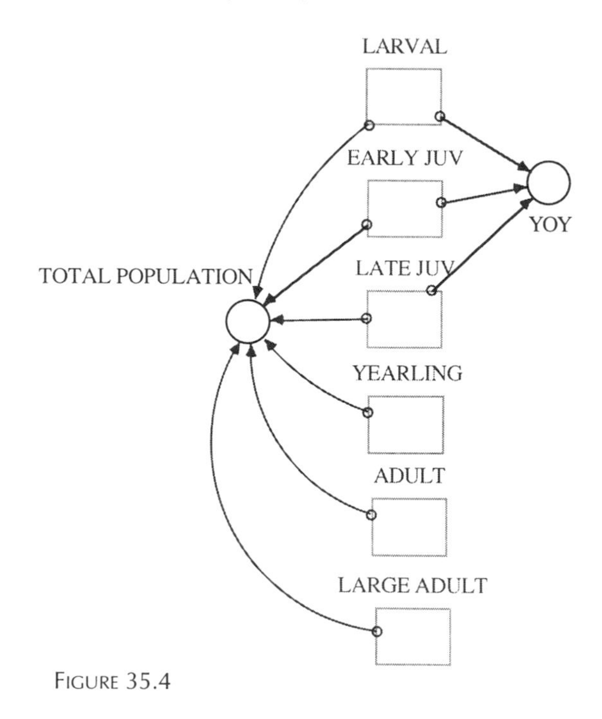

FIGURE 35.4

The population dynamics of the predator are given in the following module in Figure 35.9. Here, the death of small predators depends on the availability of gizzard shad larvae and early juveniles (Fig. 35.6).

Similarly, the effects of biomass of prey for small and adult predators is specified through control variables set up as graphs (Figs. 35.7 and 35.8).

The modules capturing algal and zooplankton growth are set up as in Figure 35.10.

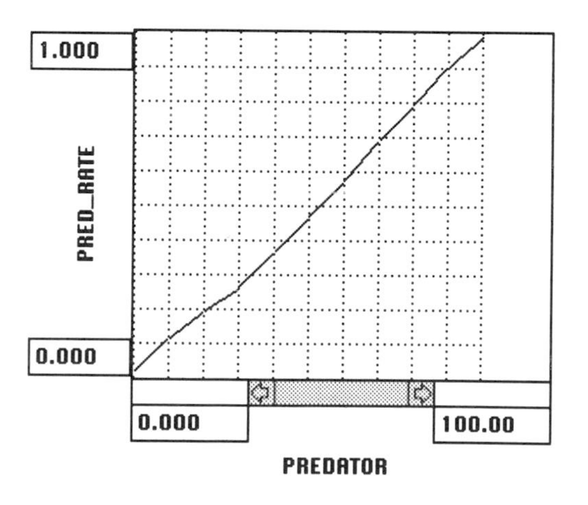

FIGURE 35.5

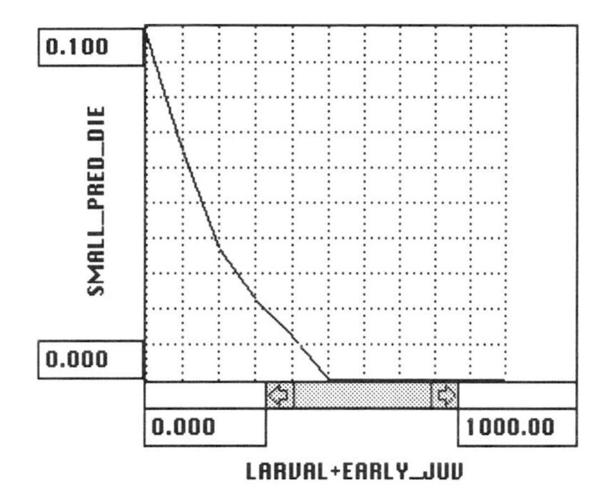

FIGURE 35.6

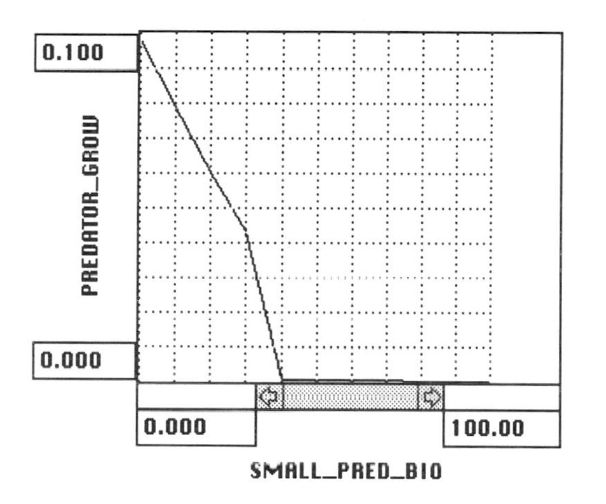

FIGURE 35.7

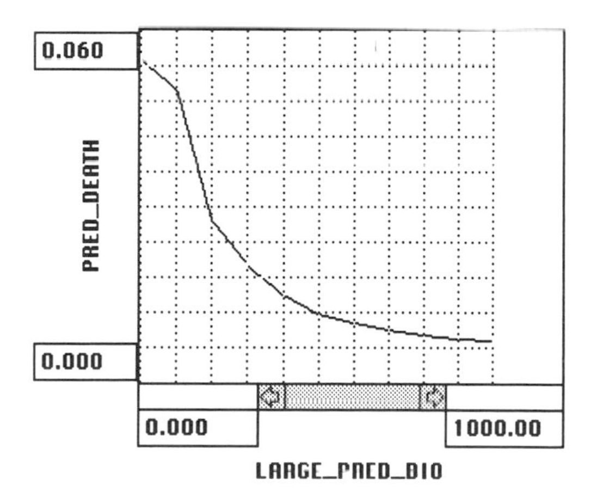

FIGURE 35.8

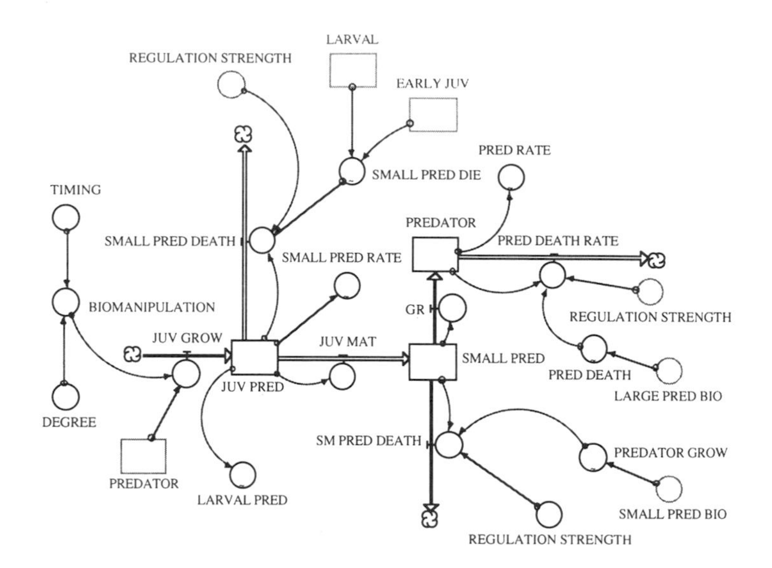

FIGURE 35.9

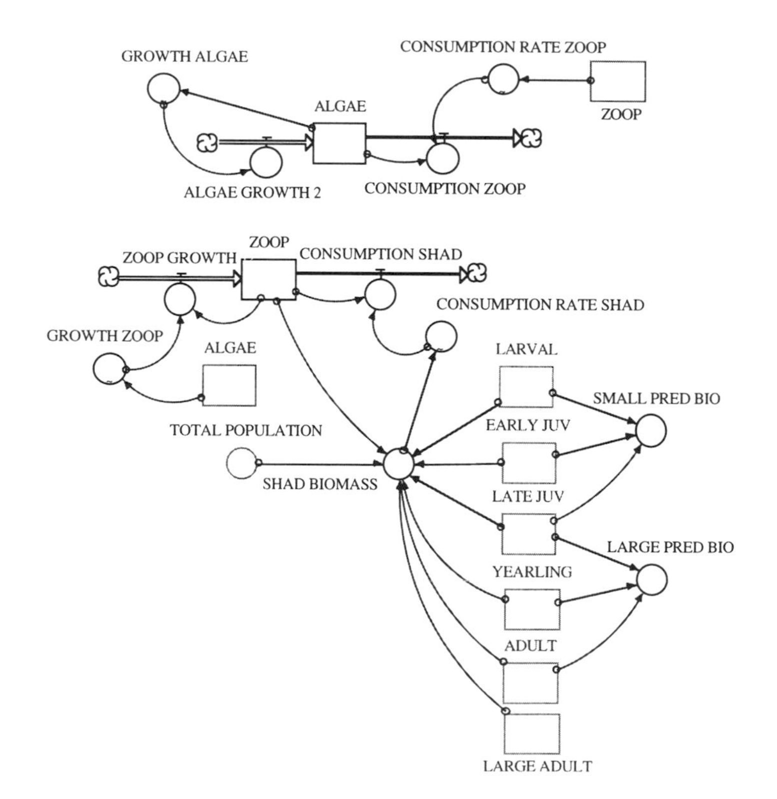

FIGURE 35.10

The growth rates for algae and zooplankton are specified, respectively, as in Figure 35.11.

Consumption rates of algae by zooplankton and of zooplankton by gizzard shad are also specified graphically in Figure 35.12.

Now all the pieces of the model are in place, and we are ready to investigate the impacts of biomanipulation. Our models show that the top-down manipulations are effective in controlling populations of gizzard shad (Figs. 35.13 and 35.14).

Conduct sensitivity analyses for the impact of the number of predators and the regulation strength on algal bloom. You should find that by increasing the number of predators that are introduced into a system, the peak of algal blooms (which occur without these predators) is reduced and the number of algal blooms is also reduced. Regulation strength is effective

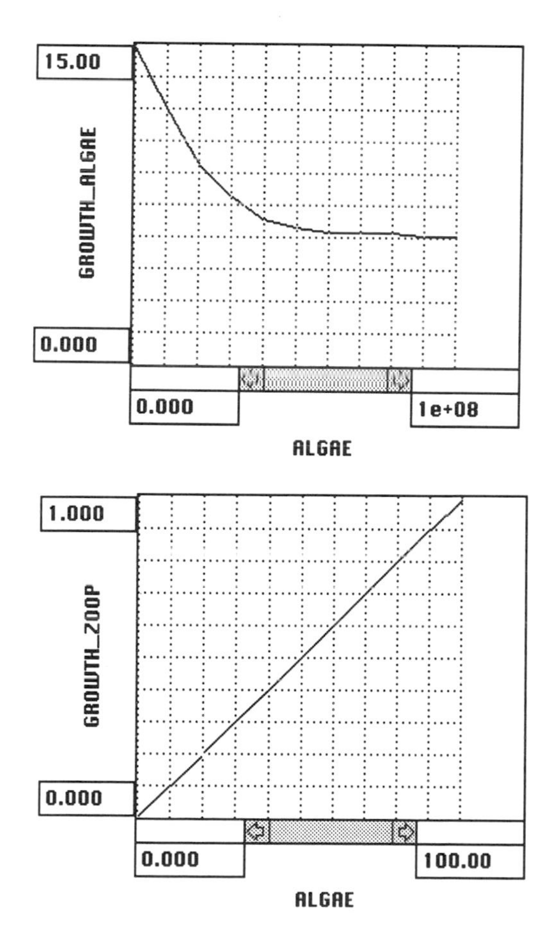

FIGURE 35.11

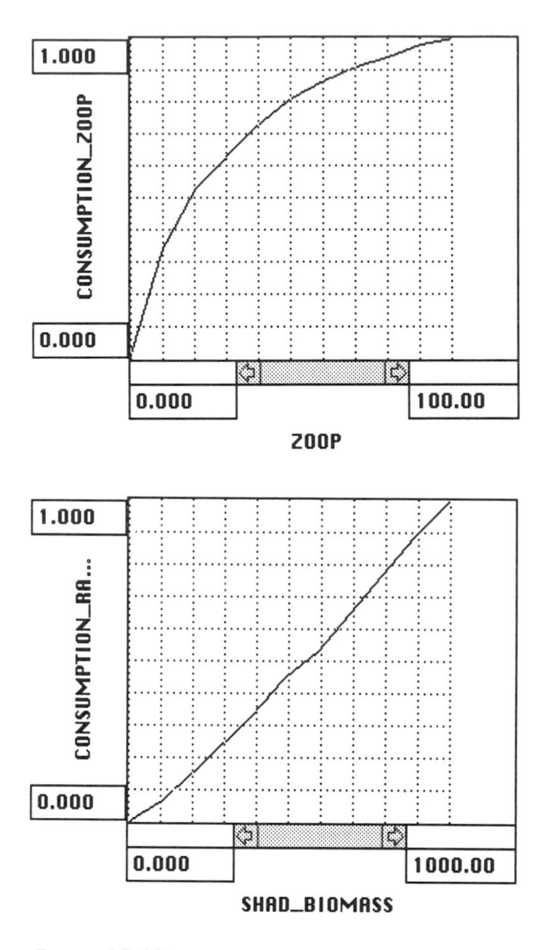

FIGURE 35.12

in controlling algal biomass and gizzard shad numbers as well. By reducing the regulation strength, the amount of algae in the system is substantially enhanced, increasing the amount of regulation strength diminishes the amount of algae in the system.

In a separate set of sensitivity analyses, assess the impacts of egg survival rates and winterkill on the system's dynamics. In the case of winterkill, reduction of the population does not matter, because reproduction of gizzard shad is so great that a few adults could fill the system. Thus the peak numbers of larval gizzard shad is slightly reduced, but the overall population patterns remain unchanged. Egg survival also does not greatly affect the population or algal biomass.

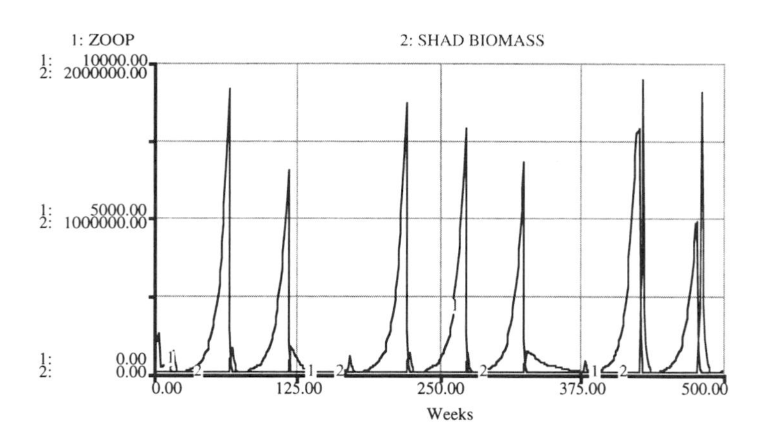

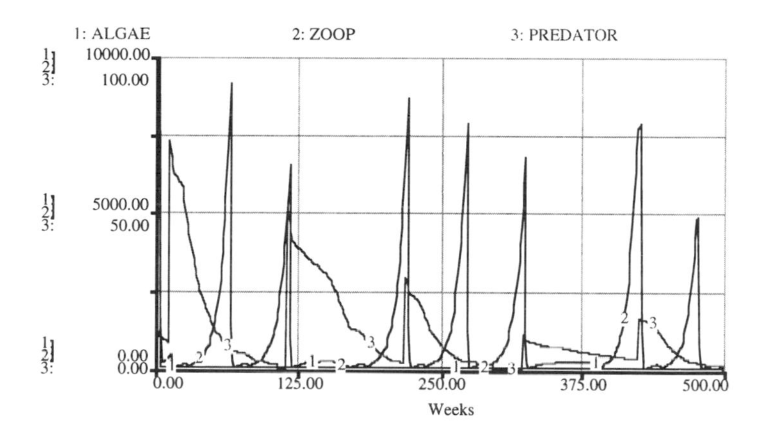

FIGURE 35.13

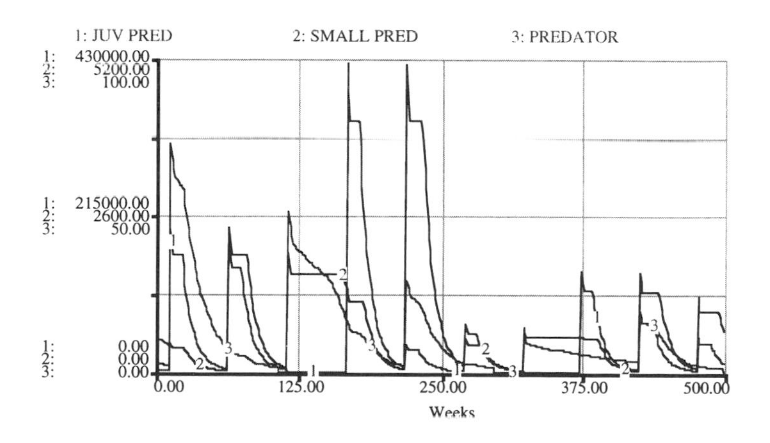

FIGURE 35.14

From the results of sensitivity testing, it appears that manipulations of predator populations may still have success in controlling gizzard shad populations despite the fact of rapid gizzard shad growth and high fecundity. Lake managers should take care to conserve and possibly enhance predator stocks (taking care not to impact local stocks or environments) in lakes where water quality is important.

GIZZARD SHAD

```
ADULT(t) = ADULT(t - dt) + (YEARLING_GROW - ADULT_GROW
- ADULT_DIE) * dt
INIT ADULT = 100 {Individuals}
INFLOWS:
YEARLING_GROW = PULSE(YEARLING,7,52) {Individuals per
Week}
OUTFLOWS:
ADULT_GROW = ADULT*ADULT_GROWTH {Individuals per Week}
ADULT_DIE = IF ADULT <1 THEN 0 ELSE
(ADULT*SIZE_RATE_5)+WINTERKILL_5 {Individuals per Week}

ALGAE(t) = ALGAE(t - dt) + (ALGAE_GROWTH_2 -
CONSUMPTION_ZOOP) * dt
INIT ALGAE = 10000 {Units of Biomass}
INFLOWS:
ALGAE_GROWTH_2 = GROWTH_ALGAE {Units of Biomass per
Week}
OUTFLOWS:
CONSUMPTION_ZOOP = (ALGAE*CONSUMPTION_RATE_ZOOP) {Units
of Biomass per Week}

EARLY_JUV(t) = EARLY_JUV(t - dt) + (LARVAE_GROW -
EARLY_JUV_GROW - EARLY_JUV_DIE) * dt
INIT EARLY_JUV = 100 {Individuals}
INFLOWS:
LARVAE_GROW = LARVAL*LARVAL_GROWTH {Individuals per
Week}
OUTFLOWS:
EARLY_JUV_GROW = EARLY_JUV*EARLY_JUV_GROWTH
{Individuals per Week}
EARLY_JUV_DIE = IF EARLY_JUV < 1 THEN 0 ELSE
(EARLY_JUV*SIZE_RATE_2)+WINTERKILL_2 {Individuals per
Week}
```

```
JUV_PRED(t) = JUV_PRED(t - dt) + (JUV_GROW -
SMALL_PRED_DEATH - JUV_MAT) * dt
INIT JUV_PRED = 1000 {Individuals}
INFLOWS:
JUV_GROW = PULSE(PREDATOR*25000,8,52)+BIOMANIPULATION
{Individuals per Week}
OUTFLOWS:
SMALL_PRED_DEATH =
(SMALL_PRED_DIE*REGULATION_STRENGTH)*JUV_PRED
{Individuals per Week}
JUV_MAT = PULSE(JUV_PRED,8,52) {Individuals per Week}

LARGE_ADULT(t) = LARGE_ADULT(t - dt) + (ADULT_GROW -
LARGE_ADULT_DIE) * dt
INIT LARGE_ADULT = 40 {Individuals}
INFLOWS:
ADULT_GROW = ADULT*ADULT_GROWTH {Individuals per Week}
OUTFLOWS:
LARGE_ADULT_DIE = IF LARGE_ADULT <1 THEN 0 ELSE
(.001*(LARGE_ADULT))+WINTERKILL_6 {Individuals per
Week}

LARVAL(t) = LARVAL(t - dt) + (EGG_DEPOSITION -
LARVAE_GROW - LARVAL_DIE) * dt
INIT LARVAL = 100 {Individuals}
INFLOWS:
EGG_DEPOSITION =
(PULSE(REPROD*20000*EGG_SURVIVAL,10,52))+(PULSE(REPROD*
160000*EGG_SURVIVAL,12,52))+(PULSE(REPROD*2000*EGG_SURV
IVAL,16,52)) {Individuals per Week}
OUTFLOWS:
LARVAE_GROW = LARVAL*LARVAL_GROWTH {Individuals per
Week}
LARVAL_DIE = IF LARVAL <1 THEN 0 ELSE
(LARVAL*SIZE_RATE)+WINTERKILL {Individuals per Week}

LATE_JUV(t) = LATE_JUV(t - dt) + (EARLY_JUV_GROW -
LATE_JUV_GROW - LATE_JUV_DIE) * dt
INIT LATE_JUV = 75 {Individuals}
INFLOWS:
EARLY_JUV_GROW = EARLY_JUV*EARLY_JUV_GROWTH
{Individuals per Week}
OUTFLOWS:
```

```
LATE_JUV_GROW = PULSE(LATE_JUV,9,52) {Individuals per
Week}
LATE_JUV_DIE = IF LATE_JUV <1 THEN 0 ELSE
((LATE_JUV*SIZE_RATE_3)*SEASONAL_SHIFT)+WINTERKILL_3
{Individuals per Week}

PREDATOR(t) = PREDATOR(t - dt) + (GR - PRED_DEATH_RATE)
* dt
INIT PREDATOR = 10 {Individuals}
INFLOWS:
GR = PULSE(SMALL_PRED,8,104) {Individuals per Week}
OUTFLOWS:
PRED_DEATH_RATE = IF PREDATOR <1 THEN 1 ELSE
(PRED_DEATH*REGULATION_STRENGTH)*PREDATOR {Individuals
per Week}

SMALL_PRED(t) = SMALL_PRED(t - dt) + (JUV_MAT - GR -
SM_PRED_DEATH) * dt
INIT SMALL_PRED = 100 {Individuals}
INFLOWS:
JUV_MAT = PULSE(JUV_PRED,8,52) {Individuals per Week}
OUTFLOWS:
GR = PULSE(SMALL_PRED,8,104) {Individuals per Week}
SM_PRED_DEATH =
(PREDATOR_GROW*REGULATION_STRENGTH)*SMALL_PRED
{Individuals per Week}

YEARLING(t) = YEARLING(t - dt) + (LATE_JUV_GROW -
YEARLING_GROW - YEAR_DIE) * dt
INIT YEARLING = 60 {Individuals}
INFLOWS:
LATE_JUV_GROW = PULSE(LATE_JUV,9,52) {Individuals per
Week}
OUTFLOWS:
YEARLING_GROW = PULSE(YEARLING,7,52) {Individuals per
Week}
YEAR_DIE = IF YEARLING <1 THEN 0 ELSE
(YEARLING*SIZE_RATE_4)+WINTERKILL_4 {Individuals per
Week}

ZOOP(t) = ZOOP(t - dt) + (ZOOP_GROWTH -
CONSUMPTION_SHAD) * dt
INIT ZOOP = 1000 {Units of Biomass}
```

```
INFLOWS:
ZOOP_GROWTH = ZOOP*GROWTH_ZOOP {Units of Biomass per
Week}
OUTFLOWS:
CONSUMPTION_SHAD = IF ZOOP <1 THEN 0 ELSE
(ZOOP*CONSUMPTION_RATE_SHAD) {Units of Biomass per Week}

ADULT_GROWTH = ADULT/ZOOP*.0009
BIOMANIPULATION = PULSE(DEGREE,50,TIMING) {Individuals
per Week}
DEGREE = 0 {Week}
EGG_SURVIVAL = ABS(RANDOM(SURVIVAL,.009))
LARGE_PRED_BIO = ADULT+LATE_JUV+YEARLING {Prey for
adults}
REGULATION_STRENGTH = 1
REPROD = ADULT+(LARGE_ADULT*1.2)
SEASONAL_SHIFT = 1-PULSE(1,8,52)
SEVERITY = .1 {Reflects year-to-year variation in
winter severity and subsequent winter kill}
SEVERITY_6 = RANDOM(.1,.01)
SHAD_BIOMASS = IF ZOOP > 2*TOTAL_POPULATION THEN
((LARVAL*.05)+(EARLY_JUV*3)+(LATE_JUV*10)+((YEARLING*30
)+(ADULT*100)+(LARGE_ADULT*200))*.001)*.7 ELSE
(LARVAL*.05)+(EARLY_JUV*3)+(LATE_JUV*3)
SIZE_RATE = LARVAL_PRED*.5
SIZE_RATE_2 = SMALL_PRED_RATE*.5
SIZE_RATE_3 = (SMALL_PRED_RATE*.5)+(PRED_RATE*.5)
SIZE_RATE_4 = PRED_RATE*.125
SIZE_RATE_5 = PRED_RATE*.01
SMALL_PRED_BIO = LARVAL+EARLY_JUV+(LATE_JUV*.5){prey
for small predators}
SURVIVAL = .008 {reflects year to year variation in egg
survival due to flooding}
TIMING = 52 {Weeks}
TOTAL_POPULATION =
ADULT+LARVAL+LARGE_ADULT+LATE_JUV+EARLY_JUV+YEARLING
{Individuals}
WINTERKILL = PULSE(LARVAL*SEVERITY,50,52) {Individuals
per Week}
WINTERKILL_2 = PULSE(EARLY_JUV*SEVERITY,50,52)
{Individuals per Week}
WINTERKILL_3 = PULSE(LATE_JUV*SEVERITY,50,52)
{Individuals per Week}
```

```
WINTERKILL_4 = PULSE(YEARLING*SEVERITY,50,52)
{Individuals per Week}
WINTERKILL_5 = PULSE(ADULT*SEVERITY,50,52) {Individuals
per Week}
WINTERKILL_6 = PULSE(LARGE_ADULT*SEVERITY_6,50,52)
{Individuals per Week}
YOY = EARLY_JUV+LARVAL+LATE_JUV {Individuals}
CONSUMPTION_RATE_SHAD = GRAPH(SHAD_BIOMASS)
(0.00, 0.00), (100, 0.065), (200, 0.155), (300, 0.25),
(400, 0.345), (500, 0.46), (600, 0.54), (700, 0.66),
(800, 0.775), (900, 0.895), (1000, 1.00)
CONSUMPTION_RATE_ZOOP = GRAPH(ZOOP )
(0.00, 0.01), (10.0, 0.34), (20.0, 0.52), (30.0,
0.625), (40.0, 0.73), (50.0, 0.805), (60.0, 0.865),
(70.0, 0.905), (80.0, 0.94), (90.0, 0.98), (100, 1.00)
EARLY_JUV_GROWTH = GRAPH((EARLY_JUV+LARVAL)/ZOOP)
(0.00, 0.99), (100, 0.995), (200, 1.00), (300, 0.99),
(400, 0.965), (500, 0.705), (600, 0.435), (700, 0.34),
(800, 0.31), (900, 0.29), (1000, 0.3)
GROWTH_ALGAE = GRAPH(ALGAE)
(0.00, 14.8), (1e+07, 12.0), (2e+07, 9.30), (3e+07,
7.95), (4e+07, 6.83), (5e+07, 6.45), (6e+07, 6.22),
(7e+07, 6.30), (8e+07, 6.22), (9e+07, 6.30), (1e+08,
6.30)
GROWTH_ZOOP = GRAPH(ALGAE )
(0.00, 0.015), (10.0, 0.115), (20.0, 0.195), (30.0,
0.295), (40.0, 0.38), (50.0, 0.475), (60.0, 0.565),
(70.0, 0.675), (80.0, 0.765), (90.0, 0.895), (100,
0.99)
LARVAL_GROWTH = GRAPH(LARVAL/ZOOP)
(0.00, 0.995), (10.0, 0.845), (20.0, 0.755), (30.0,
0.65), (40.0, 0.575), (50.0, 0.475), (60.0, 0.365),
(70.0, 0.295), (80.0, 0.255), (90.0, 0.215), (100,
0.155)
LARVAL_PRED = GRAPH(JUV_PRED)
(0.00, 2.00), (10.0, 4.50), (20.0, 15.5), (30.0, 19.5),
(40.0, 27.5), (50.0, 39.0), (60.0, 48.5), (70.0, 62.0),
(80.0, 75.0), (90.0, 87.0), (100, 97.5)
PREDATOR_GROW = GRAPH(SMALL_PRED_BIO)
(0.00, 0.098), (10.0, 0.0785), (20.0, 0.06), (30.0,
0.043), (40.0, 0.0005), (50.0, 0.0005), (60.0, 0.0005),
(70.0, 0.0005), (80.0, 0.00), (90.0, 0.00), (100, 0.00)
PRED_DEATH = GRAPH(LARGE_PRED_BIO)
```

```
(0.00, 0.055), (100, 0.05), (200, 0.0275), (300, 0.02),
(400, 0.015), (500, 0.015), (600, 0.01), (700, 0.01),
(800, 0.01), (900, 0.0075), (1000, 0.01)
PRED_RATE = GRAPH(PREDATOR )
(0.00, 0.02), (10.0, 0.115), (20.0, 0.195), (30.0,
0.26), (40.0, 0.365), (50.0, 0.465), (60.0, 0.57),
(70.0, 0.685), (80.0, 0.785), (90.0, 0.9), (100, 0.99)
SMALL_PRED_DIE = GRAPH(LARVAL+EARLY_JUV )
(0.00, 0.0975), (100, 0.0645), (200, 0.037), (300,
0.0225), (400, 0.0165), (500, 0.00), (600, 0.00), (700,
0.00), (800, 0.00), (900, 0.00), (1000, 0.00)
SMALL_PRED_RATE = GRAPH(JUV_PRED)
(0.00, 0.015), (10.0, 0.1), (20.0, 0.175), (30.0,
0.275), (40.0, 0.375), (50.0, 0.485), (60.0, 0.61),
(70.0, 0.715), (80.0, 0.83), (90.0, 0.92), (100, 1.00)
```

36

Modeling Spatial Dynamics of Predator–Prey Interactions in a Changing Environment*

It is a sad fact that several of our most noble birds of prey can no longer be studied in what were once their native haunts.

D.A. Bannerman, *Birds of the British Isles,* 1956

In Chapter 32 we modeled the spatial dynamics of two species that compete with each other for the same parcel of land. Let us now model the case of spatial predator–prey interactions. To model the spatial aspect of these interactions, we define four subdivisions of the landscape as laid out in Figure 36.1.

Cell 1 is occupied with 1000 predators and 10,000 prey. There are no predators or prey in the other three cells. The birth rates for predators and prey are given exogenously, yet the number of births depends on predator–prey interaction and the carrying capacity of their ecosystem. Similarly, the deaths of predators and prey depend on their interaction—the prey consumed by predators. For example, the births and deaths of predators and prey in cell 1 are defined as

$$\text{BIRTH PRED 1} = \text{BR PRED*(PRED 1-DEATH PRED 1)} \quad (1)$$

$$\text{DEATH PRED 1} = \text{PRED 1-CONSUME 1/CONSUME RATE} \quad (2)$$

$$\text{BIRTH PREY 1} = \text{(PREY 1-DEATH PREY 1)*BR PREY 1 *(1-PREY 1/CC PREY 1)} \quad (3)$$

1	2
3	4

FIGURE 36.1

*We wish to thank Jim Westervelt for his contributions to this chapter.

$$\text{DEATH PREY 1} = \text{CONSUME 1}, \tag{4}$$

where PREY 1 and PRED 1 are the stock of the two populations in cell 1, BR PREY 1 is the birth rate of prey in cell 1, and CONSUME 1 is the number of prey consumed by predators in cell 1. The number of prey consumed in cell 1 cannot exceed the number of prey in that cell and is at least zero and at most the consumption rate times the number of predators in that cell. The amount of prey consumed by predators in cell 1 is defined as

$$\text{CONSUME 1} = \text{MIN(PREY 1,CONSUME RATE*PRED 1)} \tag{5}$$

CC PREY 1 is the carrying capacity for prey in cell 1. The latter is assumed to change along a sinewave with an exogenously defined CYCLE TIME:

$$\text{CC PREY 1} = \text{2250*SIN(2*PI*TIME/CYCLE TIME)+2500}. \tag{6}$$

In addition, we assumed that the carrying capacity for each individual patch was out of phase by 90 degrees with the "previous" patch, that is,

$$\text{CC PREY 2} = \text{2250*SIN(2*PI*TIME/CYCLE TIME–PI/2)+2500} \tag{7}$$

$$\text{CC PREY 3} = \text{2250*SIN(2*PI*TIME/CYCLE TIME–3*PI/2)+2500} \tag{8}$$

$$\text{CC PREY 4} = \text{2250*SIN(2*PI*TIME/CYCLE TIME–PI)+2500} \tag{9}$$

Prey migrate routinely regardless of their population in the starting or the receiving cells, and the predators migrate to a new cell when they begin to starve in their current cell. Once the migration quantity is established, a random process determines its distribution to adjacent cells. This simple idea enables the prey to "escape" to a neighboring cell, where the predator population may be at a relatively low level.

The model for cell 1 of the 4 cell predator–prey model is shown in Figure 36.2. Predators and prey can move to the right (RT) and down (DN). The fraction of predators and prey that move is a random number between zero and one. For example, the migration of predators from cell one to any of the other cells is defined as

$$\text{MIG PRED 1} = \text{MIN(PRED 1–DEATH PRED 1,MIG RATE PRED*} \\ \text{DEATH PRED 1)} \tag{10}$$

The migration rate of predators, MIG RATE PRED, depends on the availability of prey in the cell. For simplicity, we assume that the number of predators that migrate is the product of the number of starvation deaths and the migration rate of predators. In the model, we define this number as a constant.

How do the predator and prey populations vary as the cycle time is varied on the carrying capacity? For what values of the carrying capacity will the predator and prey populations have the lowest standard deviation? Run several trials with cycle time being varied; you may want to make use of the

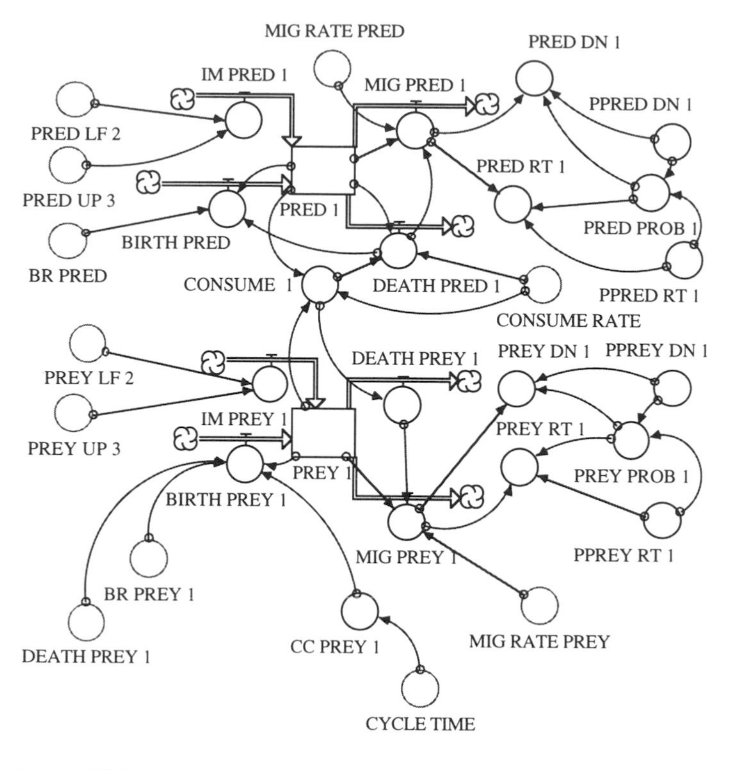

FIGURE 36.2

MADONNA software to speed up the sensitivity analysis and calculate the means and standard deviations for a series of model runs with different cycle times. To evaluate which cycle time produced the most stable populations, you must look for the least value of the standard deviation. For the predator population, a cycle time of 8 produced a low standard deviation of approximately 630. Similarly, for the prey population, a cycle time of 1 produced a low standard deviation of about 2242. Are these the lowest standard deviations one can find? Run the model many times for the same cycle time and observe the results.

The two cases of CYCLE TIME = 1 and CYCLE TIME = 8 are shown, respectively, in the graphs in Figures 36.3 and 36.4. Admittedly, even in the best cases, the population still varies significantly over time. This variation can best be minimized by increasing the number of patches that describe the ecosystem.

By holding the phase difference of the sinusoidal carrying capacity constant between patches, will a variation of frequency between patches create

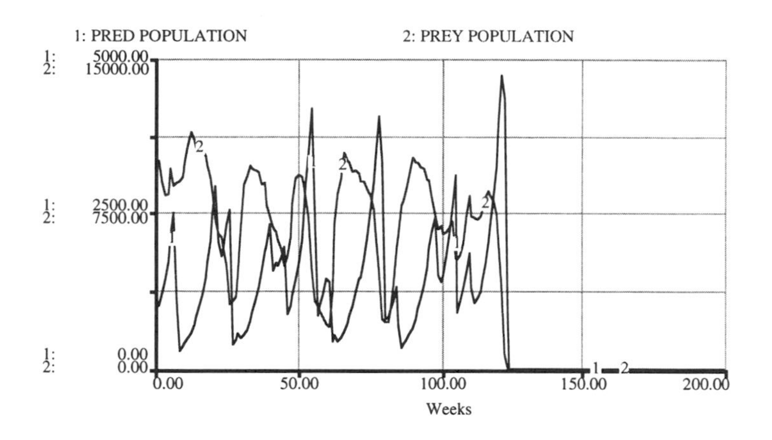

FIGURE 36.3

a more stable ecosystem? Can a similar phase difference that was used in this model be applied to a spatial model with nine or more patches? Can variations in other parameters, including birth rate, migration rate, and consumption rate, create the desired effects of this spatial predator–prey model? Try to find answers to these questions.

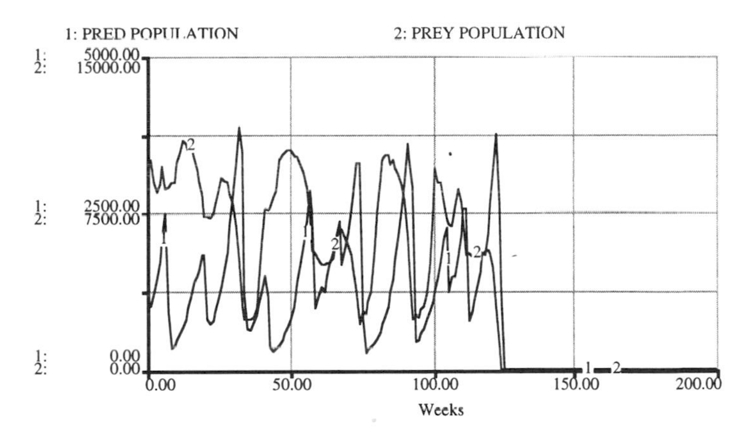

FIGURE 36.4

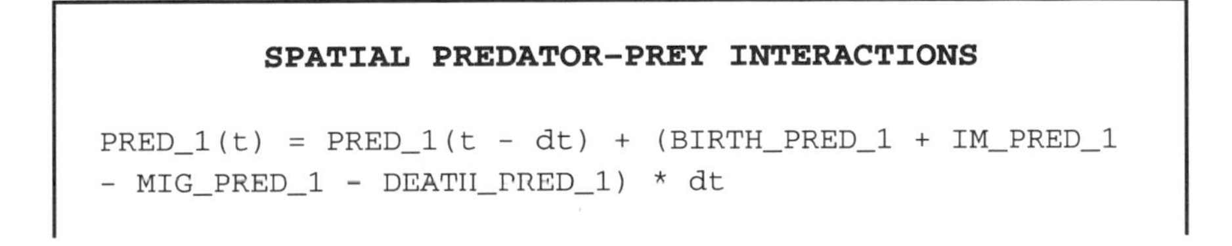

SPATIAL PREDATOR-PREY INTERACTIONS

```
PRED_1(t) = PRED_1(t - dt) + (BIRTH_PRED_1 + IM_PRED_1
- MIG_PRED_1 - DEATH_PRED_1) * dt
```

```
INIT PRED_1 = 1000
INFLOWS:
BIRTH_PRED_1 = BR_PRED*(PRED_1-DEATH_PRED_1)
IM_PRED_1 = PRED_LF_2+PRED_UP_3
OUTFLOWS:
MIG_PRED_1 = MIN(PRED_1-
DEATH_PRED_1,MIG_RATE_PRED*DEATH_PRED_1)
DEATH_PRED_1 = PRED_1-CONSUME__1/CONSUME_RATE

PRED_2(t) = PRED_2(t - dt) + (BIRTH_PRED_2 + IM_PRED_2
- MIG_PRED_2 - DEATH_PRED_2) * dt
INIT PRED_2 = 0
INFLOWS:
BIRTH_PRED_2 = BR_PRED*(PRED_2-DEATH_PRED_2)
IM_PRED_2 = PRED_RT_1+PRED_UP_4
OUTFLOWS:
MIG_PRED_2 = MIN(PRED_2-
DEATH_PRED_2,MIG_RATE_PRED*DEATH_PRED_2)
DEATH_PRED_2 = PRED_2-CONSUME_2/CONSUME_RATE

PRED_3(t) = PRED_3(t - dt) + (BIRTH_PRED_7 + IM_PRED_7
- MIG_PRED_3 - DEATH_PRED_3) * dt
INIT PRED_3 = 0
INFLOWS:
BIRTH_PRED_7 = BR_PRED*(PRED_3-DEATH_PRED_3)
IM_PRED_7 = PRED_DN_1+PRED_LF_4
OUTFLOWS:
MIG_PRED_3 = MIN(PRED_3-
DEATH_PRED_3,DEATH_PRED_3*MIG_RATE_PRED)
DEATH_PRED_3 = PRED_3-CONSUME_3/CONSUME_RATE

PRED_4(t) = PRED_4(t - dt) + (BIRTH_PRED_4 + IM_PRED_4
- MIG_PRED_4 - DEATH_PRED_4) * dt
INIT PRED_4 = 0
INFLOWS:
BIRTH_PRED_4 = BR_PRED*(PRED_4-DEATH_PRED_4)
IM_PRED_4 = PRED_DN_2+PRED_RT_3
OUTFLOWS:
MIG_PRED_4 = MIN(PRED_4-
DEATH_PRED_4,DEATH_PRED_4*MIG_RATE_PRED)
DEATH_PRED_4 = PRED_4-CONSUME_4/CONSUME_RATE

PREY_1(t) = PREY_1(t - dt) + (BIRTH_PREY_1 + IM_PREY_1
- DEATH_PREY_1 - MIG_PREY_1) * dt
```

```
INIT PREY_1 = 10000
INFLOWS:
BIRTH_PREY_1 = (PREY_1-DEATH_PREY_1)*BR_PREY_1*(1-
PREY_1/CC_PREY_1)
IM_PREY_1 = PREY_LF_2+PREY_UP_3
OUTFLOWS:
DEATH_PREY_1 = CONSUME__1
MIG_PREY_1 = MIG_RATE_PREY*(PREY_1-DEATH_PREY_1)

PREY_2(t) = PREY_2(t - dt) + (BIRTH_PREY_2 + IM_PREY_2
- DEATH_PREY_2 - MIG_PREY_2) * dt
INIT PREY_2 = 0
INFLOWS:
BIRTH_PREY_2 = (PREY_2-DEATH_PREY_2)*BR_PREY_1*(1-
PREY_2/CC_PREY_2)
IM_PREY_2 = PREY_RT_1+PREY_UP_4
OUTFLOWS:
DEATH_PREY_2 = CONSUME_2
MIG_PREY_2 = MIG_RATE_PREY*(PREY_2-DEATH_PREY_2)

PREY_3(t) = PREY_3(t - dt) + (BIRTH_PREY_3 + IM_PREY_3
- DEATH_PREY_3 - MIG_PREY_3) * dt
INIT PREY_3 = 0
INFLOWS:
BIRTH_PREY_3 = (PREY_3-DEATH_PREY_3)*BR_PREY_1*(1-
PREY_3/CC_PREY_3)
IM_PREY_3 = PREY_DN_1+PREY_LF_4
OUTFLOWS:
DEATH_PREY_3 = CONSUME_3
MIG_PREY_3 = MIG_RATE_PREY*(PREY_3-DEATH_PREY_3)

PREY_4(t) = PREY_4(t - dt) + (IM_PREY_4 + BIRTH_PREY_4
- DEATH_PREY_4 - MIG_PREY_4) * dt
INIT PREY_4 = 0
INFLOWS:
IM_PREY_4 = PREY_DN_2+PREY_RT_3
BIRTH_PREY_4 = (PREY_4-DEATH_PREY_4)*BR_PREY_1*(1-
PREY_4/CC_PREY_4)
OUTFLOWS:
DEATH_PREY_4 = CONSUME_4
MIG_PREY_4 = MIG_RATE_PREY*(PREY_4-DEATH_PREY_4)

BR_PRED = .2
BR_PREY_1 = 2
```

```
CC_PREY_1 = 2250*SIN(2*PI*TIME/CYCLE_TIME)+2500
CC_PREY_2 = 2250*SIN(2*PI*TIME/CYCLE_TIME-PI/2)+2500
CC_PREY_3 = 2250*SIN(2*PI*TIME/CYCLE_TIME-3*PI/2)+2500
CC_PREY_4 = 2250*SIN(2*PI*TIME/CYCLE_TIME-PI)+2500
CONSUME_2 = MIN(PREY_2,CONSUME_RATE*PRED_2)
CONSUME_3 = MIN(PREY_3,CONSUME_RATE*PRED_3)
CONSUME_4 = MIN(PREY_4,CONSUME_RATE*PRED_4)
CONSUME_RATE = 1
CONSUME__1 = MIN(PREY_1,CONSUME_RATE*PRED_1)
CYCLE_TIME = 8
MIG_RATE_PRED = .05 {This times the number of
starvation deaths is the number that migrate}
MIG_RATE_PREY = .1 {This is the proportion of the prey
that migrate}
PPRED_DN_1 = RANDOM(0,1)
PPRED_DN_2 = RANDOM(0,1)
PPRED_LF_2 = RANDOM(0,1)
PPRED_LF_4 = RANDOM(0,1)
PPRED_RT_1 = RANDOM(0,1)
PPRED_RT_3 = RANDOM(0,1)
PPRED_UP_3 = RANDOM(0,1)
PPRED_UP_4 = RANDOM(0,1)
PPREY_DN_1 = RANDOM(0,1)
PPREY_DN_2 = RANDOM(0,1)
PPREY_LF_2 = RANDOM(0,1)
PPREY_LF_4 = RANDOM(0,1)
PPREY_RT_1 = RANDOM(0,1)
PPREY_RT_3 = RANDOM(0,1)
PPREY_UP_3 = RANDOM(0,1)
PPREY_UP_4 = RANDOM(0,1)
PRED_DN_1 = MIG_PRED_1*PPRED_DN_1/PRED_PROB_1
PRED_DN_2 = MIG_PRED_2*PPRED_DN_2/PRED_PROB_2
PRED_LF_2 = MIG_PRED_2*PPRED_LF_2/PRED_PROB_2
PRED_LF_4 = MIG_PRED_4*PPRED_LF_4/PRED_PROB_4
PRED_POPULATION = PRED_1+PRED_2+PRED_3+PRED_4
PRED_PROB_1 = PPRED_RT_1+PPRED_DN_1
PRED_PROB_2 = PPRED_LF_2+PPRED_DN_2
PRED_PROB_3 = PPRED_RT_3+PPRED_UP_3
PRED_PROB_4 = PPRED_LF_4+PPRED_UP_4
PRED_RT_1 = MIG_PRED_1*PPRED_RT_1/PRED_PROB_1
PRED_RT_3 = MIG_PRED_3*PPRED_RT_3/PRED_PROB_3
PRED_UP_2 = MIG_PRED_4*PPRED_UP_4/PRED_PROB_4
PRED_UP_3 = MIG_PRED_3*PPRED_UP_3/PRED_PROB_3
```

```
PREY_DN_1 = MIG_PREY_1*PPREY_DN_1/PREY_PROB_1
PREY_DN_2 = MIG_PREY_2*PPREY_DN_2/PREY_PROB_2
PREY_LF_2 = MIG_PREY_2*PPREY_LF_2/PREY_PROB_2
PREY_LF_4 = MIG_PREY_4*PPREY_LF_4/PREY_PROB_4
PREY_POPULATION = PREY_1+PREY_3+PREY_2+PREY_4
PREY_PROB_1 = PPREY_RT_1+PPREY_DN_1
PREY_PROB_2 = PPREY_LF_2+PPREY_DN_2
PREY_PROB_3 = PPREY_RT_3+PPREY_UP_3
PREY_PROB_4 = PPREY_LF_4+PPREY_UP_4
PREY_RT_1 = MIG_PREY_1*PPREY_RT_1/PREY_PROB_1
PREY_RT_3 = MIG_PREY_3*PPREY_RT_3/PREY_PROB_3
PREY_UP_3 = MIG_PREY_3*PPREY_UP_3/PREY_PROB_3
PREY_UP_4 = MIG_PREY_4*PPREY_UP_4/PREY_PROB_4
```

Part 7

Catastrophe and Self-Organization

37

Catastrophe

> The choice of the name, catastrophe theory, is unfortunate as it denotes abnormal nasty events. What we have come to realize is that such events are normal and necessary for the continued smooth functioning of many systems.
>
> E.D. Schneider and J.J. Kay,
> Complexity and Thermodynamics, *Futures*, 26: 641, 1994.

If a large number of real systems exhibit dynamics that bear the potential for chaos, why do we not see more chaos in real-world processes? Fortunately, the domains over which stability of the system occurs can be releatively large. But once in a while, systems may move "towards the edge of stability" and little nudges to the system may move it from stability to instability, that is, into a castastrophe. Subsequently, reorganization of system components may occur to bring the system back into a stable domain, a kind of evolutionary process. This stable domain, however, may not be the same as the one prior to the disturbance.

The system undergoes a catastrophic event in the sense that it is moved from an initial state of stability through a dramatic phase of reorganization and back to some degree of stability. Examples for such catastrophic events include landslides, avalanches, earthquakes, and pest outbreaks in ecosystems. In each case, small changes in the system occur that individually may not be critical to the system's behavior. Collectively, however, they lead to the evolution of the system towards a critical state. This is apparent, for example, in the case of avalanches. Each individual snowflake potentially adds to the instability of the system. When a critical point is reached, the next snowflake may trigger an avalanche that affects a large part of the system. Temporary stability is quickly reached if the avalanche was not too dramatic. Even if not of a large scale, the avalanche adds to the "stress" of the system downhill, making it more susceptible to further avalanches as more snow falls at those regions or as additional small avalanches are received from higher on the hill. Ultimately, a large-scale, catastrophic event may occur that affects the entire system, not just individual regions. The system components regroup and finally enter a phase of new, temporary stability.

So, evolutionary processes are at work, making the system more "efficient" in some sense. This is evolution toward catastrophe. A system in such a state can remerge to a stable state by another process of evolution, likely faster than the first kind, and this new stable state may not be very efficient. Large living natural systems are likely constrained from operating at

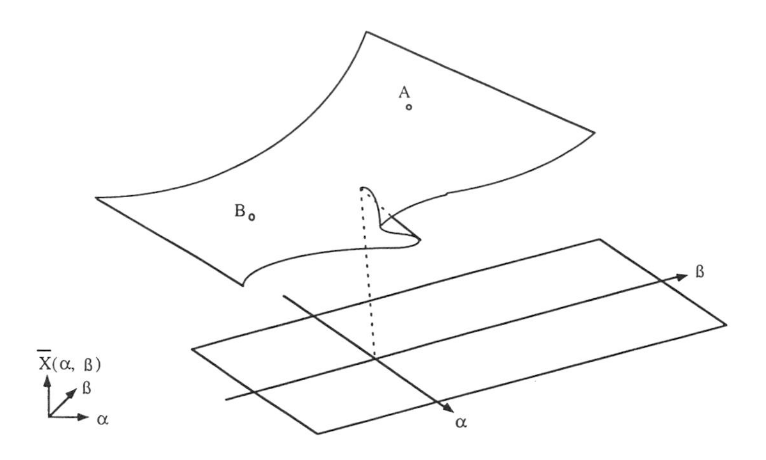

FIGURE 37.1

or near peak efficiency by random intervention of uncoordinated external processes at the regional levels.

In this chapter we develop a simple model of catastrophe and then go on to models of sandpiles, earthquakes, and pest outbreakes. Before we develop these models, consider the Figure 37.1, which illustrates the surface defined by the following equation[1]:

$$X^3 - ALPHA*X - BETA = 0 \quad (1)$$

Imagine a ball lying at the top of this surface such as point A. The ball may lie still, and very small nudges away from its equilibrium point A lead to a new equilibrium. After a series of such small perturbations, however, the ball will roll off the top part of the surface, and a priori it is difficult for us to determine exactly where it will end up. All we know for certain is that the new equilibrium position is somewhere at the bottom of the surface, say, point B.

Small nudges to the ball in B will again move it slightly away from B. And if we kick it hard enough, we can propel the ball through the fold, or "cusp," to the upper part of the surface again. Where exactly will it end up? To give a precise answer requires exact knowledge of the shape of the surface, the properties of the ball, and the magnitude and direction of the force exerted on the ball. In more complicated, real-life systems, not all the

[1]See E. Beltrami, *Mathematics for Dynamic Modeling*, Boston: Academic Press, Inc. 1987.

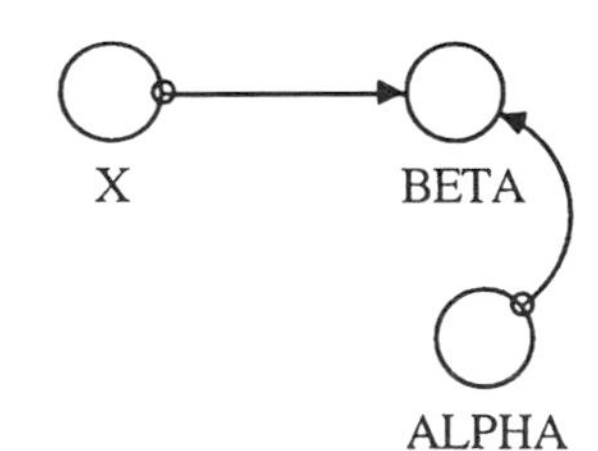

FIGURE 37.2

variables to describe the system and the forces incident upon them are well enough known. As a result, we may only know stability domains rather than specific locations.

The STELLA model for equation (1) is given in Figure 37.2. We slightly vary X with each simulation time step. Solve equation (1) for BETA. Set DT = .0025 and define X, for example, as

$$X = \text{TIME} - 2; \tag{2}$$

then run the model.

For positive ALPHA, the cusp or fold appears in the X versus BETA plot. For ALPHA = 0 and negative ALPHA, the S curve appears. The cases of positive, zero, and negative values for ALPHA are shown in the graphs in Figures 37.3, 37.4, and 37.5, respectively.

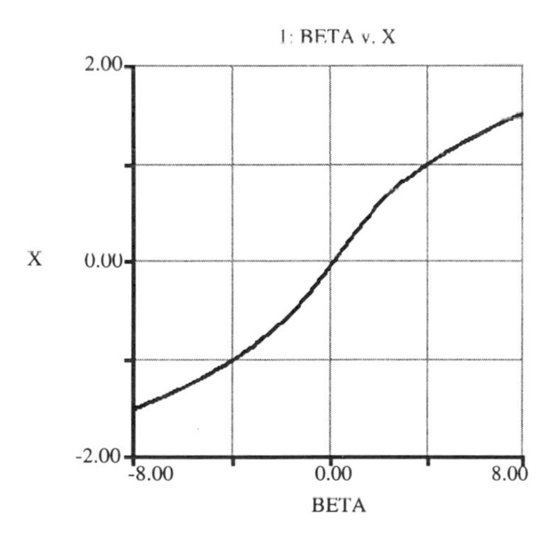

FIGURE 37.3

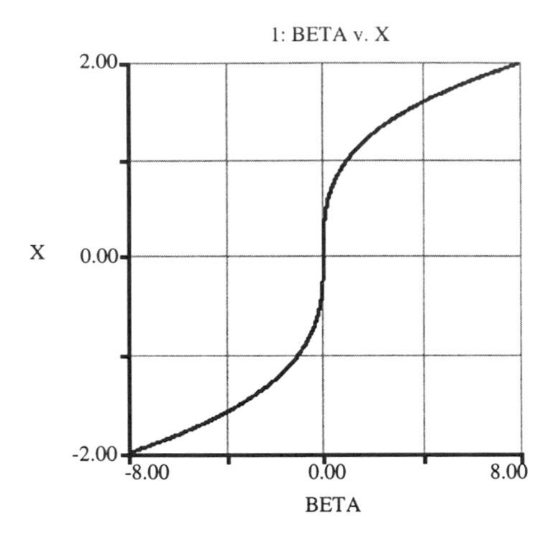

FIGURE 37.4

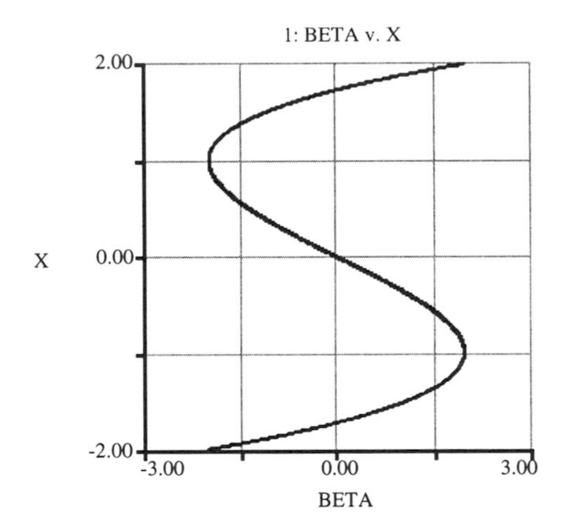

FIGURE 37.5

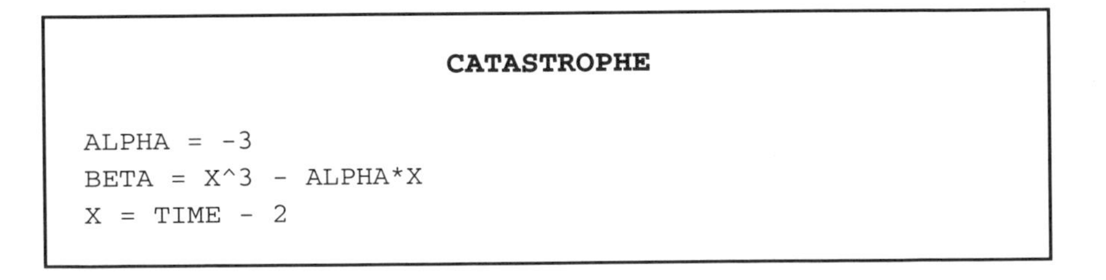

CATASTROPHE

```
ALPHA = -3
BETA = X^3 - ALPHA*X
X = TIME - 2
```

38

Spruce Budworm Dynamics

Books are subject among other Chances to fire,
and the Worme.

Richard Whitlock,
1654 *Zootomia*, 230

A classical example for the implications of catastrophes for ecosystem management is spruce budworm dynamics. Spruce budworm is a caterpillar that feeds on spruce and fir forests in the northeastern United States and eastern Canada. For many years, population sizes of spruce budworm are low and have little impact on trees. When forest stands reach maturity, however, spruce budworm populations explode, seriously affecting the forest by defoliating the trees. As a result of defoliation, trees are weakened and ultimately may die. With the death of trees comes a loss of the food source for spruce budworm and a consequent population crash.

The cycle of low spruce budworm population densities, followed by population explosions and catastrophic collapse, tend to repeat themselves over the course of years. The resulting damage and death of trees affects negatively the timber and paper pulp industries of the region. Frequently, forest managers decided to spray forest stands to control budworm populations. The dynamics inherent in the system, however, lead it to follow its own path, making ever more extensive pest control necessary. If those controls fail, outbreaks are more severe and devastating than if the system had been left to control itself, as recent experiences in the United States and Canada show.

One of the natural system controls not now used in forestry relates to the idea of patch size. Natural systems no doubt avoid large catastophes because they operate in patches, where the degree of maturity of adjacent patches is nearly always different. Consequently, pests and fires find difficulty in spreading beyond a patch, and the size of the catastrophe is kept small. Current forestry practice seems to be disconnected from such natural system behavior. Our model is not a regional one and so such interpatch dynamics are not captured. However, we have modeled such patch dynamics in Chapter 32, and you may want to combine the models of Chapter 32 with the one developed here.

To model the spruce budworm catastrophe,[1] let us denote B as the budworm population density, K as the carrying capacity, S as habitat density, and GR as the budworm's natural rate of increase. Thus,

---

[1]This model follows closely the model laid out in E. Beltrami, *Mathematics for Dynamic Modeling*, Boston: Academic Press, Inc., 1987, pp. 189–196.

$$\frac{dB}{dt} = GR * B * \left(1 - \frac{B}{K * S}\right) \tag{1}$$

would describe the population dynamics for a fixed carrying capacity and no predatory influences on population growth. This is the logistic growth equation that we have used in this book many times before. Let us introduce the effects of predation with a maximum predation rate C, which is assumed to be constant. At small population densities, predation has only little effects on the budworm population because they are well hidden in a relatively dense canopy. As population densities increase, however, predators may increasingly feed on budworm that partially or totally defoliated the trees and are then easy prey. A predation term that captures such interactions is

$$\frac{C * B^2}{A^2 + B^2}, \tag{2}$$

with A as a scalar that captures the effectiveness of the predators to spot and prey on spruce budworm. In an immature forest, predation is easier than in a mature forest with a diverse and dense canopy. Thus, A may be assumed to increase with increased maturity of the forest, that is, habitat size S

$$A = K1 * S, \tag{3}$$

and thus

$$\frac{C * B^2}{A^2 + B^2} = \frac{C * B^2}{(K1 * S)^2 + B^2}, \tag{4}$$

with K1 as a constant.

Combining predation with the logistic growth function yields

$$\Delta B = \frac{dB}{dt} = GR * B * \left(1 - \frac{B}{K * S}\right) - \frac{C * B^2}{(K1 * S)^2 + B^2}, \tag{5}$$

which is the equation used in the model to drive spruce budworm population changes, ΔB.

Changes in habitat size are assumed to also follow the logistic growth curve, with RS as the natural rates of increase and KS as carrying capacity:

$$\Delta S = \frac{dS}{dt} = RS * S * \left(1 - \frac{S}{KS * E}\right). \tag{6}$$

E is the percentage of foliage on trees. The more healthy the forest, the higher E. The percentage of foliage on trees is assumed to decrease as the average budworm density per habitat size B/S increases. To model diminishing stress as budworm populations decrease, we multiply B/S by E^2. The combined effect of logistic growth in foliage and budworm-induced foliage losses is

$$\Delta E = \frac{dE}{dt} = RE * E * (1 - E) \frac{P * B * E^2}{S}, \tag{7}$$

with RE the rate of foliage increase and P a proportionality factor.

Let us consider the case of $B \neq 0$ and introduce the following notation:

$$R = \frac{GR * K1 * S}{C} \tag{8}$$

$$Q = \frac{K}{K1}, \tag{9}$$

and rewrite

$$B = K1 * S * X. \tag{10}$$

It can be shown[2] that the nontrivial equilibria of equation (5) fulfill

$$R\left(1 - \frac{X}{Q}\right) = F(X) = G(X), \tag{11}$$

with

$$G(X) = \frac{X}{1 + X^2}. \tag{12}$$

The left side of equation (11) is a straight line F(X) with slope –R/Q. Equilibria occur where this line intersects with G(X).

S and R increase with increases in Q. At first, there is a single equilibrium, corresponding to the situation shown in Figures 38.1, left. After some time, the line becomes tangent to the curve, as in Figure 38.1, middle. With further increases in the slope, two points that "attract" system behavior emerge (Figure 38.1, right). These are the two outer points in Figure 38.1, middle.

Note that in each of the following graphs, X is the horizontal axis, the two sides of equation (11) are plotted on the vertical axis, and F has the slope R/Q. As such, F will be equal to or intersect G at equilibrium points (per equations [11] and [12]) of varying stability. The relative slopes of F and G at the points of intersection determine the stability. If F is less than G, to the right of the intersection, the point is stable. Thus, between the right graph in Figure 38.1 and the left graph in Figure 38.2 the left intersection switches suddenly from stable to unstable, the only stable point becoming the right intersection. This sudden shift in stable points is what causes the sudden shift or explosion in the budworm population. The right graph in

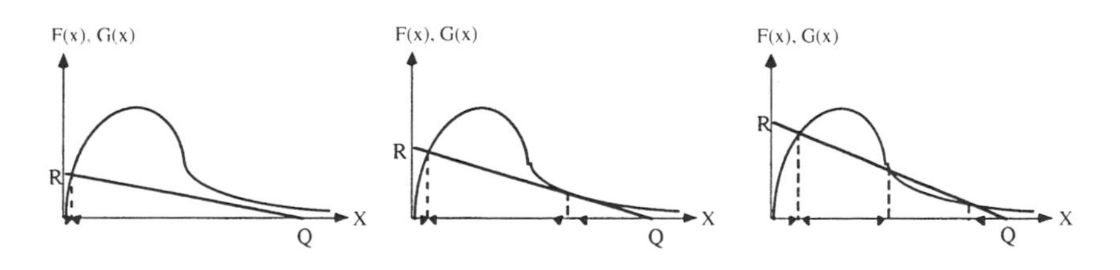

FIGURE 38.1

[2]E. Beltrami, *Mathematics for Dynamic Modeling*, Boston: Academic Press, 1987.

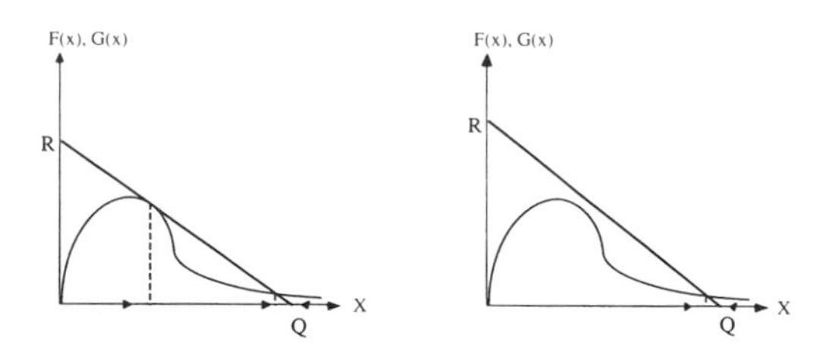

FIGURE 38.2

Figure 38.2 shows the case in which only one point of intersection exists. This point is a stable attractor.

The cusp of the spruce budworm dynamics is shown schematically in Figure 38.3 in the R-Q plane. The upper part of the surface corresponds to an outbreak level, while the lower part corresponds to a subsistence level.

The modules to solve for the dynamics of the spruce budworm population are shown in Figures 38.4 and 38.5.

We drive changes in the model by setting X = TIME. The functions G(X) and F(X) generated by our STELLA model are shown in the graphs in Figures 38.6 and 38.7.

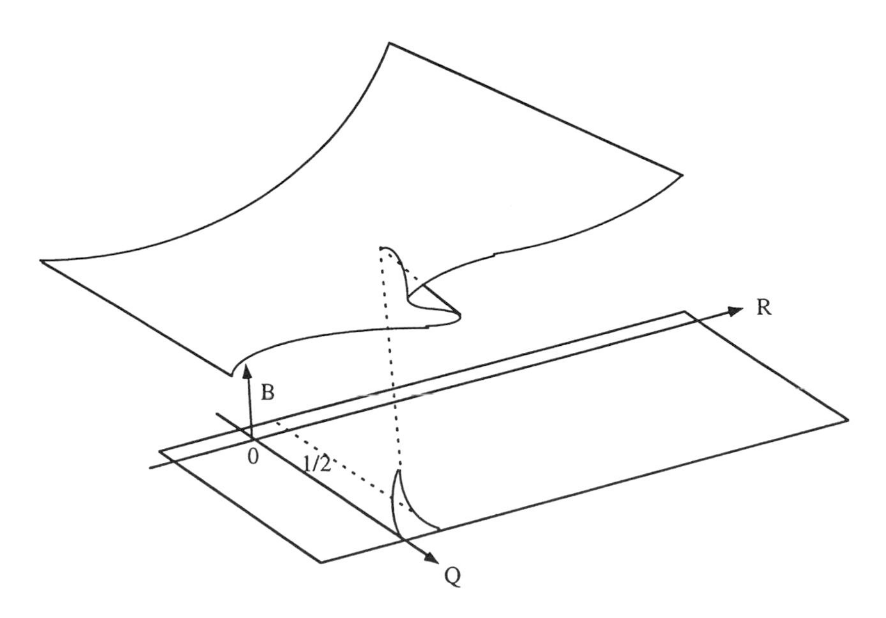

FIGURE 38.3

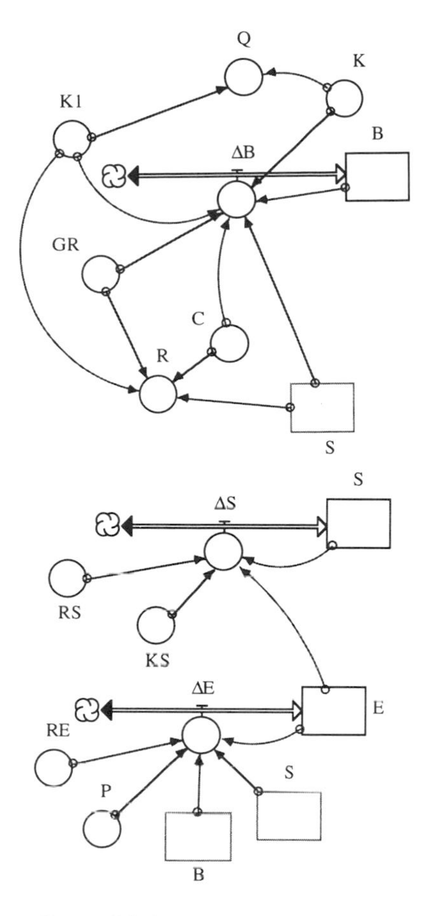

FIGURE 38.4

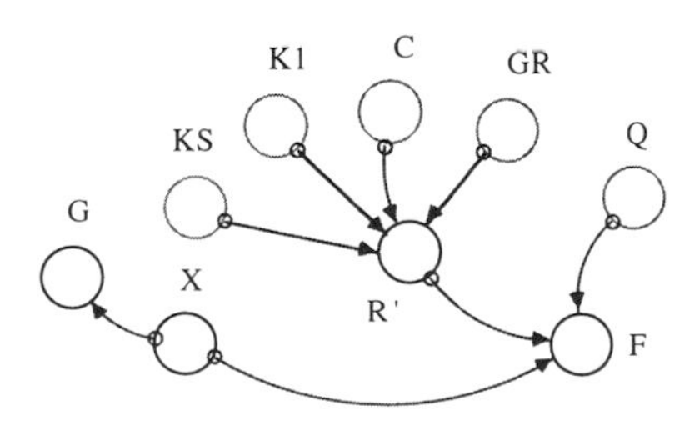

FIGURE 38.5

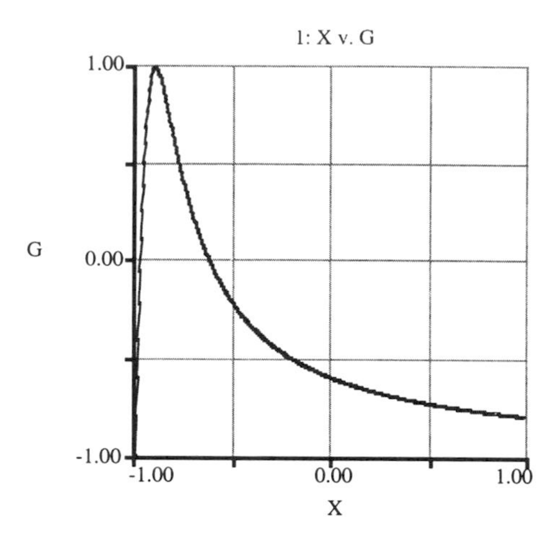

FIGURE 38.6

For KS = 2.5, the corresponding values of X are 0.69, 2.0, and 7.32. These correspond to B = 0.173, 0.500, and 1.83, respectively [see equation (10)]. These B values represent the steady states of B. However, only two of these extrema are stable; the middle value represents an unstable extremum. Within a given range of initial values of B the same steady state is reached. A different set of initial values lead to another steady state. There are two

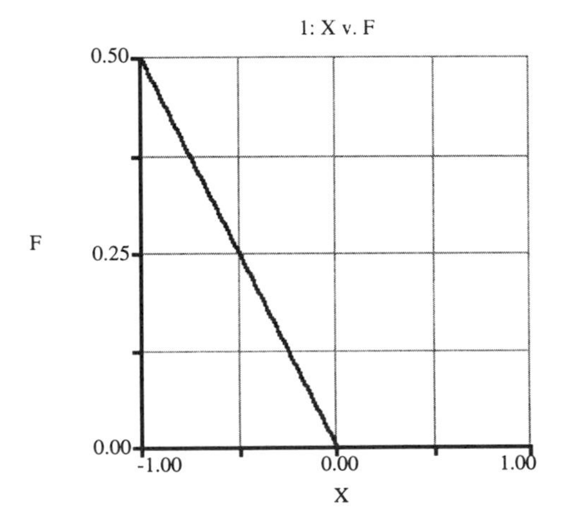

FIGURE 38.7

such ranges for initial Bs, given the way that the main model is set up, which are initial values of B greater than and less than 1/2 (Figs. 38.8 and 38.9).

Recognize that the choice of KS is a crucial one. So is the choice of Q. They determine whether there are one, two, or three extremal solutions for B. Run a series of sensitivity tests on KS and Q to explore their impact on the system's dynamics.

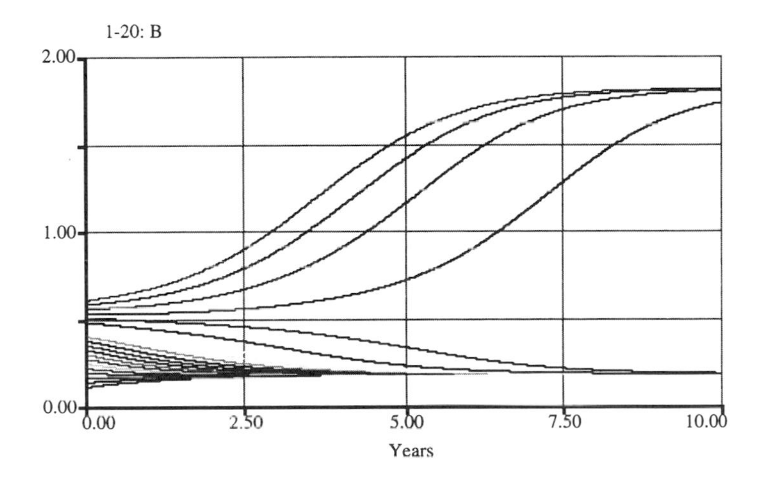

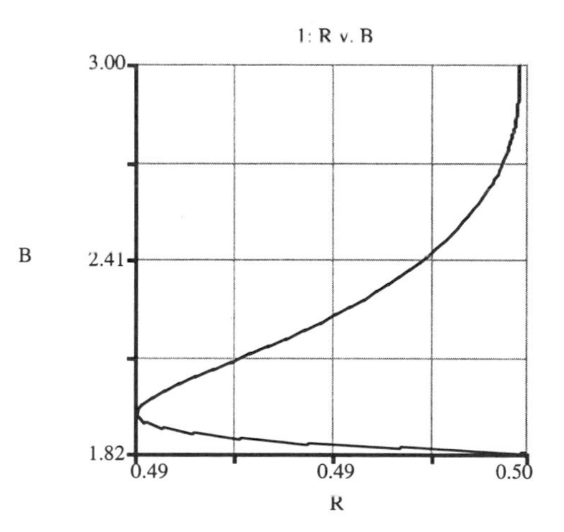

FIGURE 38.8

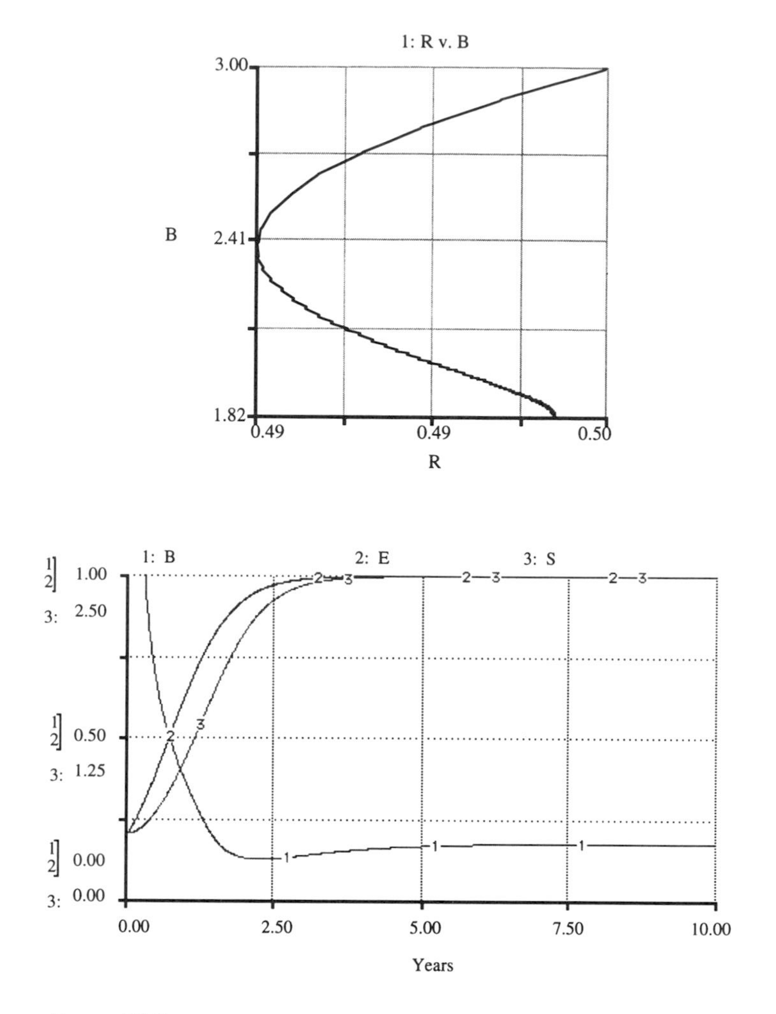

FIGURE 38.9

SPRUCE BUDWORM DYNAMICS

```
B(t) = B(t - dt) + (ΔB) * dt
INIT B = 3 {Spruce Budworm per Unit Area}
INFLOWS:
ΔB = GR*B*(1-B/K/S) - C*B^2/((K1*S)^2 + B^2) {Spruce
Budworm per Unit Area per Time Period}

E(t) = E(t - dt) + (ΔE) * dt
INIT E = .95 {Percentage of Foliage Cover}
```

```
INFLOWS:
ΔE = RE*E*(1-E) - P*B*E^2/S {Change in Percentage of
Foilage Cover per Time Period}

S(t) = S(t - dt) + (ΔS) * dt
INIT S = 2.5 {Habitat Density}
INFLOWS:
ΔS = RS*S*(1 - S/KS/E) {Habitat Density Change per Time
Period}

C = 1
F = R'*(1-X/Q)
G = X/(1+X^2)
GR = 2 {Spruce Budworm per Unit Area per Spruce Budworm
per Unit Area per Time Period}
K = 1
K1 = .1
KS = 2.5
P = .01
Q = K/K1
R = GR*K1*S/C
R' = GR*K1*KS/C
RE = 2 {Change in Percentage of Foilage Cover per
Percentage Foilage Cover Time Period}
RS = 3 {Habitat Density Change per Habitat Density
Change per Time Period}
X = TIME
```

39

Sandpile

Their error—That sand on which thy crumbling power is built.
Shelley to Lord Chancellor, 1817

One phenomenon exhibiting catastrophic events is a pile of sand to which grains of sand are continuously added from the top. As the sandpile grows higher and higher, the pile geometry approaches a critical state, such that at some point one additional grain of sand will cause a landslide. The addition of that last grain of sand leads to a catastrophe in the system. The phenomenon of a system evolving to a state in which a small perturbation can set off a large chain reaction is called *self-organized criticality.*[1]

Let us model the dynamics of the sandpile as more and more sand is added from the top. Additions of sand occur randomly, following the (arbitrary) rule

SAND DROP = If RANDOM(0,5)>4 THEN 1 ELSE 0. (1)

The model consists of five state variables that represent the amount of sand at each layer of the pile. Between each layer, a value called the *slope* is determined by dividing the amount of sand in the upper layer by the amount of sand in the lower layer. If the slope exceeds a critical value, say, 0.4, then a random amount of sand from the top layer is dumped onto the lower layer. Dumping sand onto lower layers may be enough to start another reaction on the layers below.

For example, the amount of sand dumped from LAYER A to LAYER B is specified as

```
A TO B = IF LAYER B = 0 THEN LAYER A
         ELSE IF LAYER B≥ LAYER A THEN LAYER A–LAYER B
         ELSE IF SLOPE AB>CRITICAL SLOPE THEN RAND 1
         ELSE 0                                          (2)
```

The first expression in equation (2), IF LAYER B = 0 THEN LAYER A, captures the fact of a freefall of all materials from LAYER A into the region of LAYER B. The second part of the conditional statement represents the slow

[1]For a nice discussion with examples see P. Bak, and K. Chen, Self-Organized Criticality, *Scientific American* January: 46–53, 1991.

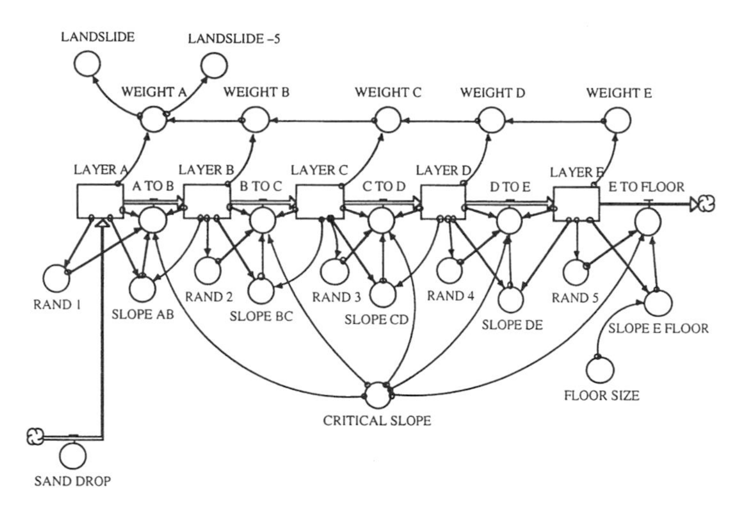

FIGURE 39.1

filling up of the lower layer by excess material from LAYER A. The third part of the statement is where a random amount of sand from the top layer gets dumped onto the lower layer. Similar statements determine the dynamics of sand movement between other layers of the sand pile.

The results of this simple landslide model (Fig. 39.1) show how the weight of sand builds up and is occasionally released in the catastrophic events of a landslide, to be built up again thereafter until the next release. The graph for the sand pile through time is shown in Figure 39.2. Notice how the bottom

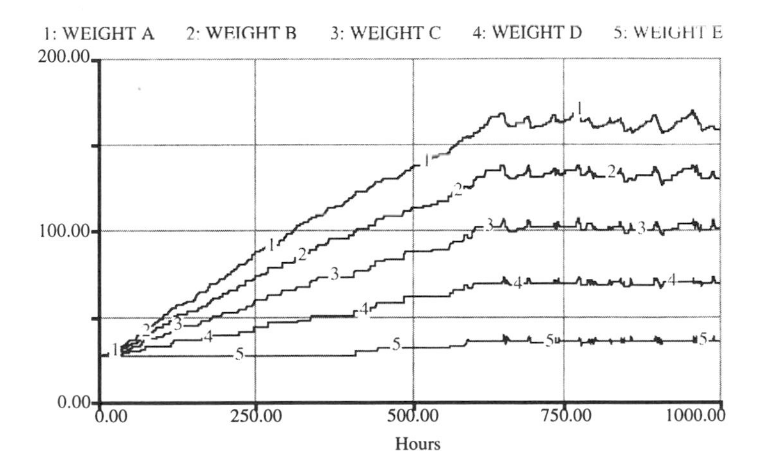

FIGURE 39.2

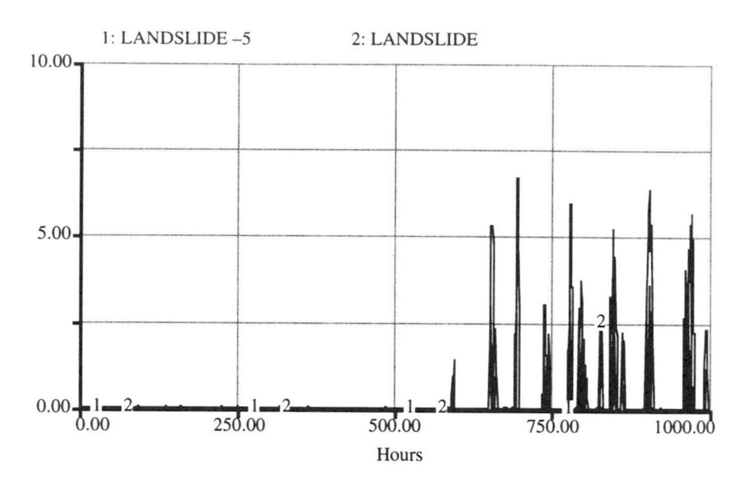

FIGURE 39.3

layer started out with a healthy amount of sand to begin with, and the other layers quietly piled themselves on grain by grain until around time period 400, when small avalanches started to add to the bottom layer. A series of larger avalanches began to occur after approximately 650 time periods. Thereafter, these catastrophic events are more frequent and severe in the second half of the model run.

The graph in Figure 39.3 shows landslide sizes by comparing the size of a layer with its size either 1 or 5 units of time ago. Again, we see how the landslide increased in severity and frequency in the second half of the model run. Change the rate at which sand grains are added to a rate that follows a sine wave. What are the effects of "erosion control" implemented at the middle layer under random rates of additions of sand and under rates that follow a sine wave?

SANDPILE

```
LAYER_A(t) = LAYER_A(t - dt) + (SAND_DROP - A_TO_B) *
dt
INIT LAYER_A = 0 {Grains of Sand}
INFLOWS:
SAND_DROP = If RANDOM(0,5)>4 THEN 1 ELSE 0 {Grains of
Sand per Time Period}
OUTFLOWS:
A_TO_B = IF LAYER_B=0 THEN LAYER_A
ELSE IF LAYER_B>=LAYER_A THEN LAYER_A-LAYER_B
```

```
ELSE IF SLOPE_AB>CRITICAL_SLOPE THEN RAND_1
ELSE 0 {Grains of Sand per Time Period}

LAYER_B(t) = LAYER_B(t - dt) + (A_TO_B - B_TO_C) * dt
INIT LAYER_B = 0 {Grains of Sand}
INFLOWS:
A_TO_B = IF LAYER_B=0 THEN LAYER_A
ELSE IF LAYER_B>=LAYER_A THEN LAYER_A-LAYER_B
ELSE IF SLOPE_AB>CRITICAL_SLOPE THEN RAND_1
ELSE 0 {Grains of Sand per Time Period}
OUTFLOWS:
B_TO_C = IF LAYER_C=0 THEN LAYER_B
ELSE IF LAYER_C>=LAYER_B THEN LAYER_B-LAYER_C
ELSE IF SLOPE_BC>CRITICAL_SLOPE THEN RAND_2
ELSE 0 {Grains of Sand per Time Period}

LAYER_C(t) = LAYER_C(t - dt) + (B_TO_C - C_TO_D) * dt
INIT LAYER_C = 0 {Grains of Sand}
INFLOWS:
B_TO_C = IF LAYER_C=0 THEN LAYER_B
ELSE IF LAYER_C>=LAYER_B THEN LAYER_B-LAYER_C
ELSE IF SLOPE_BC>CRITICAL_SLOPE THEN RAND_2
ELSE 0 {Grains of Sand per Time Period}
OUTFLOWS:
C_TO_D = IF LAYER_D=0 THEN LAYER_C
ELSE IF LAYER_D>=LAYER_C THEN LAYER_C-LAYER_D
ELSE IF SLOPE_CD>CRITICAL_SLOPE THEN RAND_3
ELSE 0 {Grains of Sand per Time Period}

LAYER_D(t) = LAYER_D(t - dt) + (C_TO_D - D_TO_E) * dt
INIT LAYER_D = 0 {Grains of Sand}
INFLOWS:
C_TO_D = IF LAYER_D=0 THEN LAYER_C
ELSE IF LAYER_D>=LAYER_C THEN LAYER_C-LAYER_D
ELSE IF SLOPE_CD>CRITICAL_SLOPE THEN RAND_3
ELSE 0 {Grains of Sand per Time Period}
OUTFLOWS:
D_TO_E = IF LAYER_E=0 then LAYER_D
ELSE IF LAYER_E>=LAYER_D THEN LAYER_D-LAYER_E
ELSE IF SLOPE_DE>CRITICAL_SLOPE THEN RAND_4
ELSE 0 {Grains of Sand per Time Period}

LAYER_E(t) = LAYER_E(t - dt) + (D_TO_E - E_TO_FLOOR) *
dt
```

```
INIT LAYER_E = 25 {Grains of Sand}
INFLOWS:
D_TO_E = IF LAYER_E=0 then LAYER_D
ELSE IF LAYER_E>=LAYER_D THEN LAYER_D-LAYER_E
ELSE IF SLOPE_DE>CRITICAL_SLOPE THEN RAND_4
ELSE 0 {Grains of Sand per Time Period}
OUTFLOWS:
E_TO_FLOOR = IF SLOPE_E_FLOOR>CRITICAL_SLOPE THEN
RAND_5 ELSE 0 {Grains of Sand per Time Period}

CRITICAL_SLOPE = 0.7 {Grains of Sand per Grains of
Sand}
FLOOR_SIZE = 50 {Grains of Sand}
LANDSLIDE = Max(0,-(WEIGHT_A-Delay(WEIGHT_A,1)))
LANDSLIDE_-5 = Max(0,-WEIGHT_A+Delay(WEIGHT_A,5))
RAND_1 = RANDOM(0,LOGN(LAYER_A))
RAND_2 = RANDOM(0,LogN(LAYER_B))
RAND_3 = RANDOM(0,LogN(LAYER_C))
RAND_4 = RANDOM(0,LogN(LAYER_D))
RAND_5 = RANDOM(0,LogN(LAYER_E))
SLOPE_AB = IF LAYER_B>0 THEN LAYER_A/LAYER_B ELSE 0
{Grains of Sand per Grains of Sand}
SLOPE_BC = IF LAYER_C>0 THEN LAYER_B/LAYER_C ELSE 0
{Grains of Sand per Grains of Sand}
SLOPE_CD = IF LAYER_D>0 THEN LAYER_C/LAYER_D ELSE 0
{Grains of Sand per Grains of Sand}
SLOPE_DE = IF LAYER_E>0 THEN LAYER_D/LAYER_E ELSE 0
{Grains of Sand per Grains of Sand}
SLOPE_E_FLOOR = LAYER_E/FLOOR_SIZE {Grains of Sand per
Grains of Sand}
WEIGHT_A = LAYER_A+WEIGHT_B {Grains of Sand}
WEIGHT_B = LAYER_B+WEIGHT_C {Grains of Sand}
WEIGHT_C = LAYER_C+WEIGHT_D {Grains of Sand}
WEIGHT_D = LAYER_D+WEIGHT_E {Grains of Sand}
WEIGHT_E = LAYER_E {Grains of Sand}
```

40

Earthquake*

All its banded anarchs fled, Like vultures frighted. Before an earthquake's tread.

P. B. Shelley, *Hellas,* 1821
in: Percy Bysshe Shelley,
Selected Poetry and Pose,
Alasdair D.F. Macrae (ed.), Routledge, London (1991)

In this chapter we take up again the issue of catastrophe and self-organized criticality. Here we develop a model of an earthquake, following the example given by Bak and Chen.[1] The model consists of three major modules: the quake region, a monitor, and a selector. The quake region is where all the excitement happens in this model; it is where the earthquakes are actually simulated and can be watched. The monitor keeps track of the strength of the earthquake, and the selector enables you to easilty adjust for subsequent runs the key model conditions, such as the location of the epicenter and quake size.

The quake region module consists of 25 state variables arranged in a square array (5 × 5) of epicenter nodes. These state variables are named CELL A1, CELL A2, . . . CELL A5, . . . CELL E5. Each node is connected to its neighboring nodes in the four compass directions. The units in each stock variable are considered to be units of stress. Each node has a capacity of 10 stress units and a minimum of 0 stress units, that is, there is no negative stress. In the event that the amount of stress in a node is exceeded, the entire amount of stress is shifted in equal parts to its neighboring nodes. The tremors are passed on in this way from node to node throughout the system. If a node happens to be on the edge, then the stress that would have been transfered to another node is dissipated from the system. The layout of this module is schematically shown in Figure 40.1.

Whenever stress is transferred out of a node, the monitor section accumulates the amount of stress transfered in the model into a quake strength stock until there is no more activity in the quake region. Next, the stock is compared against several powers of alpha (set to 4 in this model), and

*We wish to thank Mattox Beckman for his contributions to this chapter.

[1]P. Bak, and K. Chen, Self-Organized Criticality, *Scientific American* January: 46–53, 1991.

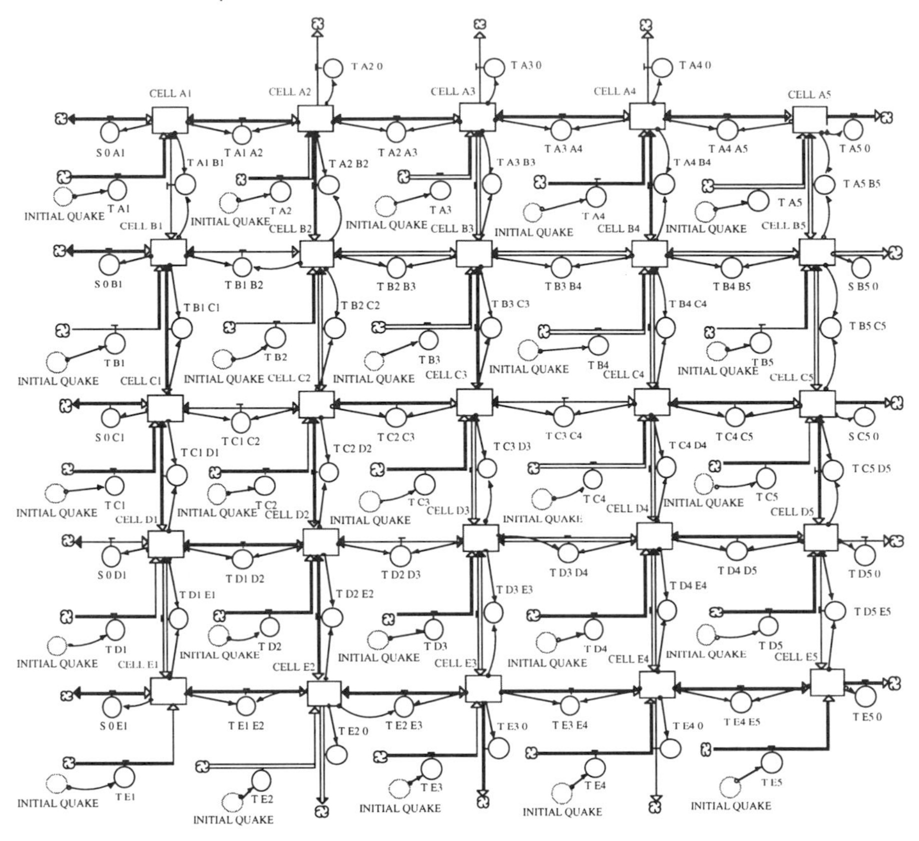

FIGURE 40.1

then an appropriate stock representing the number of quakes of a certain magnitude is incremented. This is very much like the Richter scale used with real earthquakes. Earthquakes of magnitude 1, 2, 3, 4, and >4 are tabulated (Fig. 40.2).

To better understand the mechanism of an earthquake in this model and the measurement of its strength, suppose an epicenter is selected and it causes a chain reaction leading 10 nodes to discharge. A node may discharge more than once in a single quake if the surrounding nodes discharge also and rebuild the stress levels in it. This will cause approximately 100 stress units to be transferred. Since 4^3 is 64 and 4^4 is 256, this would be classified as a magnitude 3 earthquake.

The simulation is controlled by the selector module. The selector does nothing until signaled by a counter routine that there is no activity in the quake region. Then, it selects a random epicenter and a quake size (usually from 1 to 3), and updates that node in the quake region (Fig. 40.3). If it sets off a chain reaction, the counter will prevent further perturbations until the quake region is again stable. Otherwise, the selection routine repeats its process.

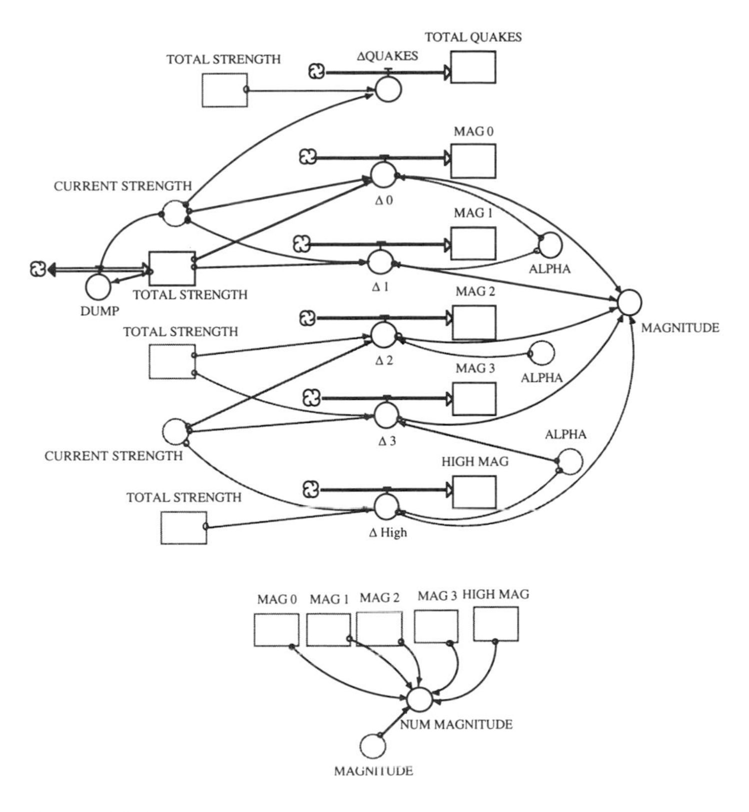

FIGURE 40.2

To do an experiment using this model, you need only to select an initial stress level for the grid and the size of each perturbation. Both of these variables are found in the selector routine. The calculation of the CURRENT STRENGTH of the quake requires summing the tremors by columns and rows of the nodes in the 5 x 5 array (Figs. 40.4 and 40.5).

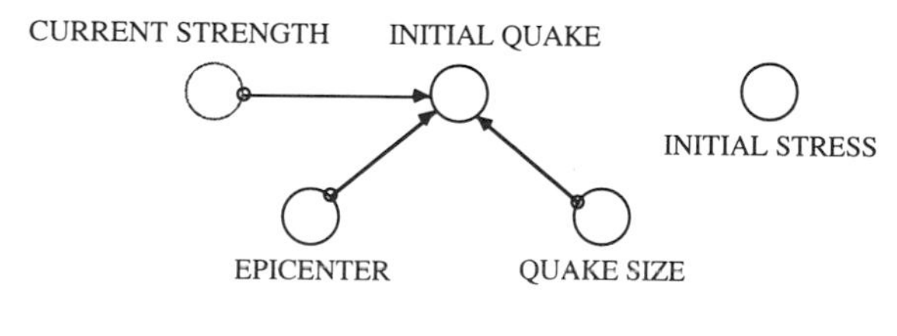

FIGURE 40.3

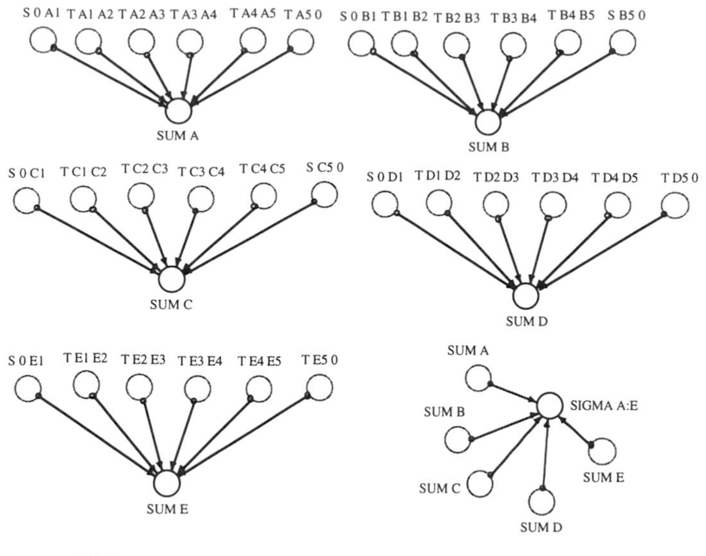

FIGURE 40.4

The results of this experiment exhibit the phenomenon of self-organized criticality. The stresses build up until it takes only a small spike to set the whole system off (Fig. 40.6). Turn on the animation of stocks for that model. The animation slows down the execution of the model a great deal but is useful to watch, and it is entertaining.

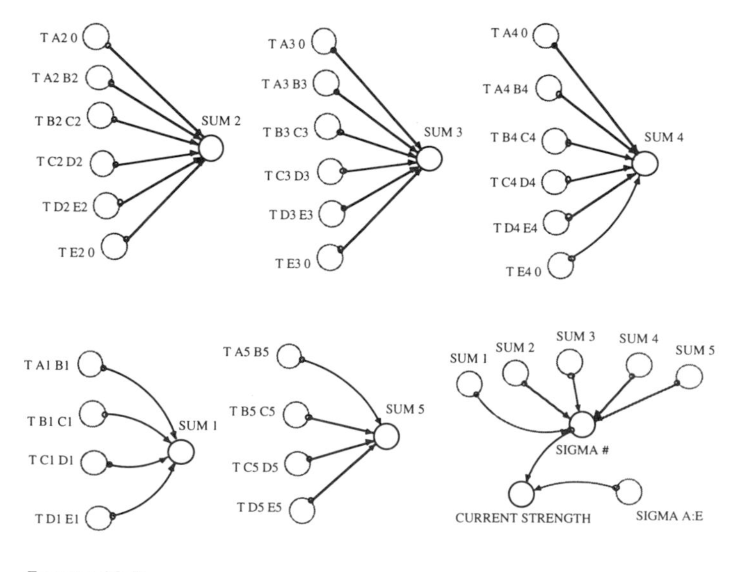

FIGURE 40.5

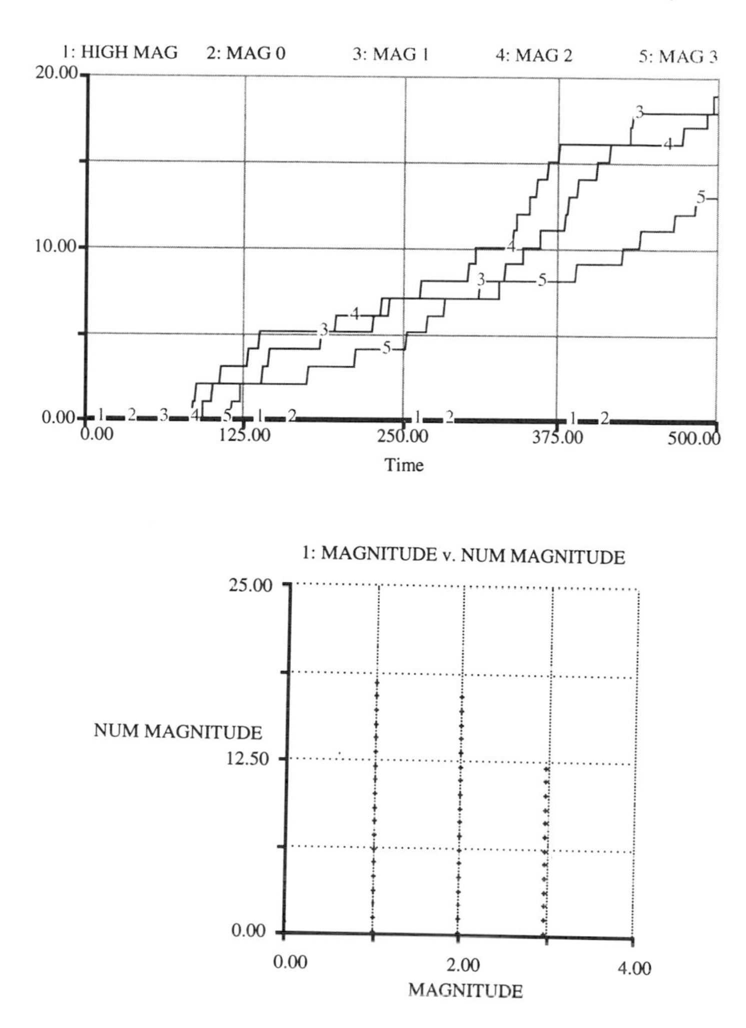

FIGURE 40.6

EARTHQUAKE

```
CELL_A1(t) = CELL_A1(t - dt) + (S_0_A1 + T_A1 - T_A1_A2
- T_A1_B1) * dt
INIT CELL_A1 = INITIAL_STRESS
INFLOWS:
S_0_A1 = if CELL_A1>10 then -CELL_A1/2 else 0
T_A1 = If Mod(INITIAL_QUAKE,100)=1 then
Int(INITIAL_QUAKE/100) else 0
```

```
OUTFLOWS:
T_A1_A2 = if CELL_A1>10 then CELL_A1/4 else
if CELL_A2>10 then -CELL_A2/4 else 0
T_A1_B1 = if CELL_A1>10 then CELL_A1/4 else
if CELL_B1>10 then -CELL_B1/4 else 0

CELL_A2(t) = CELL_A2(t - dt) + (T_A1_A2 + T_A2 -
T_A2_A3 - T_A2_B2 - T_A2_0) * dt
INIT CELL_A2 = INITIAL_STRESS
INFLOWS:
T_A1_A2 = if CELL_A1>10 then CELL_A1/4 else
if CELL_A2>10 then -CELL_A2/4 else 0
T_A2 = If Mod(INITIAL_QUAKE,100)=2 then
Int(INITIAL_QUAKE/100) else 0
OUTFLOWS:
T_A2_A3 = if CELL_A2>10 then CELL_A2/4 else
if CELL_A3>10 then -CELL_A3/4 else 0
T_A2_B2 = if CELL_A2>10 then CELL_A2/4 else
if CELL_B2>10 then -CELL_B2/4 else 0
T_A2_0 = If CELL_A2>10 then CELL_A2/4 else 0

CELL_A3(t) = CELL_A3(t - dt) + (T_A2_A3 + T_A3 -
T_A3_A4 - T_A3_B3 - T_A3_0) * dt
INIT CELL_A3 = INITIAL_STRESS
INFLOWS:
T_A2_A3 = if CELL_A2>10 then CELL_A2/4 else
if CELL_A3>10 then -CELL_A3/4 else 0
T_A3 = If Mod(INITIAL_QUAKE,100)=3 then
Int(INITIAL_QUAKE/100) else 0
OUTFLOWS:
T_A3_A4 = if CELL_A3>10 then CELL_A3/4 else
if CELL_A4>10 then -CELL_A4/4 else 0
T_A3_B3 = if CELL_A3>10 then CELL_A3/4 else
if CELL_B3>10 then -CELL_B3/4 else 0
T_A3_0 = If CELL_A3>10 then CELL_A3/4 else 0

CELL_A4(t) = CELL_A4(t - dt) + (T_A3_A4 + T_A4 -
T_A4_A5 - T_A4_B4 - T_A4_0) * dt
INIT CELL_A4 = INITIAL_STRESS
INFLOWS:
T_A3_A4 = if CELL_A3>10 then CELL_A3/4 else
if CELL_A4>10 then -CELL_A4/4 else 0
```

```
T_A4 = If Mod(INITIAL_QUAKE,100)=4 then
Int(INITIAL_QUAKE/100) else 0
OUTFLOWS:
T_A4_A5 = if CELL_A4>10 then CELL_A4/4 else
if CELL_A5>10 then -CELL_A5/4 else 0
T_A4_B4 = if CELL_A4>10 then CELL_A4/4 else
if CELL_B4>10 then -CELL_B4/4 else 0
T_A4_0 = If CELL_A4>10 then CELL_A4/4 else 0

CELL_A5(t) = CELL_A5(t - dt) + (T_A4_A5 + T_A5 - T_A5_0
- T_A5_B5) * dt
INIT CELL_A5 = INITIAL_STRESS
INFLOWS:
T_A4_A5 = if CELL_A4>10 then CELL_A4/4 else
if CELL_A5>10 then -CELL_A5/4 else 0
T_A5 = If Mod(INITIAL_QUAKE,100)=5 then
Int(INITIAL_QUAKE/100) else 0
OUTFLOWS:
T_A5_0 = If CELL_A5>10 then CELL_A5/2 else 0
T_A5_B5 = if CELL_A5>10 then CELL_A5/4 else
if CELL_B5>10 then -CELL_B5/4 else 0

CELL_B1(t) = CELL_B1(t - dt) + (T_A1_B1 + S_0_B1 + T_B1
- T_B1_B2 - T_B1_C1) * dt
INIT CELL_B1 = INITIAL_STRESS
INFLOWS:
T_A1_B1 = if CELL_A1>10 then CELL_A1/4 else
if CELL_B1>10 then -CELL_B1/4 else 0
S_0_B1 = if CELL_B1>10 then -CELL_B1/4 else 0
T_B1 = If Mod(INITIAL_QUAKE,100)=6 then
Int(INITIAL_QUAKE/100) else 0
OUTFLOWS:
T_B1_B2 = if CELL_B1>10 then CELL_B1/4 else
if CELL_B2>10 then -CELL_B2/4 else 0
T_B1_C1 = if CELL_B1>10 then CELL_B1/4 else
if CELL_C1>10 then -CELL_C1/4 else 0

CELL_B2(t) = CELL_B2(t - dt) + (T_A2_B2 + T_B1_B2 +
T_B2 - T_B2_B3 - T_B2_C2) * dt
INIT CELL_B2 = INITIAL_STRESS
INFLOWS:
T_A2_B2 = if CELL_A2>10 then CELL_A2/4 else
if CELL_B2>10 then -CELL_B2/4 else 0
```

```
T_B1_B2 = if CELL_B1>10 then CELL_B1/4 else
if CELL_B2>10 then -CELL_B2/4 else 0
T_B2 = If Mod(INITIAL_QUAKE,100)=7 then
Int(INITIAL_QUAKE/100) else 0
OUTFLOWS:
T_B2_B3 = if CELL_B2>10 then CELL_B2/4 else
if CELL_B3>10 then -CELL_B3/4 else 0
T_B2_C2 = if CELL_B2>10 then CELL_B2/4 else
if CELL_C2>10 then -CELL_C2/4 else 0

CELL_B3(t) = CELL_B3(t - dt) + (T_B2_B3 + T_A3_B3 +
T_B3 - T_B3_B4 - T_B3_C3) * dt
INIT CELL_B3 = INITIAL_STRESS
INFLOWS:
T_B2_B3 = if CELL_B2>10 then CELL_B2/4 else
if CELL_B3>10 then -CELL_B3/4 else 0
T_A3_B3 = if CELL_A3>10 then CELL_A3/4 else
if CELL_B3>10 then -CELL_B3/4 else 0
T_B3 = If Mod(INITIAL_QUAKE,100)=8 then
Int(INITIAL_QUAKE/100) else 0
OUTFLOWS:
T_B3_B4 = if CELL_B3>10 then CELL_B3/4 else
if CELL_B4>10 then -CELL_B4/4 else 0
T_B3_C3 = if CELL_B3>10 then CELL_B3/4 else
if CELL_C3>10 then -CELL_C3/4 else 0

CELL_B4(t) = CELL_B4(t - dt) + (T_B3_B4 + T_A4_B4 +
T_B4 - T_B4_B5 - T_B4_C4) * dt
INIT CELL_B4 = INITIAL_STRESS
INFLOWS:
T_B3_B4 = if CELL_B3>10 then CELL_B3/4 else
if CELL_B4>10 then -CELL_B4/4 else 0
T_A4_B4 = if CELL_A4>10 then CELL_A4/4 else
if CELL_B4>10 then -CELL_B4/4 else 0
T_B4 = If Mod(INITIAL_QUAKE,100)=9 then
Int(INITIAL_QUAKE/100) else 0
OUTFLOWS:
T_B4_B5 = if CELL_B4>10 then CELL_B4/4 else
if CELL_B5>10 then -CELL_B5/4 else 0
T_B4_C4 = if CELL_B4>10 then CELL_B4/4 else
if CELL_C4>10 then -CELL_C4/4 else 0

CELL_B5(t) = CELL_B5(t - dt) + (T_B4_B5 + T_A5_B5 +
T_B5 - S_B5_0 - T_B5_C5) * dt
```

```
INIT CELL_B5 = INITIAL_STRESS
INFLOWS:
T_B4_B5 = if CELL_B4>10 then CELL_B4/4 else
if CELL_B5>10 then -CELL_B5/4 else 0
T_A5_B5 = if CELL_A5>10 then CELL_A5/4 else
if CELL_B5>10 then -CELL_B5/4 else 0
T_B5 = If Mod(INITIAL_QUAKE,100)=10 then
Int(INITIAL_QUAKE/100) else 0
OUTFLOWS:
S_B5_0 = If CELL_B5>10 then CELL_B5/4 else 0
T_B5_C5 = if CELL_B5>10 then CELL_B5/4 else
if CELL_C5>10 then -CELL_C5/4 else 0

CELL_C1(t) = CELL_C1(t - dt) + (S_0_C1 + T_B1_C1 + T_C1
- T_C1_C2 - T_C1_D1) * dt
INIT CELL_C1 = INITIAL_STRESS
INFLOWS:
S_0_C1 = if CELL_C1>10 then -CELL_C1/4 else 0
T_B1_C1 = if CELL_B1>10 then CELL_B1/4 else
if CELL_C1>10 then -CELL_C1/4 else 0
T_C1 = If Mod(INITIAL_QUAKE,100)=11 then
Int(INITIAL_QUAKE/100) else 0
OUTFLOWS:
T_C1_C2 = if CELL_C1>10 then CELL_C1/4 else
if CELL_C2>10 then -CELL_C2/4 else 0
T_C1_D1 = if CELL_C1>10 then CELL_C1/4 else
if CELL_D1>10 then -CELL_D1/4 else 0

CELL_C2(t) = CELL_C2(t - dt) + (T_C1_C2 + T_B2_C2 +
T_C2 - T_C2_C3 - T_C2_D2) * dt
INIT CELL_C2 = INITIAL_STRESS
INFLOWS:
T_C1_C2 = if CELL_C1>10 then CELL_C1/4 else
if CELL_C2>10 then -CELL_C2/4 else 0
T_B2_C2 = if CELL_B2>10 then CELL_B2/4 else
if CELL_C2>10 then -CELL_C2/4 else 0
T_C2 = If Mod(INITIAL_QUAKE,100)=12 then
Int(INITIAL_QUAKE/100) else 0
OUTFLOWS:
T_C2_C3 = if CELL_C2>10 then CELL_C2/4 else
if CELL_C3>10 then -CELL_C3/4 else 0
T_C2_D2 = if CELL_C2>10 then CELL_C2/4 else
if CELL_D2>10 then -CELL_D2/4 else 0
```

```
CELL_C3(t) = CELL_C3(t - dt) + (T_C2_C3 + T_B3_C3 +
T_C3 - T_C3_C4 - T_C3_D3) * dt
INIT CELL_C3 = INITIAL_STRESS
INFLOWS:
T_C2_C3 = if CELL_C2>10 then CELL_C2/4 else
if CELL_C3>10 then -CELL_C3/4 else 0
T_B3_C3 = if CELL_B3>10 then CELL_B3/4 else
if CELL_C3>10 then -CELL_C3/4 else 0
T_C3 = If Mod(INITIAL_QUAKE,100)=13 then
Int(INITIAL_QUAKE/100) else 0
OUTFLOWS:
T_C3_C4 = if CELL_C3>10 then CELL_C3/4 else
if CELL_C4>10 then -CELL_C4/4 else 0
T_C3_D3 = if CELL_C3>10 then CELL_C3/4 else
if CELL_D3>10 then -CELL_D3/4 else 0

CELL_C4(t) = CELL_C4(t - dt) + (T_C3_C4 + T_B4_C4 +
T_C4 - T_C4_C5 - T_C4_D4) * dt
INIT CELL_C4 = INITIAL_STRESS
INFLOWS:
T_C3_C4 = if CELL_C3>10 then CELL_C3/4 else
if CELL_C4>10 then -CELL_C4/4 else 0
T_B4_C4 = if CELL_B4>10 then CELL_B4/4 else
if CELL_C4>10 then -CELL_C4/4 else 0
T_C4 = If Mod(INITIAL_QUAKE,100)=14 then
Int(INITIAL_QUAKE/100) else 0
OUTFLOWS:
T_C4_C5 = if CELL_C4>10 then CELL_C4/4 else
if CELL_C5>10 then -CELL_C5/4 else 0
T_C4_D4 = if CELL_C4>10 then CELL_C4/4 else
if CELL_D4>10 then -CELL_D4/4 else 0

CELL_C5(t) = CELL_C5(t - dt) + (T_C4_C5 + T_B5_C5 +
T_C5 - S_C5_0 - T_C5_D5) * dt
INIT CELL_C5 = INITIAL_STRESS
INFLOWS:
T_C4_C5 = if CELL_C4>10 then CELL_C4/4 else
if CELL_C5>10 then -CELL_C5/4 else 0
T_B5_C5 = if CELL_B5>10 then CELL_B5/4 else
if CELL_C5>10 then -CELL_C5/4 else 0
T_C5 = If Mod(INITIAL_QUAKE,100)=15 then
Int(INITIAL_QUAKE/100) else 0
OUTFLOWS:
```

```
S_C5_0 = If CELL_C5>10 then CELL_C5/4 else 0
T_C5_D5 = if CELL_C5>10 then CELL_C5/4 else
if CELL_D5>10 then -CELL_D5/4 else 0

CELL_D1(t) = CELL_D1(t - dt) + (T_C1_D1 + S_0_D1 + T_D1
- T_D1_D2 - T_D1_E1) * dt
INIT CELL_D1 = INITIAL_STRESS
INFLOWS:
T_C1_D1 = if CELL_C1>10 then CELL_C1/4 else
if CELL_D1>10 then -CELL_D1/4 else 0
S_0_D1 = if CELL_D1>10 then -CELL_D1/4 else 0
T_D1 = If Mod(INITIAL_QUAKE,100)=16 then
Int(INITIAL_QUAKE/100) else 0
OUTFLOWS:
T_D1_D2 = if CELL_D1>10 then CELL_D1/4 else
if CELL_D2>10 then -CELL_D2/4 else 0
T_D1_E1 = if CELL_D1>10 then CELL_D1/4 else
if CELL_E1>10 then -CELL_E1/4 else 0
CELL_D2(t) = CELL_D2(t - dt) + (T_D1_D2 + T_C2_D2 +
T_D2 - T_D2_D3 - T_D2_E2) * dt
INIT CELL_D2 = INITIAL_STRESS
INFLOWS:
T_D1_D2 = if CELL_D1>10 then CELL_D1/4 else
if CELL_D2>10 then -CELL_D2/4 else 0
T_C2_D2 = if CELL_C2>10 then CELL_C2/4 else
if CELL_D2>10 then -CELL_D2/4 else 0
T_D2 = If Mod(INITIAL_QUAKE,100)=17 then
Int(INITIAL_QUAKE/100) else 0
OUTFLOWS:
T_D2_D3 = if CELL_D2>10 then CELL_D2/4 else
if CELL_D3>10 then -CELL_D3/4 else 0
T_D2_E2 = if CELL_D2>10 then CELL_D2/4 else
if CELL_E2>10 then -CELL_E2/4 else 0

CELL_D3(t) = CELL_D3(t - dt) + (T_D2_D3 + T_C3_D3 +
T_D3 - T_D3_D4 - T_D3_E3) * dt
INIT CELL_D3 = INITIAL_STRESS
INFLOWS:
T_D2_D3 = if CELL_D2>10 then CELL_D2/4 else
if CELL_D3>10 then -CELL_D3/4 else 0
T_C3_D3 = if CELL_C3>10 then CELL_C3/4 else
if CELL_D3>10 then -CELL_D3/4 else 0
```

```
T_D3 = If Mod(INITIAL_QUAKE,100)=18 then
Int(INITIAL_QUAKE/100) else 0
OUTFLOWS:
T_D3_D4 = if CELL_D3>10 then CELL_D3/4 else
if CELL_D4>10 then -CELL_D4/4 else 0
T_D3_E3 = if CELL_D3>10 then CELL_D3/4 else
if CELL_E3>10 then -CELL_E3/4 else 0

CELL_D4(t) = CELL_D4(t - dt) + (T_D3_D4 + T_C4_D4 +
T_D4 - T_D4_D5 - T_D4_E4) * dt
INIT CELL_D4 = INITIAL_STRESS
INFLOWS:
T_D3_D4 = if CELL_D3>10 then CELL_D3/4 else
if CELL_D4>10 then -CELL_D4/4 else 0
T_C4_D4 = if CELL_C4>10 then CELL_C4/4 else
if CELL_D4>10 then -CELL_D4/4 else 0
T_D4 = If Mod(INITIAL_QUAKE,100)=19 then
Int(INITIAL_QUAKE/100) else 0
OUTFLOWS:
T_D4_D5 = if CELL_D4>10 then CELL_D4/4 else
if CELL_D5>10 then -CELL_D5/4 else 0
T_D4_E4 = if CELL_D4>10 then CELL_D4/4 else
if CELL_E4>10 then -CELL_E4/4 else 0

CELL_D5(t) = CELL_D5(t - dt) + (T_D4_D5 + T_C5_D5 +
T_D5 - T_D5_0 - T_D5_E5) * dt
INIT CELL_D5 = INITIAL_STRESS
INFLOWS:
T_D4_D5 = if CELL_D4>10 then CELL_D4/4 else
if CELL_D5>10 then -CELL_D5/4 else 0
T_C5_D5 = if CELL_C5>10 then CELL_C5/4 else
if CELL_D5>10 then -CELL_D5/4 else 0
T_D5 = If Mod(INITIAL_QUAKE,100)=20 then
Int(INITIAL_QUAKE/100) else 0
OUTFLOWS:
T_D5_0 = If CELL_D5>10 then CELL_D5/4 else 0
T_D5_E5 = if CELL_D5>10 then CELL_D5/4 else
if CELL_E5>10 then -CELL_E5/4 else 0

CELL_E1(t) = CELL_E1(t - dt) + (T_D1_E1 + S_0_E1 + T_E1
- T_E1_E2) * dt
INIT CELL_E1 = INITIAL_STRESS
INFLOWS:
```

```
T_D1_E1 = if CELL_D1>10 then CELL_D1/4 else
if CELL_E1>10 then -CELL_E1/4 else 0
S_0_E1 = if CELL_E1>10 then -CELL_E1/2 else 0
T_E1 = If Mod(INITIAL_QUAKE,100)=21 then
Int(INITIAL_QUAKE/100) else 0
OUTFLOWS:
T_E1_E2 = if CELL_E1>10 then CELL_E1/4 else
if CELL_E2>10 then -CELL_E2/4 else 0

CELL_E2(t) = CELL_E2(t - dt) + (T_E1_E2 + T_D2_E2 +
T_E2 - T_E2_E3 - T_E2_0) * dt
INIT CELL_E2 = INITIAL_STRESS
INFLOWS:
T_E1_E2 = if CELL_E1>10 then CELL_E1/4 else
if CELL_E2>10 then -CELL_E2/4 else 0
T_D2_E2 = if CELL_D2>10 then CELL_D2/4 else
if CELL_E2>10 then -CELL_E2/4 else 0
T_E2 = If Mod(INITIAL_QUAKE,100)=21 then
Int(INITIAL_QUAKE/100) else 0
OUTFLOWS:
T_E2_E3 = if CELL_E2>10 then CELL_E2/4 else
if CELL_E3>10 then -CELL_E3/4 else 0
T_E2_0 = IF CELL_E2>10 then CELL_E2/4 else 0

CELL_E3(t) = CELL_E3(t - dt) + (T_E2_E3 + T_D3_E3 +
T_E3 - T_E3_E4 - T_E3_0) * dt
INIT CELL_E3 = INITIAL_STRESS
INFLOWS:
T_E2_E3 = if CELL_E2>10 then CELL_E2/4 else
if CELL_E3>10 then -CELL_E3/4 else 0
T_D3_E3 = if CELL_D3>10 then CELL_D3/4 else
if CELL_E3>10 then -CELL_E3/4 else 0
T_E3 = If Mod(INITIAL_QUAKE,100)=22 then
Int(INITIAL_QUAKE/100) else 0
OUTFLOWS:
T_E3_E4 = if CELL_E3>10 then CELL_E3/4 else
if CELL_E4>10 then -CELL_E4/4 else 0
T_E3_0 = IF CELL_E3>10 then CELL_E3/4 else 0

CELL_E4(t) = CELL_E4(t - dt) + (T_E3_E4 + T_D4_E4 +
T_E4 - T_E4_E5 - T_E4_0) * dt
INIT CELL_E4 - INITIAL_STRESS
INFLOWS:
```

```
T_E3_E4 = if CELL_E3>10 then CELL_E3/4 else
if CELL_E4>10 then -CELL_E4/4 else 0
T_D4_E4 = if CELL_D4>10 then CELL_D4/4 else
if CELL_E4>10 then -CELL_E4/4 else 0
T_E4 = If Mod(INITIAL_QUAKE,100)=24 then
Int(INITIAL_QUAKE/100) else 0
OUTFLOWS:
T_E4_E5 = if CELL_E4>10 then CELL_E4/4 else
if CELL_E5>10 then -CELL_E5/4 else 0
T_E4_0 = IF CELL_E4>10 then CELL_E4/4 else 0

CELL_E5(t) = CELL_E5(t - dt) + (T_E4_E5 + T_D5_E5 +
T_E5 - T_E5_0) * dt
INIT CELL_E5 = INITIAL_STRESS
INFLOWS:
T_E4_E5 = if CELL_E4>10 then CELL_E4/4 else
if CELL_E5>10 then -CELL_E5/4 else 0
T_D5_E5 = if CELL_D5>10 then CELL_D5/4 else
if CELL_E5>10 then -CELL_E5/4 else 0
T_E5 = If Mod(INITIAL_QUAKE,100)=25 then
Int(INITIAL_QUAKE/100) else 0
OUTFLOWS:
T_E5_0 = If CELL_E5>10 then CELL_E5/2 else 0

HIGH_MAG(t) = HIGH_MAG(t - dt) + (Δ_High) * dt
INIT HIGH_MAG = 0
INFLOWS:
Δ_High = If CURRENT_STRENGTH=0 and
TOTAL_STRENGTH>ALPHA^4 then 1 else 0
MAG_0(t) = MAG_0(t - dt) + (Δ_0) * dt
INIT MAG_0 = 0
INFLOWS:
Δ_0 = If CURRENT_STRENGTH=0 and
 TOTAL_STRENGTH>0 and TOTAL_STRENGTH<\#61>ALPHA
then 1
else 0

MAG_1(t) = MAG_1(t - dt) + (Δ_1) * dt
INIT MAG_1 = 0
INFLOWS:
Δ_1 = If CURRENT_STRENGTH=0 and
 TOTAL_STRENGTH>ALPHA and TOTAL_STRENGTH<\#61>ALPHA^2
then 1
```

```
else 0
MAG_2(t) = MAG_2(t - dt) + (Δ_2) * dt
INIT MAG_2 = 0
INFLOWS:
Δ_2 = If CURRENT_STRENGTH=0 and
 TOTAL_STRENGTH>ALPHA^2 and TOTAL_STRENGTH<\#61>ALPHA^3
then 1
else 0

MAG_3(t) = MAG_3(t - dt) + (Δ_3) * dt
INIT MAG_3 = 0
INFLOWS:
Δ_3 = If CURRENT_STRENGTH=0 and
 TOTAL_STRENGTH>ALPHA^3 and TOTAL_STRENGTH<\#61>ALPHA^4
then 1
else 0

TOTAL_QUAKES(t) = TOTAL_QUAKES(t - dt) + (ΔQUAKES) * dt
INIT TOTAL_QUAKES = 0
INFLOWS:
ΔQUAKES = If CURRENT_STRENGTH=0 and TOTAL_STRENGTH>0
then 1 else 0

TOTAL_STRENGTH(t) = TOTAL_STRENGTH(t - dt) + (DUMP) *
dt
INIT TOTAL_STRENGTH = 0
INFLOWS:
DUMP = If CURRENT_STRENGTH>0 then CURRENT_STRENGTH else
-TOTAL_STRENGTH

ALPHA = 4
CURRENT_STRENGTH = SIGMA_#+SIGMA_A:E
EPICENTER = Int(Random(0,25)+1)
INITIAL_QUAKE = IF CURRENT_STRENGTH=0 or time=5 then
EPICENTER + QUAKE_SIZE * 100 else 0
INITIAL_STRESS = 1
MAGNITUDE = Δ_0+Δ_1*2+Δ_2*3+Δ_3*4+Δ_High*5-1
NUM_MAGNITUDE = IF MAGNITUDE=0 THEN MAG_0 ELSE
IF MAGNITUDE=1 THEN MAG_1 ELSE
IF MAGNITUDE=2 THEN MAG_2 ELSE
IF MAGNITUDE=3 THEN MAG_3 ELSE
IF MAGNITUDE=4 THEN HIGH_MAG ELSE 0
QUAKE_SIZE = Int(Random(.5,1.5)+1)
```

```
SIGMA_# = SUM_1+SUM_2+SUM_3+SUM_4+SUM_5
SIGMA_A:E = SUM_A+SUM_B+SUM_C+SUM_D+SUM_E
SUM_1 =
ABS(T_A1_B1)+ABS(T_B1_C1)+ABS(T_C1_D1)+ABS(T_D1_E1)
SUM_2 =
ABS(T_A2_0)+ABS(T_A2_B2)+ABS(T_B2_C2)+ABS(T_C2_D2)+ABS(
T_D2_E2)+ABS(T_E2_0)
SUM_3 =
ABS(T_A3_0)+ABS(T_A3_B3)+ABS(T_B3_C3)+ABS(T_C3_D3)+ABS(
T_D3_E3)+ABS(T_E3_0)
SUM_4 =
ABS(T_A4_0)+ABS(T_A4_B4)+ABS(T_B4_C4)+ABS(T_C4_D4)+ABS(
T_D4_E4)+ABS(T_E4_0)
SUM_5 =
ABS(T_A5_B5)+ABS(T_B5_C5)+ABS(T_C5_D5)+ABS(T_D5_E5)
SUM_A =
ABS(S_0_A1)+ABS(T_A1_A2)+ABS(T_A2_A3)+ABS(T_A3_A4)+ABS(
T_A4_A5)+ABS(T_A5_0)
SUM_B =
ABS(S_0_B1)+ABS(T_B1_B2)+ABS(T_B2_B3)+ABS(T_B3_B4)+ABS(
T_B4_B5)+ABS(S_B5_0)
SUM_C =
ABS(S_0_C1)+ABS(T_C1_C2)+ABS(T_C2_C3)+ABS(T_C3_C4)+ABS(
T_C4_C5)+ABS(S_C5_0)
SUM_D =
ABS(S_0_D1)+ABS(T_D1_D2)+ABS(T_D2_D3)+ABS(T_D3_D4)+ABS(
T_D4_D5)+ABS(T_D5_0)
SUM_E =
ABS(S_0_E1)+ABS(T_E1_E2)+ABS(T_E2_E3)+ABS(T_E3_E4)+ABS(
T_E4_E5)+ABS(T_E5_0)
```

41

Game of Life

Now, gentlemen, I have another game to play.

D'Israeli, 1826

In the previous two chapters we modeled physical systems whose structure built over time and then changed drastically in response to some random external force, such as an additional grain of sand falling onto a sandpile, or a random shock in a system, leading to an earthquake. Following drastic change in the structure of the system, the system organized itself to a state at which it was relatively stable again, at least over a short period of time. Analogies can easily be drawn between such self-organizing physical systems and biological and ecological systems. But rather than pursue analogies in the context of the sandpile and earthquake models, let us set up a new model that focuses on two features of natural systems that are closely related to self-organized criticality, and essential for the maintenance of order in ecological systems.

Two of the most striking features of nature are the ability of these systems to evolve and reproduce. For example, gradients in material concentrations and temperature in the primordial soup enabled the formation of simple structures that were not only distinguishable from their environment but eventually also possessed the ability to change that environment. The emergence of catalytic RNA in a heterogeneous environment, in turn, not only led to change in the environment of these simple structures but, with chance, to the change of these structures themselves.[1] Increasingly complex molecular structures evolved, reproduced, and continuously transformed the environment in which they lived.

The evolution and reproduction of life forms was long interpreted as the result of some "vital force." It is only since the beginning of this century that we have gained a better understanding of evolution and reproduction from a physical perspective.[2] In 1966, John von Neumann created a simple model of a reproducing machine[3] that sparked the development of numerous

[1]L.E. Orgel, The Origin of Life on the Earth, *Scientific American*, October: 77–83, 1994.
[2]E. Schrödinger, *What is Life*, Cambridge: Cambridge University Press, 1944.
[3]J. von Neumann, *Theory of Self-Reproducing Automata*, Chicago: University of Illinois Press, 1966.

efforts to boil down rules for replication to simple logical statements that take on the form of mechanistic rules. Collectively, these models are known as *cellular automata.* In general, a cellular automaton is any set of individuals with a finite number of characteristics that change over time based on previous characteristics and the characteristics of individuals with which it interacts. Typically, individuals are represented as cells in a grid. The features of each cell are determined by its features in the past and those of the neighboring cells.

The model of this chapter is set up for 100 cells arranged in a square. It is an extremely simple cellular automaton that has been developed by the British mathematician John H. Conway. Conway[4] called this model "Life," perhaps because the displays of the model looks remarkably like a petri dish full of microscopic organisms, moving back and forth, merging with each other, replicating themselves, or generating new forms.

The rules for the game of life are as follows:

1. Each cell on the grid can only take on two characteristics, for example, the colors black or white. Black cells are "alive," white ones have no life.
2. If, for a given cell, the number of alive neighbors is exactly two, the value of the square does not change at the next time step.
3. If the number of black neighbors is exactly three, the square will be black in the next time step.
4. If the number of black neighbors is neither two nor three, the square will be white at the next time step.

We have named the cells in our model with letters A, B, . . . , J for rows and numbers 1, 2, . . . , 10 for columns. The state of each cell is captured by a state variable that either takes on a value of one if its is black or zero if it is white. Changes in the state of a cell are captured by biflows and are calculated on the basis of its own state and the value its neighbors take on. If a cell is black (i.e., its state variable takes on a value of one) and if exactly three of its neighbors are black (the sum of the state variables of all neighbors is three), then, according to the third rule, there is no change in the state of the cell and the biflow is zero. If the cell is white but exactly three of its neighbors are black, then, according to the third rule, the biflow takes on a value of one, changing the state of the respective cell from zero to one. Rules 2 and 4 can be accommodated analogously in the model. An example for the application of these rules to one of the cells in the model is given in the module in Figure 41.1.

The parameter BIRTH # is the number of neighbor cells that must be black to "give birth" to a cell. According to our rules we set BIRTH # = 3. Similarly, NEIGHBOR MAX is the number of cells being black above which

---

[4]W. Poundstone, *The Recursive Universe,* Chicago: Contemporary Books, 1985.

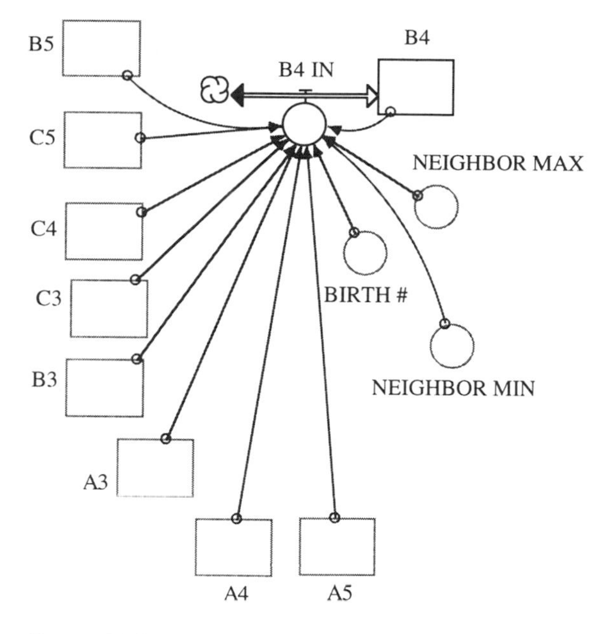

FIGURE 41.1

the cell dies from crowding, and NEIGHBOR MIN is the minimum number of neighboring cells below which the cell is no longer viable. NEIGHBOR MAX and NEIGHBOR MIN in our model are set to 3 and 2, respectively. The rule for the change in cell B4 is

```
B4 IN = IF (A3+A4+A5+B3+B5+C3+C4+C5)=BIRTH # AND B4=0
        THEN 1
        ELSE IF (A3+A4+A5+B3+B5+C3+C4+C5)<NEIGHBOR MIN
        OR (A3+A4+A5+B3+B5+C3+C4+C5)>NEIGHBOR MAX AND
        B4−1 THEN −1
        ELSE 0
```

Randomly initialize the value for each cell and observe the system's dynamics. You will find that some initial conditions lead to more interesting dynamics than others. Among the more interesting initial conditions for the model are the ones shown in the set-ups in Figure 41.2, where a black cell is one with initial value of one, and a white cell has initial value of zero. Start the model with one of these initial conditions, and observe its spatio-temporal dynamics. Find your own constellations of initial conditions that lead to self-replicating structures. With this model you can construct "factories" that produce gliding swarms of "alive" cells, or intricate structures of squares within squares.

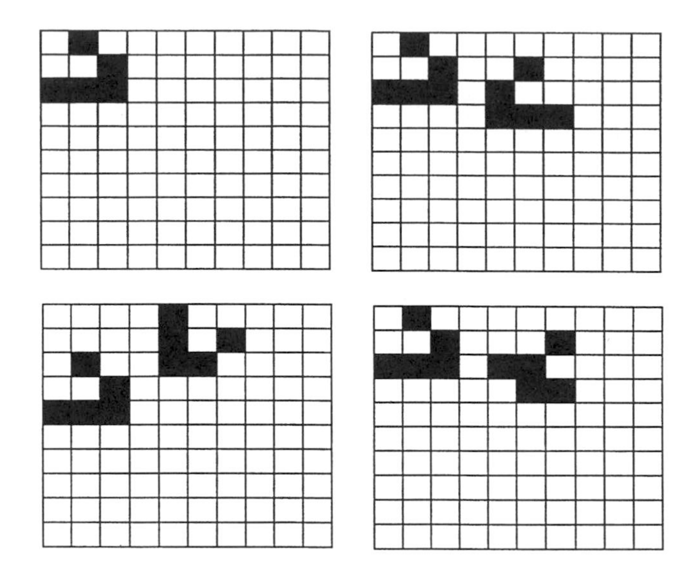

FIGURE 41.2

We have animated the stocks in the STELLA model and placed ghosts of the stocks close together to set up a checkerboard-like pattern. The results of the model with the initial conditions shown at the top left in Figure 41.2 are represented in Figure 41.3.

The initial "glider" moves downward in the grid and to the right, repeatedly replicating itself until it reaches the lower right-hand corner of the model, where it gets stuck. If you keep enlarging the number of cells, you can keep the slider going. Can you set up the decision rules such that the glider moves out of the cells, rather than gets stuck? The other initial conditions shown in Figure 41.3 yield situations in which two gliders annihilate and in which new, stable patterns emerge.

Cellular automata not only generate interesting-looking spatio-temporal patterns but have been proven to be quite informative in physics when modeling the behavior of fluids[5] and in landscape ecology, where they have been used, for example, to model sediment transport on coastal landscapes[6] or the spread of forest fires.[7]

---

[5]P. Bak, K. Chen, and M. Creutz, Self-Organized Criticality in the "Game of Life," *Nature* 342:780–782, 1989.
[6]M. Ruth, and F. Pieper, Modeling Spatial Dynamics of Sea Level Rise in a Coastal Area, *System Dynamics Review* 10:375–389, 1994.
[7]P. Bak, K. Chen, and C. Tang, A Forest Fire Model and Some Thoughts on Turbulence, *Physics Letters*, 147:297–300, 1990.

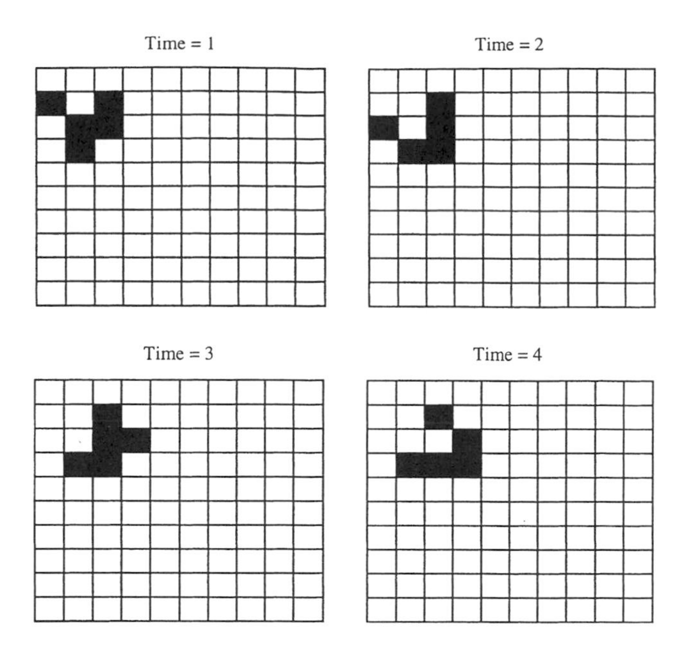

FIGURE 41.3

In the model in Figure 41.3, a black cell indicated the presence of an organism, white meant that there were no organisms in a particular cell. Can you set up the model to capture competition of two species with each other? When introducing a second type of organism, make only as few changes to the rules as possible to model that organism and its competitive behavior.

GAME OF LIFE

```
A1(t) = A1(t - dt) + (A1_IN) * dt
INIT A1 = 0
INFLOWS:
A1_IN = IF (A2+B1+B2)=BIRTH_# and A1=0 THEN 1 ELSE
IF (A2+B1+B2)<NEIGHBOR_MIN OR (A2+B1+B2)>NEIGHBOR_MAX
AND A1=1 THEN -1 ELSE 0

A10(t) = A10(t - dt) + (A10_IN) * dt
INIT A10 = 0
INFLOWS:
```

```
A10_IN = IF (A9+B10+B9)=BIRTH_# AND A10=0 THEN 1 ELSE
IF (A9+B10+B9)<NEIGHBOR_MIN OR
(A9+B10+B9)>NEIGHBOR_MAX AND A10=1 THEN -1 ELSE 0

A2(t) = A2(t - dt) + (A2_IN) * dt
INIT A2 = 1
INFLOWS:
A2_IN = IF (A1+A3+B1+B2+B3)=BIRTH_# AND A2=0 THEN 1
ELSE
IF (A1+A3+B1+B2+B3)<NEIGHBOR_MIN OR
(A1+A3+B1+B2+B3)>NEIGHBOR_MAX AND A2=1 THEN -1 ELSE 0

A3(t) = A3(t - dt) + (A3_IN) * dt
INIT A3 = 0
INFLOWS:
A3_IN = IF (A2+A4+B2+B3+B4)=BIRTH_# AND A3=0 THEN 1
ELSE
IF (A2+A4+B2+B3+B4)<NEIGHBOR_MIN OR
(A2+A4+B2+B3+B4)>NEIGHBOR_MAX AND A3=1 THEN -1 ELSE 0

A4(t) = A4(t - dt) + (A4_IN) * dt
INIT A4 = 0
INFLOWS:
A4_IN = IF (A3+A5+B3+B4+B5)=BIRTH_# AND A4=0 THEN 1
ELSE
IF (A3+A5+B3+B4+B5)<NEIGHBOR_MIN OR
(A3+A5+B3+B4+B5)>NEIGHBOR_MAX AND A4=1 THEN -1 ELSE 0

A5(t) = A5(t - dt) + (A5_IN) * dt
INIT A5 = 0
INFLOWS:
A5_IN = IF (A6+B4+B5+B6+A4)=BIRTH_# AND A5=0 THEN 1
ELSE
IF (A6+B4+B5+B6+A4)<NEIGHBOR_MIN OR
(A6+B4+B5+B6+A4)>NEIGHBOR_MAX AND A5=1 THEN -1 ELSE 0

A6(t) = A6(t - dt) + (A6_IN) * dt
INIT A6 = 0
INFLOWS:
A6_IN = IF (A7+B5+B6+B7+A5)=BIRTH_# AND A6=0 THEN 1
ELSE
IF (A7+B5+B6+B7+A5)<NEIGHBOR_MIN OR
(A7+B5+B6+B7+A5)>NEIGHBOR_MAX AND A6=1 THEN -1 ELSE 0
```

```
A7(t) = A7(t - dt) + (A7_IN) * dt
INIT A7 = 0
INFLOWS:
A7_IN = IF (A6+A8+B6+B7+B8)=BIRTH_# AND A7=0 THEN 1
ELSE
IF (A6+A8+B6+B7+B8)<NEIGHBOR_MIN OR
(A6+A8+B6+B7+B8)>NEIGHBOR_MAX AND A7=1 THEN -1 ELSE 0

A8(t) = A8(t - dt) + (A8_IN) * dt
INIT A8 = 0
INFLOWS:
A8_IN = IF (A7+A9+B7+B8+B9)=BIRTH_# AND A8=0 THEN 1
ELSE
IF (A7+A9+B7+B8+B9)<NEIGHBOR_MIN OR
(A7+A9+B7+B8+B9)>NEIGHBOR_MAX AND A8=1 THEN -1 ELSE 0

A9(t) = A9(t - dt) + (A9_IN) * dt
INIT A9 = 0
INFLOWS:
A9_IN = IF (A10+A8+B10+B8+B9)=BIRTH_# AND A9=0 THEN 1
ELSE
IF (A10+A8+B10+B8+B9)<NEIGHBOR_MIN OR
(A10+A8+B10+B8+B9)>NEIGHBOR_MAX AND A9=1 THEN -1 ELSE
0

B1(t) = B1(t - dt) + (B1_IN) * dt
INIT B1 = 0
INFLOWS:
B1_IN = IF (A1+A2+B2+C1+C2)=BIRTH_# AND B1=0 THEN 1
ELSE
IF (A1+A2+B2+C1+C2)<NEIGHBOR_MIN OR
(A1+A2+B2+C1+C2)>NEIGHBOR_MAX AND B1=1 THEN -1 ELSE 0

B10(t) = B10(t - dt) + (B10_IN) * dt
INIT B10 = 0
INFLOWS:
B10_IN = IF (A10+A9+B9+C10+C9)=BIRTH_# AND B10=0 THEN 1
ELSE IF (A10+A9+B9+C10+C9)<NEIGHBOR_MIN OR
(A10+A9+B9+C10+C9)>NEIGHBOR_MAX AND B10=1 THEN -1 ELSE
0

B2(t) = B2(t - dt) + (B2_IN) * dt
INIT B2 = 0
```

```
INFLOWS:
B2_IN = IF (A1+A2+A3+B1+B3+C1+C2+C3)=BIRTH_# AND B2=0
THEN 1 ELSE IF (A1+A2+A3+B1+B3+C1+C2+C3)<NEIGHBOR_MIN
OR (A1+A2+A3+B1+B3+C1+C2+C3)>NEIGHBOR_MAX AND B2=1
THEN -1 ELSE 0

B3(t) = B3(t - dt) + (B3_IN) * dt
INIT B3 = 1
INFLOWS:
B3_IN = IF (A2+A3+A4+B2+B4+C2+C3+C4)=BIRTH_# AND B3=0
THEN 1 ELSE IF (A2+A3+A4+B2+B4+C2+C3+C4)<NEIGHBOR_MIN
OR (A2+A3+A4+B2+B4+C2+C3+C4)>NEIGHBOR_MAX AND B3=1
THEN -1 ELSE 0

B4(t) = B4(t - dt) + (B4_IN) * dt
INIT B4 = 0
INFLOWS:
B4_IN = IF (A3+A4+A5+B3+B5+C3+C4+C5)=BIRTH_# AND B4=0
THEN 1 ELSE IF (A3+A4+A5+B3+B5+C3+C4+C5)<NEIGHBOR_MIN
OR (A3+A4+A5+B3+B5+C3+C4+C5)>NEIGHBOR_MAX AND B4=1
THEN -1 ELSE 0

B5(t) = B5(t - dt) + (B5_IN) * dt
INIT B5 = 0
INFLOWS:
B5_IN = IF (A4+A5+A6+B4+B6+C4+C5+C6)=BIRTH_# AND B5=0
THEN 1 ELSE IF (A4+A5+A6+B4+B6+C4+C5+C6)<NEIGHBOR_MIN
OR (A4+A5+A6+B4+B6+C4+C5+C6)>NEIGHBOR_MAX AND B5=1
THEN -1 ELSE 0

B6(t) = B6(t - dt) + (B6_IN) * dt
INIT B6 = 0
INFLOWS:
B6_IN = IF (A5+A6+A7+B5+B7+C5+C6+C7)=BIRTH_# AND B6=0
THEN 1 ELSE IF (A5+A6+A7+B5+B7+C5+C6+C7)<NEIGHBOR_MIN
OR (A5+A6+A7+B5+B7+C5+C6+C7)>NEIGHBOR_MAX AND B6=1
THEN -1 ELSE 0

B7(t) = B7(t - dt) + (B7_IN) * dt
INIT B7 = 0
INFLOWS:
B7_IN = IF (A6+A7+A8+B6+B8+C6+C7+C8)=BIRTH_# AND B7=0
THEN 1 ELSE IF (A6+A7+A8+B6+B8+C6+C7+C8)<NEIGHBOR_MIN
```

```
OR (A6+A7+A8+B6+B8+C6+C7+C8)>NEIGHBOR_MAX AND B7=1
THEN -1 ELSE 0

B8(t) = B8(t - dt) + (B8_IN) * dt
INIT B8 = 0
INFLOWS:
B8_IN = IF (A7+A8+A9+B7+B9+C7+C8+C9)=BIRTH_# AND B8=0
THEN 1 ELSE IF (A7+A8+A9+B7+B9+C7+C8+C9)<NEIGHBOR_MIN
OR (A7+A8+A9+B7+B9+C7+C8+C9)>NEIGHBOR_MAX AND B8=1
THEN -1 ELSE 0

B9(t) = B9(t - dt) + (B9_IN) * dt
INIT B9 = 0
INFLOWS:
B9_IN = IF (A10+A8+A9+B10+B8+C10+C8+C9)=BIRTH_# AND
B9=0 THEN 1 ELSE IF
(A10+A8+A9+B10+B8+C10+C8+C9)<NEIGHBOR_MIN OR
(A10+A8+A9+B10+B8+C10+C8+C9)>NEIGHBOR_MAX AND B9=1
THEN -1 ELSE 0

C1(t) = C1(t - dt) + (C1_IN) * dt
INIT C1 = 1
INFLOWS:
C1_IN = IF (B1+B2+C2+D1+D2)=BIRTH_# AND C1=0 THEN 1
ELSE
IF (B1+B2+C2+D1+D2)<NEIGHBOR_MIN OR
(B1+B2+C2+D1+D2)>NEIGHBOR_MAX AND C1=1 THEN -1 ELSE 0

C10(t) = C10(t - dt) + (C10_IN) * dt
INIT C10 = 0
INFLOWS:
C10_IN = IF (B10+B9+C9+D10+D9)=BIRTH_# AND C10=0 THEN 1
ELSE IF (B10+B9+C9+D10+D9)<NEIGHBOR_MIN OR
(B10+B9+C9+D10+D9)>NEIGHBOR_MAX AND C10=1 THEN -1 ELSE
0

C2(t) = C2(t - dt) + (C2_IN) * dt
INIT C2 = 1
INFLOWS:
C2_IN = IF (B1+B2+B3+C1+C3+D1+D2+D3)=BIRTH_# AND C1=0
THEN 1 ELSE IF (B1+B2+B3+C1+C3+D1+D2+D3)<NEIGHBOR_MIN
OR (B1+B2+B3+C1+C3+D1+D2+D3)>NEIGHBOR_MAX AND C2=1
THEN -1 ELSE 0
```

```
C3(t) = C3(t - dt) + (C3_IN) * dt
INIT C3 = 1
INFLOWS:
C3_IN = IF (B2+B3+B4+C2+C4+D2+D3+D4)=BIRTH_# AND C3=0
THEN 1 ELSE IF (B2+B3+B4+C2+C4+D2+D3+D4)<NEIGHBOR_MIN
OR (B2+B3+B4+C2+C4+D2+D3+D4)>NEIGHBOR_MAX AND C3=1
THEN -1 ELSE 0

C4(t) = C4(t - dt) + (C4_IN) * dt
INIT C4 = 0
INFLOWS:
C4_IN = IF (B3+B4+B5+C3+C5+D3+D4+D5)=BIRTH_# AND C4=0
THEN 1 ELSE IF (B3+B4+B5+C3+C5+D3+D4+D5)<NEIGHBOR_MIN
OR (B3+B4+B5+C3+C5+D3+D4+D5)>NEIGHBOR_MAX AND C4=1
THEN -1 ELSE 0

C5(t) = C5(t - dt) + (C5_IN) * dt
INIT C5 = 0
INFLOWS:
C5_IN = IF (B4+B5+B6+C4+C6+D4+D5+D6)=BIRTH_# AND C5=0
THEN 1 ELSE IF (B4+B5+B6+C4+C6+D4+D5+D6)<NEIGHBOR_MIN
OR (B4+B5+B6+C4+C6+D4+D5+D6)>NEIGHBOR_MAX AND C5=1
THEN -1 ELSE 0

C6(t) = C6(t - dt) + (C6_IN) * dt
INIT C6 = 0
INFLOWS:
C6_IN = IF (B5+B6+B7+C5+C7+D5+D6+D7)=BIRTH_# AND C6=0
THEN 1 ELSE IF (B5+B6+B7+C5+C7+D5+D6+D7)<NEIGHBOR_MIN
OR (B5+B6+B7+C5+C7+D5+D6+D7)>NEIGHBOR_MAX AND C6=1
THEN -1 ELSE 0

C7(t) = C7(t - dt) + (C7_IN) * dt
INIT C7 = 0
INFLOWS:
C7_IN = IF (B6+B7+B8+C6+C8+D6+D7+D8)=BIRTH_# AND C7=0
THEN 1 ELSE IF (B6+B7+B8+C6+C8+D6+D7+D8)<NEIGHBOR_MIN
OR (B6+B7+B8+C6+C8+D6+D7+D8)>NEIGHBOR_MAX AND C7=1
THEN -1 ELSE 0

C8(t) = C8(t - dt) + (C8_IN) * dt
INIT C8 = 0
INFLOWS:
```

```
C8_IN = IF (B7+B8+B9+C7+C9+D7+D8+D9)=BIRTH_# AND C8=0
THEN 1 ELSE IF (B7+B8+B9+C7+C9+D7+D8+D9)<NEIGHBOR_MIN
OR (B7+B8+B9+C7+C9+D7+D8+D9)>NEIGHBOR_MAX AND C8=1
THEN -1 ELSE 0

C9(t) = C9(t - dt) + (C9_IN) * dt
INIT C9 = 0
INFLOWS:
C9_IN = IF (B10+B8+B9+C10+C8+D10+D8+D9)=BIRTH_# AND
C9=0 THEN 1 ELSE IF
(B10+B8+B9+C10+C8+D10+D8+D9)<NEIGHBOR_MIN OR
(B10+B8+B9+C10+C8+D10+D8+D9)>NEIGHBOR_MAX AND C9=1
THEN -1 ELSE 0

D1(t) = D1(t - dt) + (D1_IN) * dt
INIT D1 = 0
INFLOWS:
D1_IN = IF (C1+C2+D2+E1+E2)=BIRTH_# AND D1=0 THEN 1
ELSE IF (C1+C2+D2+E1+E2)<NEIGHBOR_MIN OR
(C1+C2+D2+E1+E2)>NEIGHBOR_MAX AND D1=1 THEN -1 ELSE 0

D10(t) = D10(t - dt) + (D10_IN) * dt
INIT D10 = 0
INFLOWS:
D10_IN = IF (C10+C9+D9+E10+E9)=BIRTH_# AND D10=0 THEN 1
ELSE IF (C10+C9+D9+E10+E9)<NEIGHBOR_MIN OR
(C10+C9+D9+E10+E9)>NEIGHBOR_MAX AND D10=1 THEN -1 ELSE
0

D2(t) = D2(t - dt) + (D2_IN) * dt
INIT D2 = 0
INFLOWS:
D2_IN = IF (C1+C2+C3+D1+D3+E1+E2+E3)=BIRTH_# AND D2=0
THEN 1 ELSE IF (C1+C2+C3+D1+D3+E1+E2+E3)<NEIGHBOR_MIN
OR (C1+C2+C3+D1+D3+E1+E2+E3)>NEIGHBOR_MAX AND D2=1
THEN -1 ELSE 0

D3(t) = D3(t - dt) + (D3_IN) * dt
INIT D3 = 0
INFLOWS:
D3_IN = IF (C2+C3+C4+D2+D4+E2+E3+E4)=BIRTH_# AND D3=0
THEN 1 ELSE IF (C2+C3+C4+D2+D4+E2+E3+E4)<NEIGHBOR_MIN
```

```
OR (C2+C3+C4+D2+D4+E2+E3+E4)>NEIGHBOR_MAX AND D3=1
THEN -1 ELSE 0

D4(t) = D4(t - dt) + (D4_IN) * dt
INIT D4 = 0
INFLOWS:
D4_IN = IF (C3+C4+C5+D3+D5+E3+E4+E5)=BIRTH_# AND D4=0
THEN 1 ELSE IF (C3+C4+C5+D3+D5+E3+E4+E5)<NEIGHBOR_MIN
OR (C3+C4+C5+D3+D5+E3+E4+E5)>NEIGHBOR_MAX AND D4=1
THEN -1 ELSE 0

D5(t) = D5(t - dt) + (D5_IN) * dt
INIT D5 = 0
INFLOWS:
D5_IN = IF (C4+C5+C6+D4+D6+E4+E5+E6)=BIRTH_# AND D5=0
THEN 1 ELSE IF (C4+C5+C6+D4+D6+E4+E5+E6)<NEIGHBOR_MIN
OR (C4+C5+C6+D4+D6+E4+E5+E6)>NEIGHBOR_MAX AND D5=1
THEN -1 ELSE 0

D6(t) = D6(t - dt) + (D6_IN) * dt
INIT D6 = 0
INFLOWS:
D6_IN = IF (C5+C6+C7+D5+D7+E5+E6+E7)=BIRTH_# AND D6=0
THEN 1 ELSE IF (C5+C6+C7+D5+D7+E5+E6+E7)<NEIGHBOR_MIN
OR (C5+C6+C7+D5+D7+E5+E6+E7)>NEIGHBOR_MAX AND D6=1
THEN -1 ELSE 0

D7(t) = D7(t - dt) + (D7_IN) * dt
INIT D7 = 0
INFLOWS:
D7_IN = IF (C6+C7+C8+D6+D8+E6+E7+E8)=BIRTH_# AND D7=0
THEN 1 ELSE IF (C6+C7+C8+D6+D8+E6+E7+E8)<NEIGHBOR_MIN
OR (C6+C7+C8+D6+D8+E6+E7+E8)>NEIGHBOR_MAX AND D7=1
THEN -1 ELSE 0

D8(t) = D8(t - dt) + (D8_IN) * dt
INIT D8 = 0
INFLOWS:
D8_IN = IF (C7+C8+C9+D7+D9+E7+E8+E9)=BIRTH_# AND D8=0
THEN 1 ELSE IF (C7+C8+C9+D7+D9+E7+E8+E9)<NEIGHBOR_MIN
OR (C7+C8+C9+D7+D9+E7+E8+E9)>NEIGHBOR_MAX AND D8=1
THEN -1 ELSE 0
```

```
D9(t) = D9(t - dt) + (D9_IN) * dt
INIT D9 = 0
INFLOWS:
D9_IN = IF (C10+C8+C9+D10+D8+E10+E8+E9)=BIRTH_# AND
D9=0 THEN 1 ELSE IF
(C10+C8+C9+D10+D8+E10+E8+E9)<NEIGHBOR_MIN OR
(C10+C8+C9+D10+D8+E10+E8+E9)>NEIGHBOR_MAX AND D9=1
THEN -1 ELSE 0

E1(t) = E1(t - dt) + (E1_IN) * dt
INIT E1 = 0
INFLOWS:
E1_IN = IF (D1+D2+E2+F1+F2)=BIRTH_# AND E1=0 THEN 1
ELSE IF (D1+D2+E2+F1+F2)<NEIGHBOR_MIN OR
(D1+D2+E2+F1+F2)>NEIGHBOR_MAX AND E1=1 THEN -1 ELSE 0

E10(t) = E10(t - dt) + (E10_IN) * dt
INIT E10 = 0
INFLOWS:
E10_IN = IF (D10+D9+E9+F10+F9)=BIRTH_# AND E10=0 THEN 1
ELSE IF (D10+D9+E9+F10+F9)<NEIGHBOR_MIN OR
(D10+D9+E9+F10+F9)>NEIGHBOR_MAX AND E10=1 THEN -1 ELSE
0

E2(t) = E2(t - dt) + (E2_IN) * dt
INIT E2 = 0
INFLOWS:
E2_IN = IF (D1+D2+D3+E1+E3+F1+F2+F3)=BIRTH_# AND E2=0
THEN 1 ELSE IF (D1+D2+D3+E1+E3+F1+F2+F3)<NEIGHBOR_MIN
OR (D1+D2+D3+E1+E3+F1+F2+F3)>NEIGHBOR_MAX AND E2=1
THEN -1 ELSE 0

E3(t) = E3(t - dt) + (E3_IN) * dt
INIT E3 = 0
INFLOWS:
E3_IN = IF (D2+D3+D4+E2+E4+F2+F3+F4)=BIRTH_# AND E3=0
THEN 1 ELSE IF (D2+D3+D4+E2+E4+F2+F3+F4)<NEIGHBOR_MIN
OR (D2+D3+D4+E2+E4+F2+F3+F4)>NEIGHBOR_MAX AND E3=1
THEN -1 ELSE 0

E4(t) = E4(t - dt) + (E4_IN) * dt
INIT E4 = 0
INFLOWS:
```

```
E4_IN = IF (D3+D4+D5+E3+E5+F3+F4+F5)=BIRTH_# AND E4=0
THEN 1 ELSE IF (D3+D4+D5+E3+E5+F3+F4+F5)<NEIGHBOR_MIN
OR (D3+D4+D5+E3+E5+F3+F4+F5)>NEIGHBOR_MAX AND E4=1
THEN -1 ELSE 0

E5(t) = E5(t - dt) + (E5_IN) * dt
INIT E5 = 0
INFLOWS:
E5_IN = IF (D4+D5+D6+E4+E6+F4+F5+F6)=BIRTH_# AND E5=0
THEN 1 ELSE IF (D4+D5+D6+E4+E6+F4+F5+F6)<NEIGHBOR_MIN
OR (D4+D5+D6+E4+E6+F4+F5+F6)>NEIGHBOR_MAX AND E5=1
THEN -1 ELSE 0

E6(t) = E6(t - dt) + (E6_IN) * dt
INIT E6 = 0
INFLOWS:
E6_IN = IF (D5+D6+D7+E5+E7+F5+F6+F7)=BIRTH_# AND E6=0
THEN 1 ELSE IF (D5+D6+D7+E5+E7+F5+F6+F7)<NEIGHBOR_MIN
OR (D5+D6+D7+E5+E7+F5+F6+F7)>NEIGHBOR_MAX AND E6=1
THEN -1 ELSE 0

E7(t) = E7(t - dt) + (E7_IN) * dt
INIT E7 = 0
INFLOWS:
E7_IN = IF (D6+D7+D8+E6+E8+F6+F7+F8)=BIRTH_# AND E7=0
THEN 1 ELSE IF (D6+D7+D8+E6+E8+F6+F7+F8)<NEIGHBOR_MIN
OR (D6+D7+D8+E6+E8+F6+F7+F8)>NEIGHBOR_MAX AND E7=1
THEN -1 ELSE 0

E8(t) = E8(t - dt) + (E8_IN) * dt
INIT E8 = 0
INFLOWS:
E8_IN = IF (D7+D8+D9+E7+E9+F7+F8+F9)=BIRTH_# AND E8=0
THEN 1 ELSE IF (D7+D8+D9+E7+E9+F7+F8+F9)<NEIGHBOR_MIN
OR (D7+D8+D9+E7+E9+F7+F8+F9)>NEIGHBOR_MAX AND E8=1
THEN -1 ELSE 0

E9(t) = E9(t - dt) + (E9_IN) * dt
INIT E9 = 0
INFLOWS:
E9_IN = IF (D10+D8+D9+E10+E8+F10+F8+F9)=BIRTH_# AND
E9=0 THEN 1 ELSE IF
(D10+D8+D9+E10+E8+F10+F8+F9)<NEIGHBOR_MIN OR
```

```
(D10+D8+D9+E10+E8+F10+F8+F9)>NEIGHBOR_MAX AND E9=1
THEN -1 ELSE 0

F1(t) = F1(t - dt) + (F1_IN) * dt
INIT F1 = 0
INFLOWS:
F1_IN = IF (E1+E2+F2+G1+G2) = BIRTH_# AND F1=0 THEN 1
ELSE IF (E1+E2+F2+G1+G2) <NEIGHBOR_MIN OR
(E1+E2+F2+G1+G2)>NEIGHBOR_MAX AND F1=1 THEN -1 ELSE 0

F10(t) = F10(t - dt) + (F10_IN) * dt
INIT F10 = 0
INFLOWS:
F10_IN = IF (E10+E9+F9+G10+G9)=BIRTH_# AND F10=0 THEN 1
ELSE IF (E10+E9+F9+G10+G9)<NEIGHBOR_MIN OR
(E10+E9+F9+G10+G9)>NEIGHBOR_MAX AND F10=1 THEN -1 ELSE
0

F2(t) = F2(t - dt) + (F2_IN) * dt
INIT F2 = 0
INFLOWS:
F2_IN = IF (E1+E2+E3+F1+F3+G1+G2+G3)=BIRTH_# AND F2=0
THEN 1 ELSE IF (E1+E2+E3+F1+F3+G1+G2+G3)<NEIGHBOR_MIN
OR (E1+E2+E3+F1+F3+G1+G2+G3)>NEIGHBOR_MAX AND F2=1
THEN -1 ELSE 0

F3(t) = F3(t - dt) + (F3_IN) * dt
INIT F3 = 0
INFLOWS:
F3_IN = IF (E2+E3+E4+F2+F4+G2+G3+G4)=BIRTH_# AND F3=0
THEN 1 ELSE IF (E2+E3+E4+F2+F4+G2+G3+G4)<NEIGHBOR_MIN
OR (E2+E3+E4+F2+F4+G2+G3+G4)>NEIGHBOR_MAX AND F3=1
THEN -1 ELSE 0

F4(t) = F4(t - dt) + (F4_IN) * dt
INIT F4 = 0
INFLOWS:
F4_IN = IF (E3+E4+E5+F3+F5+G3+G4+G5)=BIRTH_# AND F4=0
THEN 1 ELSE IF (E3+E4+E5+F3+F5+G3+G4+G5)<NEIGHBOR_MIN
OR (E3+E4+E5+F3+F5+G3+G4+G5)>NEIGHBOR_MAX AND F4=1
THEN -1 ELSE 0

F5(t) = F5(t - dt) + (F5_IN) * dt
```

```
INIT F5 = 0
INFLOWS:
F5_IN = IF (E4+E5+E6+F4+F6+G4+G5+G6)=BIRTH_# AND F5=0
THEN 1 ELSE IF (E4+E5+E6+F4+F6+G4+G5+G6)<NEIGHBOR_MIN
OR (E4+E5+E6+F4+F6+G4+G5+G6)>NEIGHBOR_MAX AND F5=1
THEN -1 ELSE 0

F6(t) = F6(t - dt) + (F6_IN) * dt
INIT F6 = 0
INFLOWS:
F6_IN = IF (E5+E6+E7+F5+F7+G5+G6+G7)=BIRTH_# AND F6=0
THEN 1 ELSE IF (E5+E6+E7+F5+F7+G5+G6+G7)<NEIGHBOR_MIN
OR (E5+E6+E7+F5+F7+G5+G6+G7)>NEIGHBOR_MAX AND F6=1
THEN -1 ELSE 0

F7(t) = F7(t - dt) + (F7_IN) * dt
INIT F7 = 0
INFLOWS:
F7_IN = IF (E6+E7+E8+F6+F8+G6+G7+G8)=BIRTH_# AND F7=0
THEN 1 ELSE IF (E6+E7+E8+F6+F8+G6+G7+G8)<NEIGHBOR_MIN
OR (E6+E7+E8+F6+F8+G6+G7+G8)>NEIGHBOR_MAX AND F7=1
THEN -1 ELSE 0

F8(t) = F8(t - dt) + (F8_IN) * dt
INIT F8 = 0
INFLOWS:
F8_IN = IF (E7+E8+E9+F7+F9+G7+G8+G9)=BIRTH_# AND F8=0
THEN 1 ELSE IF (E7+E8+E9+F7+F9+G7+G8+G9)<NEIGHBOR_MIN
OR (E7+E8+E9+F7+F9+G7+G8+G9)>NEIGHBOR_MAX AND F8=1
THEN -1 ELSE 0

F9(t) = F9(t - dt) + (F9_IN) * dt
INIT F9 = 0
INFLOWS:
F9_IN = IF (E10+E8+E9+F10+F8+G10+G8+G9)=BIRTH_# AND
F9=0 THEN 1 ELSE IF
(E10+E8+E9+F10+F8+G10+G8+G9)<NEIGHBOR_MIN OR
(E10+E8+E9+F10+F8+G10+G8+G9)>NEIGHBOR_MAX AND F9=1
THEN -1 ELSE 0

G1(t) = G1(t - dt) + (G1_IN) * dt
INIT G1 = 0
INFLOWS:
```

```
G1_IN = IF (F1+F2+G2+H1+H2)=BIRTH_# AND G1=0 THEN 1
ELSE IF (F1+F2+G2+H1+H2)<NEIGHBOR_MIN OR
(F1+F2+G2+H1+H2)>NEIGHBOR_MAX AND G1=1 THEN -1 ELSE 0

G10(t) = G10(t - dt) + (G10_IN) * dt
INIT G10 = 0
INFLOWS:
G10_IN = IF (F10+F9+G9+H10+H9)=BIRTH_# AND G10=0 THEN 1
ELSE IF (F10+F9+G9+H10+H9)<NEIGHBOR_MIN OR
(F10+F9+G9+H10+H9)>NEIGHBOR_MAX AND G10=1 THEN -1 ELSE
0

G2(t) = G2(t - dt) + (G2_IN) * dt
INIT G2 = 0
INFLOWS:
G2_IN = IF (F1+F2+F3+G1+G3+H1+H2+H3)=BIRTH_# AND G2=0
THEN 1 ELSE IF (F1+F2+F3+G1+G3+H1+H2+H3)<NEIGHBOR_MIN
OR (F1+F2+F3+G1+G3+H1+H2+H3)>NEIGHBOR_MAX AND G2=1
THEN -1 ELSE 0

G3(t) = G3(t - dt) + (G3_IN) * dt
INIT G3 = 0
INFLOWS:
G3_IN = IF (F2+F3+F4+G2+G4+H2+H3+H4)=BIRTH_# AND G3=0
THEN 1 ELSE IF (F2+F3+F4+G2+G4+H2+H3+H4)<NEIGHBOR_MIN
OR (F2+F3+F4+G2+G4+H2+H3+H4)>NEIGHBOR_MAX AND G3=1
THEN -1 ELSE 0

G4(t) = G4(t - dt) + (G4_IN) * dt
INIT G4 = 0
INFLOWS:
G4_IN = IF (F3+F4+F5+G3+G5+H3+H4+H5)=BIRTH_# AND G4=0
THEN 1 ELSE IF (F3+F4+F5+G3+G5+H3+H4+H5)<NEIGHBOR_MIN
OR (F3+F4+F5+G3+G5+H3+H4+H5)>NEIGHBOR_MAX AND G4=1
THEN -1 ELSE 0

G5(t) = G5(t - dt) + (G5_IN) * dt
INIT G5 = 0
INFLOWS:
G5_IN = IF (F4+F5+F6+G4+G6+H4+H5+H6)=BIRTH_# AND G5=0
THEN 1 ELSE IF (F4+F5+F6+G4+G6+H4+H5+H6)<NEIGHBOR_MIN
OR (F4+F5+F6+G4+G6+H4+H5+H6)>NEIGHBOR_MAX AND G5=1
THEN -1 ELSE 0
```

```
G6(t) = G6(t - dt) + (G6_IN) * dt
INIT G6 = 0
INFLOWS:
G6_IN = IF (F5+F6+F7+G5+G7+H5+H6+H7)=BIRTH_# AND G6=0
THEN 1 ELSE IF (F5+F6+F7+G5+G7+H5+H6+H7)<NEIGHBOR_MIN
OR (F5+F6+F7+G5+G7+H5+H6+H7)>NEIGHBOR_MAX AND G6=1
THEN -1 ELSE 0

G7(t) = G7(t - dt) + (G7_IN) * dt
INIT G7 = 0
INFLOWS:
G7_IN = IF (F6+F7+F8+G6+G8+H6+H7+H8)=BIRTH_# AND G7=0
THEN 1 ELSE IF (F6+F7+F8+G6+G8+H6+H7+H8)<NEIGHBOR_MIN
OR (F6+F7+F8+G6+G8+H6+H7+H8)>NEIGHBOR_MAX AND G7=1
THEN -1 ELSE 0

G8(t) = G8(t - dt) + (G8_IN) * dt
INIT G8 = 0
INFLOWS:
G8_IN = IF (F7+F8+F9+G7+G9+H7+H8+H9)=BIRTH_# AND G8=0
THEN 1 ELSE IF (F7+F8+F9+G7+G9+H7+H8+H9)<NEIGHBOR_MIN
OR (F7+F8+F9+G7+G9+H7+H8+H9)>NEIGHBOR_MAX AND G8=1
THEN -1 ELSE 0

G9(t) = G9(t - dt) + (G9_IN) * dt
INIT G9 = 0
INFLOWS:
G9_IN = IF (F10+F8+F9+G10+G8+H10+H8+H9)=BIRTH_# AND
G9=0 THEN 1 ELSE IF
(F10+F8+F9+G10+G8+H10+H8+H9)<NEIGHBOR_MIN OR
(F10+F8+F9+G10+G8+H10+H8+H9)>NEIGHBOR_MAX AND G9=1
THEN -1 ELSE 0

H1(t) = H1(t - dt) + (H1_IN) * dt
INIT H1 = 0
INFLOWS:
H1_IN = IF (G1+G2+H2+I1+I2)=BIRTH_# AND H1=0 THEN 1
ELSE IF (G1+G2+H2+I1+I2)<NEIGHBOR_MIN OR
(G1+G2+H2+I1+I2)>NEIGHBOR_MAX AND H1=1 THEN -1 ELSE 0

H10(t) = H10(t - dt) + (H10_IN) * dt
INIT H10 = 0
INFLOWS:
```

```
H10_IN = IF (G10+G9+H9+I10+I9)=BIRTH_# AND H10=0 THEN 1
ELSE IF (G10+G9+H9+I10+I9)<NEIGHBOR_MIN OR
(G10+G9+H9+I10+I9)>NEIGHBOR_MAX OR H10=1 THEN -1 ELSE
0

H2(t) = H2(t - dt) + (H2_IN) * dt
INIT H2 = 0
INFLOWS:
H2_IN = IF (G1+G2+G3+H1+H3+I1+I2+I3)=BIRTH_# AND H2=0
THEN 1 ELSE IF (G1+G2+G3+H1+H3+I1+I2+I3)<NEIGHBOR_MIN
OR (G1+G2+G3+H1+H3+I1+I2+I3)>NEIGHBOR_MAX AND H2=1
THEN -1 ELSE 0

H3(t) = H3(t - dt) + (H3_IN) * dt
INIT H3 = 0
INFLOWS:
H3_IN = IF (G2+G3+G4+H2+H4+I2+I3+I4)=BIRTH_# AND H3=0
THEN 1 ELSE IF (G2+G3+G4+H2+H4+I2+I3+I4)<NEIGHBOR_MIN
OR (G2+G3+G4+H2+H4+I2+I3+I4)>NEIGHBOR_MAX AND H3=1
THEN -1 ELSE 0

H4(t) = H4(t - dt) + (H4_IN) * dt
INIT H4 = 0
INFLOWS:
H4_IN = IF (G3+G4+G5+H3+H5+I3+I4+I5)=BIRTH_# AND H4=0
THEN 1 ELSE IF (G3+G4+G5+H3+H5+I3+I4+I5)<NEIGHBOR_MIN
OR (G3+G4+G5+H3+H5+I3+I4+I5)>NEIGHBOR_MAX AND H4=1
THEN -1 ELSE 0

H5(t) = H5(t - dt) + (H5_IN) * dt
INIT H5 = 0
INFLOWS:
H5_IN = IF (G4+G5+G6+H4+H6+I4+I5+I6)=BIRTH_# AND H5=0
THEN 1 ELSE IF (G4+G5+G6+H4+H6+I4+I5+I6)<NEIGHBOR_MIN
OR (G4+G5+G6+H4+H6+I4+I5+I6)>NEIGHBOR_MAX AND H5=1
THEN -1 ELSE 0

H6(t) = H6(t - dt) + (H6_IN) * dt
INIT H6 = 0
INFLOWS:
H6_IN = IF (G5+G6+G7+H5+H7+I5+I6+I7)=BIRTH_# AND H6=0
THEN 1 ELSE IF (G5+G6+G7+H5+H7+I5+I6+I7)<NEIGHBOR_MIN
```

```
OR (G5+G6+G7+H5+H7+I5+I6+I7)>NEIGHBOR_MAX AND H6=1
THEN -1 ELSE 0

H7(t) = H7(t - dt) + (H7_IN) * dt
INIT H7 = 0
INFLOWS:
H7_IN = IF (G6+G7+G8+H6+H8+I6+I7+I8)=BIRTH_# AND H7=0
THEN 1 ELSE IF (G6+G7+G8+H6+H8+I6+I7+I8)<NEIGHBOR_MIN
OR (G6+G7+G8+H6+H8+I6+I7+I8)>NEIGHBOR_MAX AND H7=1
THEN -1 ELSE 0

H8(t) = H8(t - dt) + (H8_IN) * dt
INIT H8 = 0
INFLOWS:
H8_IN = IF (G7+G8+G9+H7+H9+I7+I8+I9)=BIRTH_# AND H8=0
THEN 1 ELSE IF (G7+G8+G9+H7+H9+I7+I8+I9)<NEIGHBOR_MIN
OR (G7+G8+G9+H7+H9+I7+I8+I9)>NEIGHBOR_MAX AND H8=1
THEN -1 ELSE 0

H9(t) = H9(t - dt) + (H9_IN) * dt
INIT H9 = 0
INFLOWS:
H9_IN = IF (G10+G8+G9+H10+H8+I10+I8+I9)=BIRTH_# AND
H9=0 THEN 1 ELSE IF
(G10+G8+G9+H10+H8+I10+I8+I9)<NEIGHBOR_MIN OR
(G10+G8+G9+H10+H8+I10+I8+I9)>NEIGHBOR_MAX AND H9=1
THEN -1 ELSE 0

I1(t) = I1(t - dt) + (I1_IN) * dt
INIT I1 = 0
INFLOWS:
I1_IN = IF (H1+H2+I2+J1+J2)=BIRTH_# AND I1=0 THEN 1
ELSE IF (H1+H2+I2+J1+J2)<NEIGHBOR_MIN OR
(H1+H2+I2+J1+J2)>NEIGHBOR_MAX AND I1=1 THEN -1 ELSE 0

I10(t) = I10(t - dt) + (I10_IN) * dt
INIT I10 = 0
INFLOWS:
I10_IN = IF (H10+H9+I9+J10+J9)=BIRTH_# AND I10=0 THEN 1
ELSE IF (H10+H9+I9+J10+J9)<NEIGHBOR_MIN OR
(H10+H9+I9+J10+J9)>NEIGHBOR_MAX AND I10=1 THEN -1
ELSE 0
```

```
I2(t) = I2(t - dt) + (I2_IN) * dt
INIT I2 = 0
INFLOWS:
I2_IN = IF (H1+H2+H3+I1+I3+J1+J2+J3)=BIRTH_# AND I2=0
THEN 1 ELSE IF (H1+H2+H3+I1+I3+J1+J2+J3)<NEIGHBOR_MIN
OR (H1+H2+H3+I1+I3+J1+J2+J3)>NEIGHBOR_MAX AND I2=1
THEN -1 ELSE 0

I3(t) = I3(t - dt) + (I3_IN) * dt
INIT I3 = 0
INFLOWS:
I3_IN = IF (H2+H3+H4+I2+I4+J2+J3+J4)=BIRTH_# AND I3=0
THEN 1 ELSE IF (H2+H3+H4+I2+I4+J2+J3+J4)<NEIGHBOR_MIN
OR (H2+H3+H4+I2+I4+J2+J3+J4)>NEIGHBOR_MAX AND I3=1
THEN -1 ELSE 0

I4(t) = I4(t - dt) + (I4_IN) * dt
INIT I4 = 0
INFLOWS:
I4_IN = IF (H3+H4+H5+I3+I5+J3+J4+J5)=BIRTH_# AND I4=0
THEN 1 ELSE IF (H3+H4+H5+I3+I5+J3+J4+J5)<NEIGHBOR_MIN
OR (H3+H4+H5+I3+I5+J3+J4+J5)>NEIGHBOR_MAX AND I4=1
THEN -1 ELSE 0

I5(t) = I5(t - dt) + (I5_IN) * dt
INIT I5 = 0
INFLOWS:
I5_IN = IF (H4+H5+H6+I4+I6+J4+J5+J6)=BIRTH_# AND I5=0
THEN 1 ELSE IF (H4+H5+H6+I4+I6+J4+J5+J6)<NEIGHBOR_MIN
OR (H4+H5+H6+I4+I6+J4+J5+J6)>NEIGHBOR_MAX AND I5=1
THEN -1 ELSE 0

I6(t) = I6(t - dt) + (I6_IN) * dt
INIT I6 = 0
INFLOWS:
I6_IN = IF (H5+H6+H7+I5+I7+J5+J6+J7)=BIRTH_# AND I6=0
THEN 1 ELSE IF (H5+H6+H7+I5+I7+J5+J6+J7)<NEIGHBOR_MIN
OR (H5+H6+H7+I5+I7+J5+J6+J7)>NEIGHBOR_MAX AND I6=1
THEN -1 ELSE 0

I7(t) = I7(t - dt) + (I7_IN) * dt
INIT I7 = 0
INFLOWS:
```

```
I7_IN = IF (H6+H7+H8+I6+I8+J6+J7+J8)=BIRTH_# AND I7=0
THEN 1 ELSE IF (H6+H7+H8+I6+I8+J6+J7+J8)<NEIGHBOR_MIN
OR (H6+H7+H8+I6+I8+J6+J7+J8)>NEIGHBOR_MAX AND I7=1
THEN -1 ELSE 0

I8(t) = I8(t - dt) + (I8_IN) * dt
INIT I8 = 0
INFLOWS:
I8_IN = IF (H7+H8+H9+I7+I9+J7+J8+J9)=BIRTH_# AND I8=0
THEN 1 ELSE IF (H7+H8+H9+I7+I9+J7+J8+J9)<NEIGHBOR_MIN
OR (H7+H8+H9+I7+I9+J7+J8+J9)>NEIGHBOR_MAX AND I8=1
THEN -1 ELSE 0

I9(t) = I9(t - dt) + (I9_IN) * dt
INIT I9 = 0
INFLOWS:
I9_IN = IF (H10+H8+I10+I8+H9+J10+J8+J9)=BIRTH_# AND
I9=0 THEN 1 ELSE IF
(H10+H8+I10+I8+H9+J10+J8+J9)<NEIGHBOR_MIN OR
(H10+H8+I10+I8+H9+J10+J8+J9)>NEIGHBOR_MAX AND I9=1
THEN -1 ELSE 0

J1(t) = J1(t - dt) + (J1_IN) * dt
INIT J1 = 0
INFLOWS:
J1_IN = IF (I1+I2+J2)=BIRTH_# AND J1=0 THEN 1 ELSE IF
(I1+I2+J2)<NEIGHBOR_MIN OR (I1+I2+J2)>NEIGHBOR_MAX AND
J1=1 THEN -1 ELSE 0

J10(t) = J10(t - dt) + (J10_IN) * dt
INIT J10 = 0
INFLOWS:
J10_IN = IF (I10+I9+J9)=BIRTH_# AND J10=0 THEN 1 ELSE
IF (I10+I9+J9)<NEIGHBOR_MIN OR
(I10+I9+J9)>NEIGHBOR_MAX AND J10=1 THEN -1 ELSE 0

J2(t) = J2(t - dt) + (J2_IN) * dt
INIT J2 = 0
INFLOWS:
J2_IN = IF (I1+I2+I3+J1+J3)=BIRTH_# AND J2=0 THEN 1
ELSE IF (I1+I2+I3+J1+J3)<NEIGHBOR_MIN OR
(I1+I2+I3+J1+J3)>NEIGHBOR_MAX AND J2=1 THEN -1 ELSE 0
```

```
J3(t) = J3(t - dt) + (J3_IN) * dt
INIT J3 = 0
INFLOWS:
J3_IN = IF (I2+I3+I4+J2+J4)=BIRTH_# AND J3=0 THEN 1
ELSE IF (I2+I3+I4+J2+J4)<NEIGHBOR_MIN OR
(I2+I3+I4+J2+J4)>NEIGHBOR_MAX AND J3=1 THEN -1 ELSE 0

J4(t) = J4(t - dt) + (J4_IN) * dt
INIT J4 = 0
INFLOWS:
J4_IN = IF (I3+I4+I5+J3+J5)=BIRTH_# AND J4=0 THEN 1
ELSE IF (I3+I4+I5+J3+J5)<NEIGHBOR_MIN OR
(I3+I4+I5+J3+J5)>NEIGHBOR_MAX AND J4=1 THEN -1 ELSE 0

J5(t) = J5(t - dt) + (J5_IN) * dt
INIT J5 = 0
INFLOWS:
J5_IN = IF (I4+I5+I6+J4+J6)=BIRTH_# AND J5=0 THEN 1
ELSE IF (I4+I5+I6+J4+J6)<NEIGHBOR_MIN OR
(I4+I5+I6+J4+J6)>NEIGHBOR_MAX AND J5=1 THEN -1 ELSE 0

J6(t) = J6(t - dt) + (J6_IN) * dt
INIT J6 = 0
INFLOWS:
J6_IN = IF (I5+I6+I7+J5+J7)=BIRTH_# AND J6=0 THEN 1
ELSE IF (I5+I6+I7+J5+J7)<NEIGHBOR_MIN OR
(I5+I6+I7+J5+J7)>NEIGHBOR_MAX AND J6=1 THEN -1 ELSE 0

J7(t) = J7(t - dt) + (J7_IN) * dt
INIT J7 = 0
INFLOWS:
J7_IN = IF (I6+I7+I8+J6+J8)=BIRTH_# AND J7=0 THEN 1
ELSE IF (I6+I7+I8+J6+J8)<NEIGHBOR_MIN OR
(I6+I7+I8+J6+J8)>NEIGHBOR_MAX AND J7=1 THEN -1 ELSE 0

J8(t) = J8(t - dt) + (J8_IN) * dt
INIT J8 = 0
INFLOWS:
J8_IN = IF (I7+I8+I9+J7+J9)=BIRTH_# AND J8=0 THEN 1
ELSE IF (I7+I8+I9+J7+J9)<NEIGHBOR_MIN OR
(I7+I8+I9+J7+J9)>NEIGHBOR_MAX AND J8=1 THEN -1 ELSE 0

J9(t) = J9(t - dt) + (J9_IN) * dt
```

```
INIT J9 = 0
INFLOWS:
J9_IN = IF (I10+I8+I9+J10+J8)=BIRTH_# AND J9=0 THEN 1
ELSE IF (I10+I8+I9+J10+J8) <NEIGHBOR_MIN OR
(I10+I8+I9+J10+J8)>NEIGHBOR_MAX AND J9=1 THEN -1 ELSE 0

BIRTH_# = 3 {NUMBER OF NEIGHBOR CELLS REQUIRED TO GIVE
BIRTH TO THE CELL}
EAT_LAG = 3
NEIGHBOR_MAX = 3 {Number of neighbor cells above which
the cell dies from crowding}
NEIGHBOR_MIN = 2 {Minimum number of neighboring cells
below which the cell is no longer viable}
REPROD_PROB = 1
```

42

Daisyworld*

One ring to rule them all, one ring to bind them,
one ring to bring them all, and in the darkness bind them.
J.R.R. Tolkien, *The Lord of the Rings,* Part 1, 1965.

During the late 1960s and early 1970s, James Lovelock, an independent inventor and scientist, and Lynn Margulis, a professor at Boston University, worked with the NASA Jet Propulsion Laboratory to develop a means to detect life on Mars. It was noted in the progress of this work that one striking property of the Earth is that its atmosphere is far from chemical equilibrium since the biota use it as a resevoir for nutrients and waste products. In other words, the atmosphere is, in the steady state, not derived of ordinary chemistry and physics. In fact, it is derived of life.

Conventional thought is that the Earth and the life upon it evolved separately by different mechanisms. The evolution of the planet's surface features and the atmosphere is a result of chemical and physical processes to which the biota must adapt to prevent extinction. In some instances, it is very clear that life has had a significant effect on the environment, yet these feedbacks are regarded as accidental. Lovelock has criticized this view for not adequately describing the sophisticated interaction between life and its environment and has developed his own theory, known as the Gaia hypothesis.

The Gaia hypothesis states, quite simply, that "the Earth is homeostatic, with the biota actively seeking to keep the environment optimal for life."[1] Lovelock suggests that life has the capacity to regulate the environment of the Earth in order to maintain its fitness for that life. Thus, according to the Gaia hypothesis, life and the environment evolve together as a single system. The species that leaves the most progeny tend to inherit a particular environment, and in turn, the environment that favors the most progeny is itself sustained.

There has been little success in finding the feedback mechanisms that regulate Gaia, but these mechanisms are likely to be subtle and complex. Scientists have questioned the validity of the Gaia hypothesis for this reason and others. One of the most important criticisms has been that the active

---

*We wish to thank Nathan Gardner for his contributions to this chapter.
[1]J.E. Lovelock, *The Ages of Gaia,* New York: W.W. Norton & Co., 1988.

regulation of the environment that is the central property of Gaia supposes natural selection for traits that will not be beneficial for thousands of years. But the ecosystem has no capacity for conscious forethought and planning, and it is clear that the biota would not evolve altruistic planetary regulation systems in the pursuit of local self-interest. In other words, if bacteria appear on the Earth's surface, how can these bacteria regulate the atmosphere to suit their own needs when they cannot possibly have a significant effect on the atmosphere's chemistry for thousands of years? Lovelock chose to challenge this criticism with a paradigm and a numerical model: Daisyworld. Since he first proposed the model, he also asserted that Daisyworld fully exhibits all of the characteristics of a Gaian world.

In our simplified view, Daisyworld is a right cylindrical planet in Earth's orbit around a sun of Sol's luminescence. This planet has a clear atmosphere and is populated by two life forms: light and dark daisies. Here, albedo is specified as a unitless measure of the reflectivity of a surface where 1 means perfectly reflective and 0 means perfectly absorptive. Light daisies have an albedo greater than the planet surface albedo, and dark daisies have an albedo less than that of the planet's surface. If there is bare planet surface available, the daisies will grow and cover it, and change its albedo.

The model of Daisyworld is based on Lovelock's equations.[2] We briefly present them here. There are only two stocks in this model, LIGHT DAISY AREA and DARK DAISY AREA. These stocks represent the fraction of planet surface covered by each type of daisy (Fig. 42.1). The daisy populations are increased by growth, which is temperature dependent. For each daisy species, there is a temperature window where growth can occur. That window extends from 5° to 45°C. Within this window, growth is described by the parabola

LIGHT DAISY GROWTH = 1 – 0.003265(25 – LIGHT DAISY TEMP)^2. (1)

DARK DAISY GROWTH = 1 – 0.003265*(25 – DARK DAISY TEMP)^2 (2)

These functions are maximal at a temperature of 25°C. The actual daisy growth is then given by the differential equations

d (LIGHT DAISY AREA)/dt = LIGHT DAISY AREA * NO DAISY AREA * LIGHT DAISY GROWTH (3)

d (DARK DAISY AREA)/dt = DARK DAISY AREA * NO DAISY AREA * DARK DAISY GROWTH, (4)

where NO DAISY AREA is the fraction of fertile planet surface area not populated by daisies:

[2]The original equations and parameters are listed in *Tellus*, Series B, 4:284, 1983.

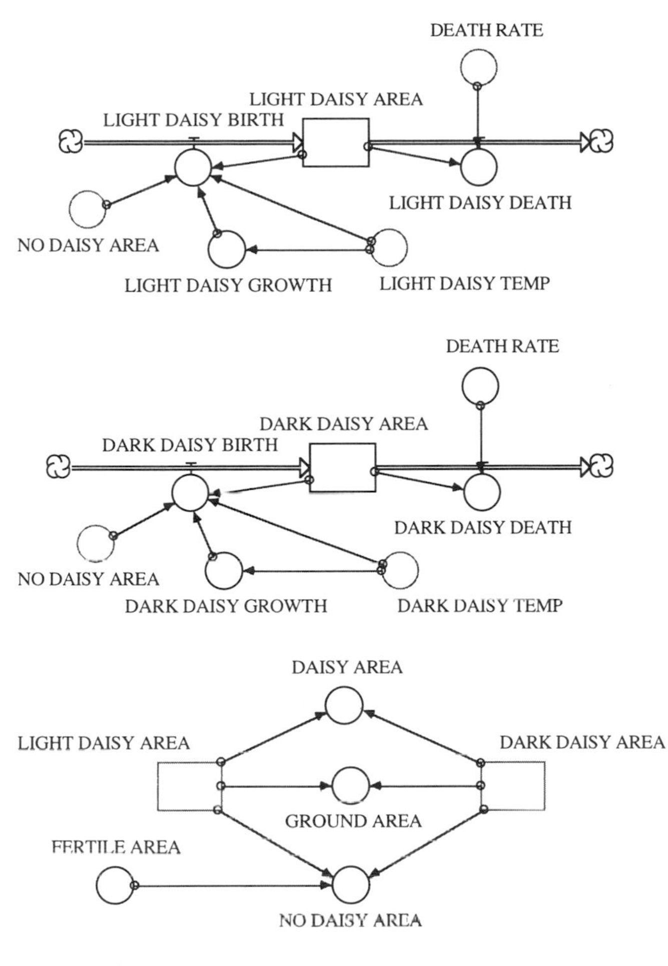

FIGURE 42.1

$$\text{NO DAISY AREA} = \text{FERTILE AREA} - \text{LIGHT DAISY AREA} - \text{DARK DAISY AREA} \quad (5)$$

FERTILE AREA is the exogenously given fraction of the planet surface that is habitable by daisies. Similarly, the area occupied by daisies is

$$\text{DAISY AREA} = \text{LIGHT DAISY AREA} + \text{DARK DAISY AREA}, \quad (6)$$

and the area not occupied by daisies is

$$\text{GROUND AREA} = 1 - \text{DARK DAISY AREA} - \text{LIGHT DAISY AREA} \quad (7)$$

The daisies are also subject to death at the rate of 30% of the population per unit time.

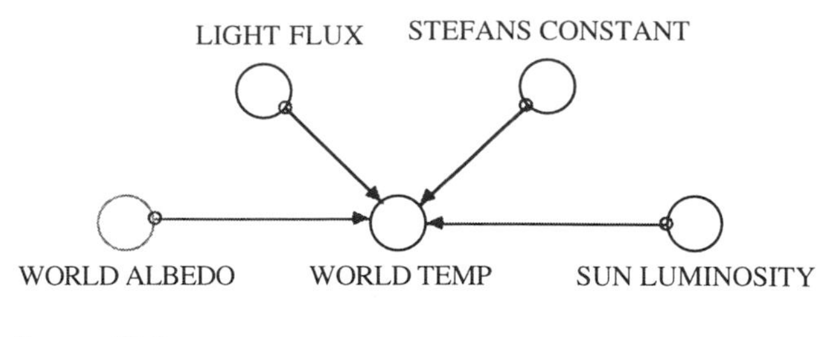

FIGURE 42.2

The most crucial aspect of the of Daisyworld model lies in the calculation of the temperature of the world. Modules in Figures 42.2 to 42.4 are devoted to this calculation. The first of these modules (Fig. 42.2) calculates world temperature. Physically, the planet surface temperature is determined by the amount of solar radiation incident upon the planet and the amount reflected by the planet. Using the Stefan–Boltzmann rule, WORLD TEMP is given by

WORLD TEMP = ((SUN LUMINOSITY*LIGHT FLUX*(1-WORLD ALBEDO))/
STEFANS CONSTANT)^0.25–273, (8)

where LIGHT FLUX is equal to the solar flux of our Sun, SUN LUMINOSITY is a dimensionless measure of the Sun's luminosity, and STEFANS CONSTANT is the proportionality between solar radiation flux and temperature.

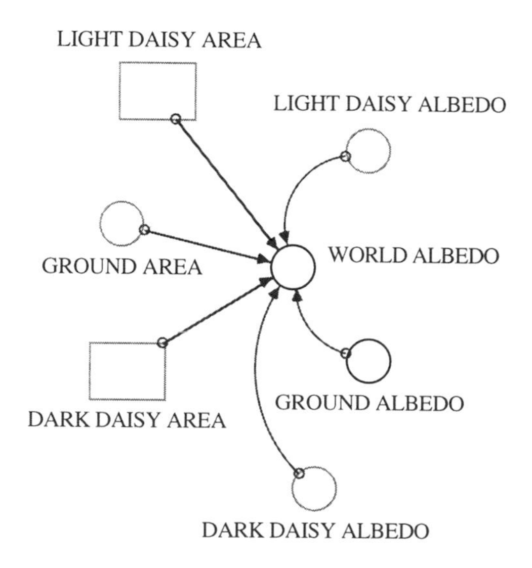

FIGURE 42.3

WORLD ALBEDO, in turn, is determined by

$$\text{WORLD ALBEDO} = \text{LIGHT DAISY AREA*LIGHT DAISY ALBEDO} + \text{DARK DAISY ALBEDO*DARK DAISY AREA} + \text{GROUND AREA*GROUND ALBEDO.} \quad (9)$$

WORLD ALBEDO is calculated in the module of Figure 42.3. In our first scenario, the Sun's luminosity is assumed to change over time:

$$\text{SUN LUMINOSITY} = \text{IF TIME} < 150 \text{ THEN } 0.01\text{*TIME ELSE } 1.5\text{–}0.01\text{*(TIME–150)} \quad (10)$$

The key to Daisyworld's ability to regulate temperature is the different albedos of the light and dark daisies. Because dark daisies absorb more solar radiation, they are warmer at a given solar flux than light daisies. Thus, they are at a different point on their growth curve. Equations (8) and (9) show how Daisyworld's temperature is regulated by the daisies. But the WORLD TEMP is not what determines the growth of the daisy species. Rather, it is the local temperature of each daisy that regulates growth. The local temperature for the two daisy species is given by

$$\text{LIGHT DAISY TEMP} = \text{LOCAL TEMP FUNC*(WORLD ALBEDO} - \text{LIGHT DAISY ALBEDO)} + \text{WORLD TEMP} \quad (11)$$

$$\text{DARK DAISY TEMP} = \text{LOCAL TEMP FUNC*(WORLD ALBEDO} - \text{DARK DAISY ALBEDO)} + \text{WORLD TEMP,} \quad (12)$$

where LOCAL TEMP FUNC is a constant that measures the degree of insulation between daisies (Fig. 42.4). A low LOCAL TEMP FUNC means that there is a great deal of heat conduction between daisy species and all local temperatures equal the mean temperature, while a high LOCAL TEMP FUNC (on the order of 10^2) indicates insulation between high and low temperature regions on the surface. For this simulation, LOCAL TEMP FUNC has been chosen to be 20, which is between the two extremes, but closer to a "conductive" planet surface.

We assume that the Sun's luminosity changes over time according to equation (10) in order to investigate how a Gaian system reacts to perturbations. Since the environment of Daisyworld is solely specified by temperature, the best way to test the Gaia hypothesis is to affect Daisyworld in such a way that temperature is the independent variable. Note, however, that most of the graphs presented here have time as the independent variable. This is somewhat misleading since time is not significant to Daisyworld; instead, the luminosity of the Sun is the independent variable used in this investigation. But SUN LUMINOSITY is varied linearly with respect to time, so the time axes in the graphs can easily be translated to a SUN LUMINOSITY axis.

First, let us consider what is expected of a Daisyworld unaffected by biology. In this case, the albedo of the planet surface is constant and equal to 0.5. As the solar luminosity increases, the temperature of the planet surface

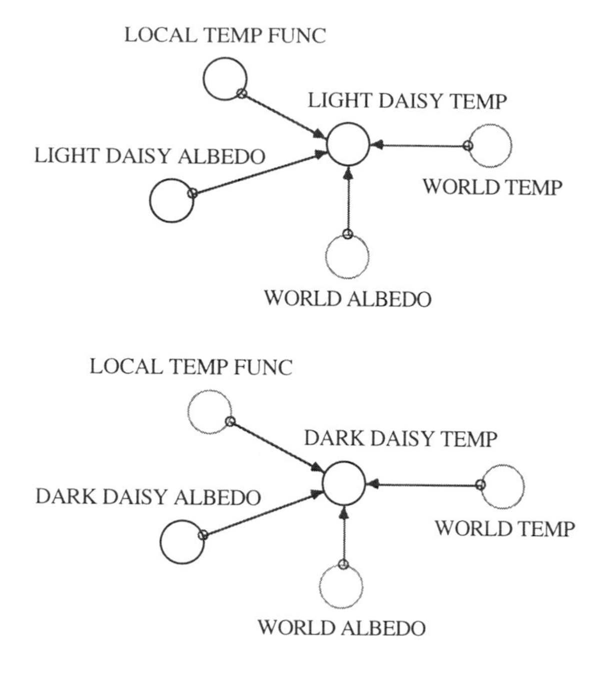

FIGURE 42.4

increases (almost) linearly by the Stefan–Boltzman rule. There is only a small range of luminosities where the planet surface temperature is suitable for daisy growth, so we expect few daisies to appear. This is clearly not a situation where life that requires a certain temperature to grow would thrive. It is this type of environment that life without Gaia faces. Whatever biota are present must be able to survive in the given environmental conditions. As we see in the graphs in Figures 42.5 and 42.6, daisies are only pre-

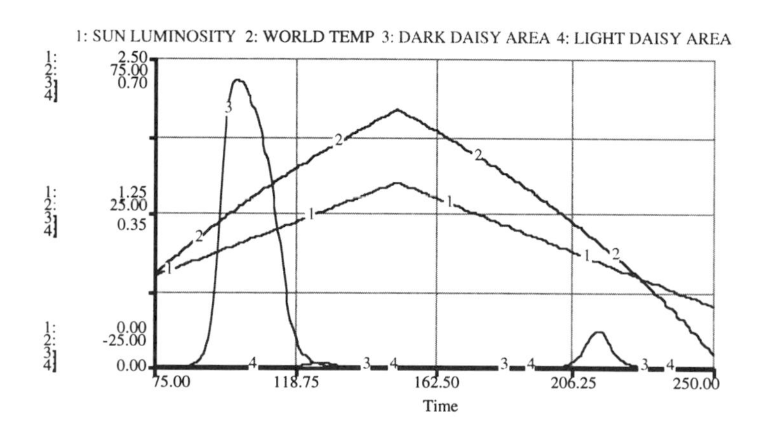

FIGURE 42.5

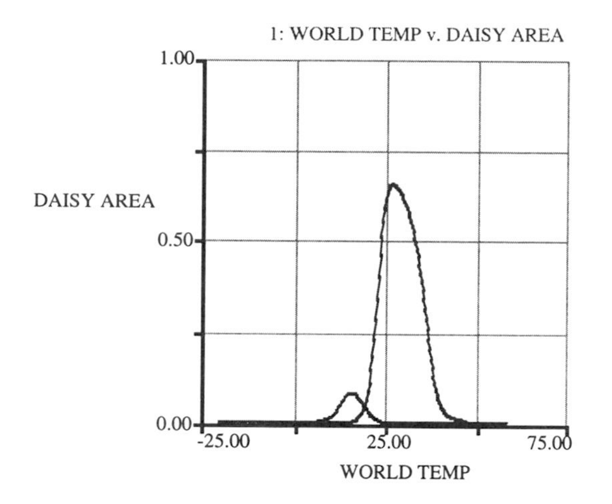

FIGURE 42.6

sent during the short interval in which the planet's surface temperature is appropriate for their growth. Daisies are entirely subject to their environment.

How should a Gaian Daisyworld respond to changing solar flux? Lovelock asserts that the daisies should regulate the temperature of Daisyworld at the optimum growth temperature, 25°C, over a range of solar luminosities. Furthermore, perturbations that are not too large should be accomodated by Daisyworld. The most important characteristic of Gaia is that Gaian feedbacks are homeostatic and should increase the stability of the system. Indeed, Lovelock claims the model always shows greater stability with daisies than it does without them. Thus, if we find that Daisyworld is not more stable with daisies than without, then we can safely assert that Daisyworld is not regulated by Gaia. The results in Figures 42.5 and 42.6 provide a baseline for our assessment.

What happens when there is feedback between the biota and the environment? The following graphs in Figures 42.7 and 42.8 show the "classic" Daisyworld that Lovelock has published numerous times. These are fascinating graphs and demonstrate quite clearly that the daisies are capable without foresight, or planning, of regulating the temperature of Daisyworld to the optimum 25°C. The graphs show the mean temperature, sun luminosity, and planet surface area covered by light and dark daisies. When the solar flux on Daisyworld is so small that the world temperature is less than 25°C, dark daisies flourish. Only when solar luminosity declines significantly can the daisies not survive.

Dark daisies, because they are dark, absorb more light than either the bare ground or light daisies, so they are warmer under the same solar flux

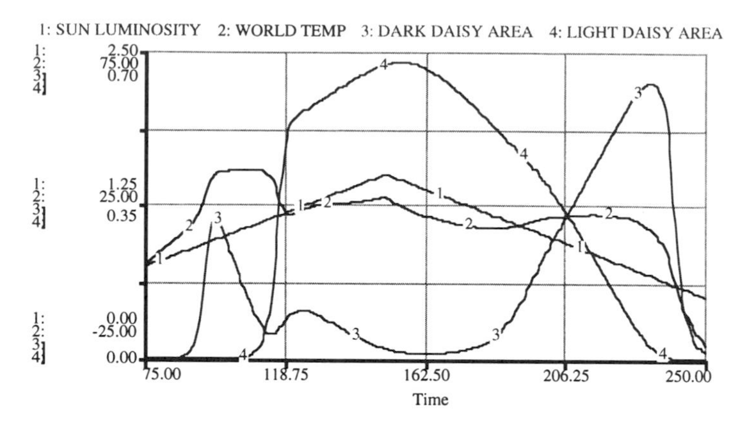

FIGURE 42.7

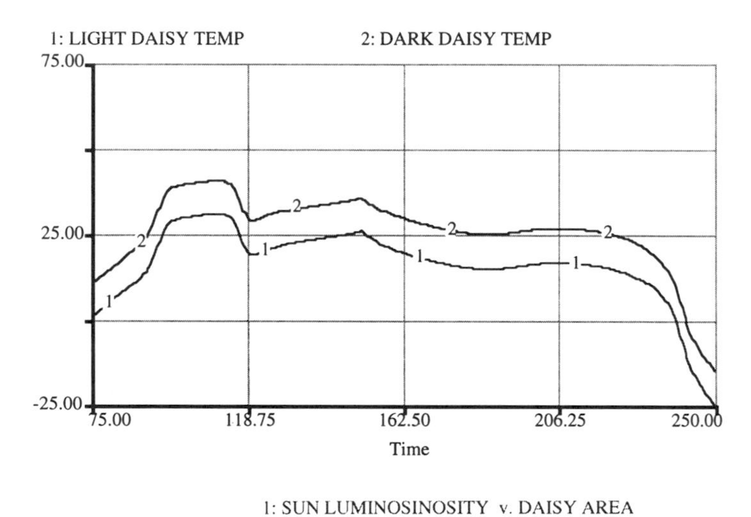

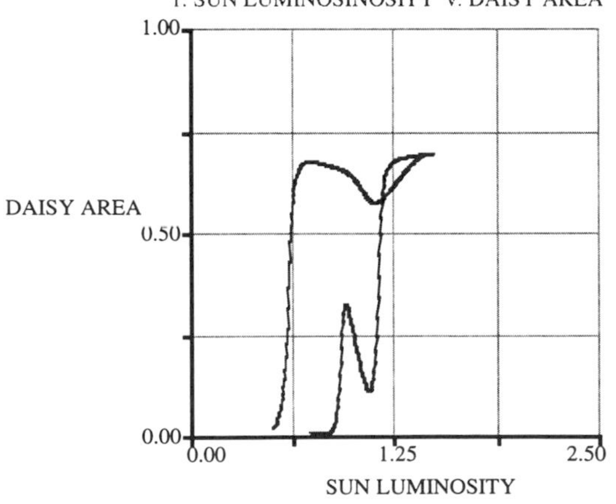

FIGURE 42.8

conditions and can grow when it is too cool for light daisies. As the dark daisy population increases, the world temperature increases, and soon it is warm enough for light daisies. But the solar luminosity is increasing at the same time, so now Daisyworld is getting too hot for the dark daisies and they start to die off. The light daisies cool off the planet surface, so mean temperature drops. As one can see, the light and dark daisies adjust their respective populations to maintain the temperature of Daisyworld near the optimal 25° C (Figs. 42.7 and 42.8)! The temperature regulation is not perfect, of course, in the case of only dark and light daisies. Increase the number of shades to improve Daisyworld.

Daisyworld must also be able to withstand perturbations more dramatic than a simple steady change of solar flux. The following shows a situation where there is a discontinuity in solar flux:

$$\text{SUN LUMINOSITY} = \text{IF TIME} < 150 \text{ THEN } 0.01{*}\text{TIME}$$

$$\text{ELSE IF TIME} \geq 150 \text{ AND TIME} \leq 200 \text{ THEN } 1$$

$$\text{ELSE } 1.5\text{-}0.01{*}(\text{TIME-150}) \tag{13}$$

The effect of the jump in solar flux is to eradicate approximately half of the light daisies. But notice that Daisyworld recovers almost immediately! The dark daisies begin to grow in response to the lower temperature of Daisyworld and the newly available surface area no longer occupied by light daisies. Consequently, the temperature of Daisyworld is quickly returned to the optimum 25°C (Figs. 42.9 and 42.10).

This is a dramatic demonstration of the ability of Daisyworld to regulate itself to an optimum point and is certainly evidence in favor of Gaia. This kind of behavior is also realistic in that life on Earth has withstood such catastrophes before and survived, witness the extinction of the dinosaurs.

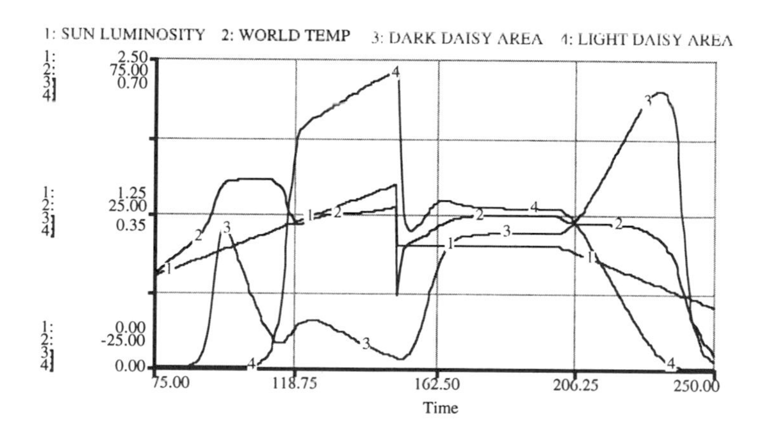

FIGURE 42.9

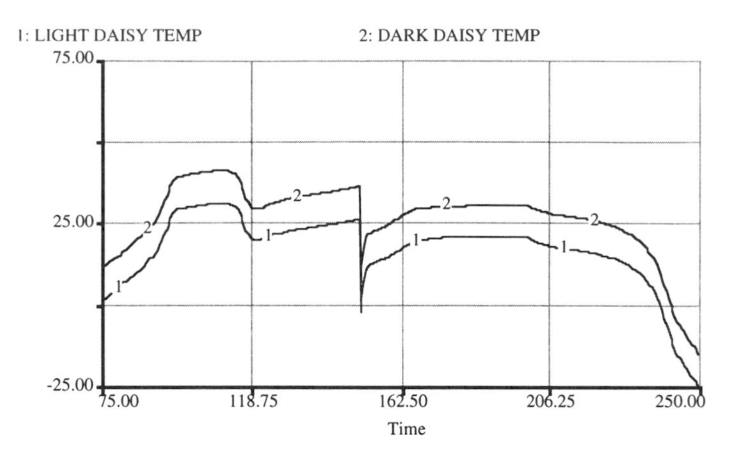

FIGURE 42.10

Granted, dinosaurs were replaced by another genus altogether rather than a new bunch of the same dinosaur species, but the point is that Gaia must be robust enough to sustain life under drastically changing conditions.

Thus far we have demonstrated two characterisitcs of Gaia: regulation of the environment to a point optimal for the growth of the regulating life forms and maintenance of that optimum, even under severe shock. But what about the last, and probably most important point of the Gaia hypothesis that there will be increased stability on the planet? In order to assess the implications of Gaia for the stability of the system, you must first define what is meant by stability. You also must specify the range of perturbations that you allow for the system. For example, find a solar luminosity that yields a steady state for the LIGHT DAISY AREA and DARK DAISY AREA. Then perturb the system and note how close the system gets to the original steady state after disturbance occurred. We have done this with

$$\text{SUN LUMINOSITY} = \text{IF TIME} \geq 120 \text{ AND TIME} \leq 180 \text{ THEN 2 ELSE 1,} \tag{14}$$

and the results are shown in the graphs in Figures 42.11 and 42.12.

How are the results changed if the optimum temperature range for the two daisies differs? You will find for some ranges that there is hysteresis in Daisyworld, implying that for a given amount of solar radiation on Daisyworld there are two stable temperatures and two stable daisy population sizes. How can this finding be reconciled with the notion of Gaia's stability?

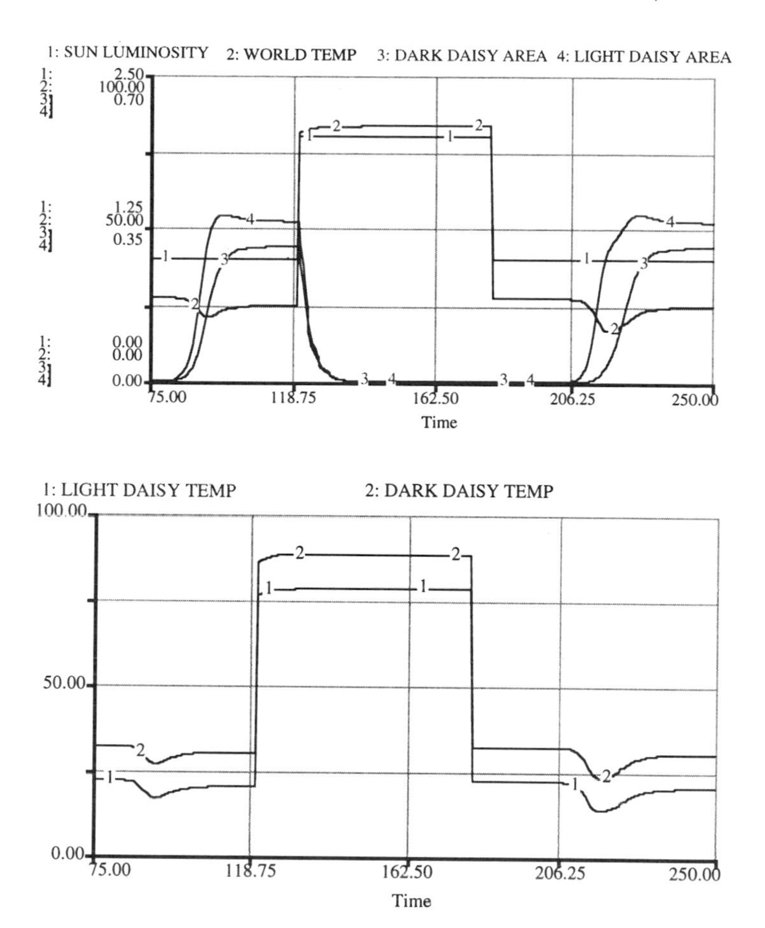

Figure 42.11

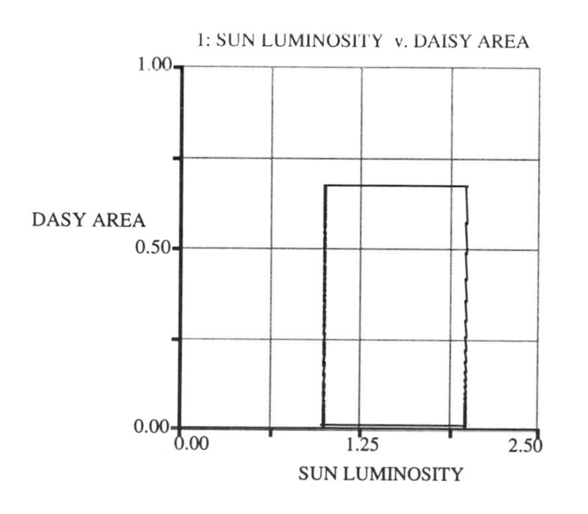

Figure 42.12

DAISYWORLD

```
DARK_DAISY_AREA(t) = DARK_DAISY_AREA(t - dt) +
(DARK_DAISY_BIRTH - DARK_DAISY_DEATH) * dt
INIT DARK_DAISY_AREA = 0.0001
INFLOWS:
DARK_DAISY_BIRTH = if ((DARK_DAISY_TEMP>5) and
(DARK_DAISY_TEMP<45)) then
DARK_DAISY_AREA*NO_DAISY_AREA*DARK_DAISY_GROWTH else 0
OUTFLOWS:
DARK_DAISY_DEATH = DARK_DAISY_AREA*DEATH_RATE
LIGHT_DAISY_AREA(t) = LIGHT_DAISY_AREA(t - dt) +
(LIGHT_DAISY_BIRTH -

LIGHT_DAISY_DEATH) * dt
INIT LIGHT_DAISY_AREA = 0.0001
INFLOWS:
LIGHT_DAISY_BIRTH = if ((LIGHT_DAISY_TEMP>5) and
(LIGHT_DAISY_TEMP<45)) then
LIGHT_DAISY_AREA*NO_DAISY_AREA*LIGHT_DAISY_GROWTH else
0
OUTFLOWS:
LIGHT_DAISY_DEATH = LIGHT_DAISY_AREA*DEATH_RATE
DAISY_AREA = LIGHT_DAISY_AREA+DARK_DAISY_AREA {Fraction
of area covered by daisies}

DARK_DAISY_ALBEDO = 0.25
DARK_DAISY_GROWTH = 1 - 0.003265*(25 -
DARK_DAISY_TEMP)^2
DARK_DAISY_TEMP = LOCAL_TEMP_FUNC*(WORLD_ALBEDO -
DARK_DAISY_ALBEDO) + WORLD_TEMP
DEATH_RATE = 0.3
FERTILE_AREA = 1.0
GROUND_ALBEDO = 0.5
GROUND_AREA = 1 - DARK_DAISY_AREA - LIGHT_DAISY_AREA
{Fraction of total area not covered by daisies}
LIGHT_DAISY_ALBEDO = 0.75
LIGHT_DAISY_GROWTH = 1 - 0.003265*(25 -
LIGHT_DAISY_TEMP)^2
LIGHT_DAISY_TEMP = LOCAL_TEMP_FUNC*(WORLD_ALBEDO -
LIGHT_DAISY_ALBEDO) + WORLD_TEMP
LIGHT_FLUX = 9.17e2 {W/(sec m^2)}
LOCAL_TEMP_FUNC = 20
```

```
NO_DAISY_AREA = FERTILE_AREA - LIGHT_DAISY_AREA -
DARK_DAISY_AREA {Fraction of fertile area not covered
by daisies}
STEFANS_CONSTANT = 5.6703e-8 {W/(m^2 K^4)}
SUN_LUMINOSITY = IF TIME < 150 THEN 0.01*TIME
ELSE 1.5-0.01*(TIME-150)
WORLD_ALBEDO = LIGHT_DAISY_AREA*LIGHT_DAISY_ALBEDO +
DARK_DAISY_ALBEDO*DARK_DAISY_AREA +
GROUND_AREA*GROUND_ALBEDO
WORLD_TEMP = ((SUN_LUMINOSITY*LIGHT_FLUX*(1 -
WORLD_ALBEDO))/STEFANS_CONSTANT)^0.25 - 273
```

Part 8

Conclusion

43

Building a Modeling Community

It was a community of studies, and a community of skill.

D'Israeli, 1841

The models and concepts that we developed in this book are powerful means to investigate the behavior of biological and ecological systems. The modeling approach that we chose is dynamic with regard to four issues. First, the systems that we modeled are dynamics ones, and we portray their dynamics, rather than use a static or comparative-static approach or a statistical model! Second, our model development process itself is dynamic. We encouraged you to start with simple models of complex systems. You will soon find that your models become increasingly complex but more representative of the system about which you are asking important questions. STELLA, through its use of graphics, and MADONNA, through its fast compilation, sensitivity methods, and statistical features, are excellent tools to organize and assess the various aspects of the system that you wish to capture. Once the model system is sufficiently understood, you can easily move on to expand on the model and capture additional features of the real system.

Third, the learning process that accompanies model development and model runs is a dynamic one. By carefully phrasing the questions that the model should answer, and by stating the assumptions that underlie the model that we develop, we structure our knowledge about a system. By running the model and observing the results, we learn about some aspect of a system. Subsequent model runs and model refinements provide more of this insight and should sharpen our focus for future model development and improve our intuition about the behavior of dynamic systems.

Fourth, through building and running computer models, we provide a basis for communicating data and model assumptions. Frequently, model efforts become large-scale multidisciplinary endeavors. STELLA and MADONNA are sufficiently versatile to enable development of complex, large-scale dynamic models. Such models can include a variety of features that are typically not dealt with by an individual modeler. Through easy incorporation of new modules into existing dynamic models and flexibility in adjusting models to specific real-world problems, STELLA and MADONNA foster dialog and collaboration among modelers. It is a superb organizing and knowledge-capturing device for model building in an interdisciplinary arena. Individuals can easily integrate their knowledge into a STELLA model without losing sight of, or influence on, their particular part of the model. We anticipate that the modeling approach presented in this book will increase interaction among modelers and will be generating new momentum for interdisciplinary and cross-cultural exchange of ideas.

With our books on dynamic modeling,[1] this book, and the others in this series, we wish to initiate a dialogue with (and among) you and other modelers. We invite you to share with us your ideas, suggestions, and criticisms of the book, its models, and its presentation format. We also encourage you to send us your best STELLA models. We intend to make the best models available to a larger audience, possibly in the form of books, acknowledging you or your group as one of the selected contributors. The models will be chosen based on their simplicity and their application to an interesting phenomenon or real-world problem.

So register now by (e-) mailing or faxing us your name and address, and possibly something about your modeling concerns. Invite your interested colleagues and students to also register with us now. We can build a modeling community only if we know how to make, and maintain, contact with you. We believe that the dynamic modeling enthusiasm, the ecolate skill, spreads by word of mouth, by people in groups of two or three sitting around a computer doing this modeling together, building a new model or reviewing one by another such group. Share your thoughts and insights with us, and through us, with other modelers. Here is how you can reach us:

Bruce Hannon, Professor
Department of Geography
220 Davenport Hall, MC 150
University of Illinois
Urbana, IL 61801, USA
Phone: (217) 333-0348
Fax: (217) 244-1785 fax
e-mail: bhannon@ncsa.uiuc.edu

Matthias Ruth, Professor
Center for Energy and Environmental Studies
Boston University
675 Commonwealth Avenue
Boston, MA 02215, USA
Phone: (617) 353-5741
Fax: (617) 353-5986
e-mail: mruth@bu.edu

[1] B. Hannon, and M. Ruth. *Dynamic Modeling*, New York: Springer-Verlag, 1994.

Appendix

A1 Installation Instructions for MacIntosh Version

1. Prepare for installation.

- Quit all applications.
- Make sure that you have at least 2.5 MG free space on your hard disk.

2. Launch the STELLA® II installer.

- Insert CD of your package.
- Double click the installer icon.

The **STELLA®** *Screen will appear.*

3. Begin installation to your hard disk.

- Click **Continue** on the **STELLA®** screen.
- Select the desired disk to install onto from the "Select destination folder:" window.

A folder named "STELLA® II Run-time" will be automatically created for you on the disk you choose.

- Click the **Install into** button.

4. Wrap up the installation.

- Wait while installation takes place.
- When installation is complete, click the **Quit** button.
- Eject the disk.

The **STELLA® II** *application and the models will be found in the "STELLA® II Run-time" folder.*

For technical support or to purchase a full version of the **STELLA®** software, call (603) 643-9636.

A2 Installation Instructions for Windows Version

1. Prepare for installation.

- Quit all applications, except for Windows.
- Make sure that you have at least 5 MB free space on your hard disk.

2. Launch the STELLA® installer.

- Insert **CD** of your package into the **a** drive.
- Choose **Run** from the **File** menu in the **Program Manager**.
- Type: **a:\setup**
- Click **OK.**
- Wait while the Installer application loads.

The Setup screen will appear, showing a pre-defined installation directory (stella2r).

- Click the **OK** button in the Setup screen.

3. Wrap up the installation.

- Wait while installation takes place.
- When installation is complete, eject the CD.

The **STELLA® II** *3.0.7 Run time application will be found in the* "**STELLA® II**" *group in the Program Manager. The application (stella2.exe) and models will be located in the stella2r directory in the File Manager.*

For technical support or to purchase a full version of the **STELLA®** software, call (603) 643-9636.

A3. Quick Help Guide[1]

A3.1 Overview of STELLA Operating Environment

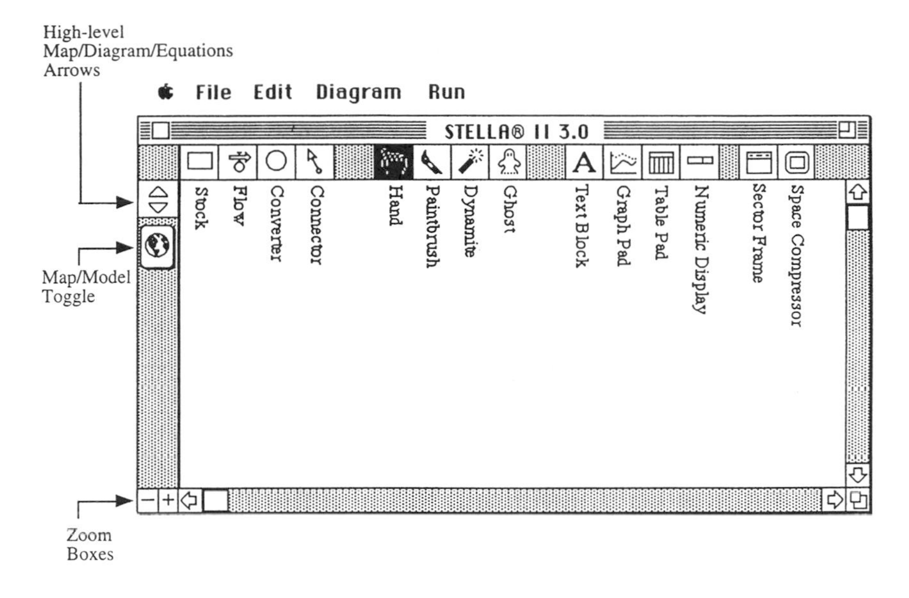

[1]By High Performance Systems, Inc.

A3.2. Drawing an Inflow to a Stock

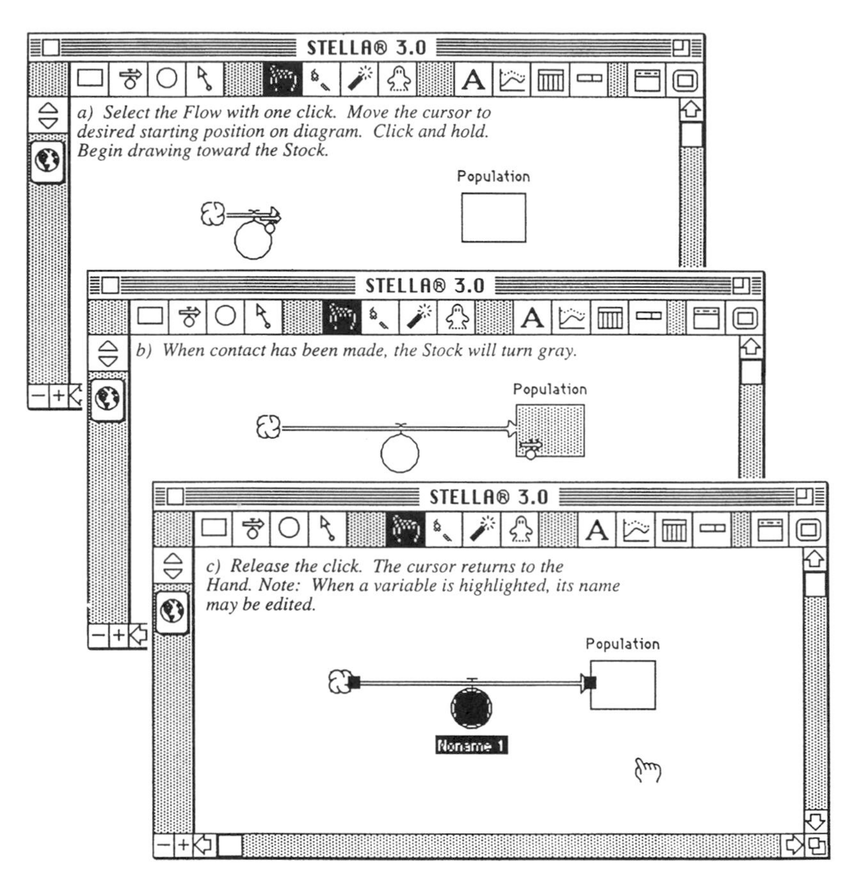

A3.3 Drawing an Outflow from a Stock

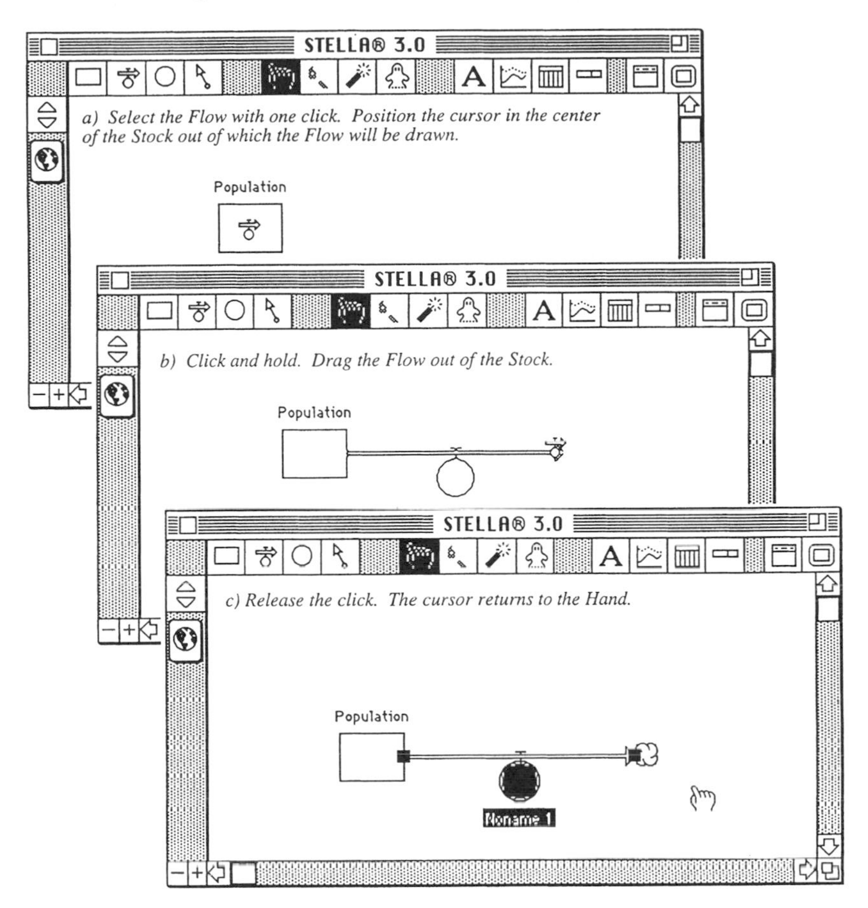

A3.4 Replacing a Cloud with a Stock

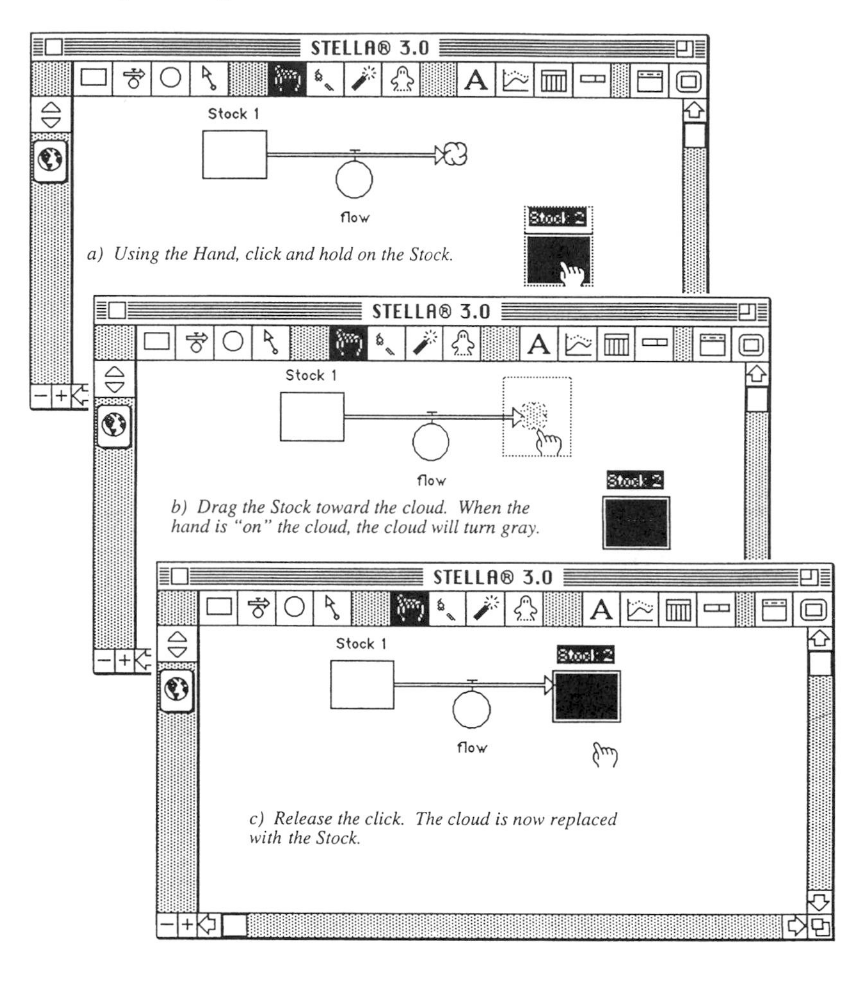

A3.5 Bending Flow Pipes

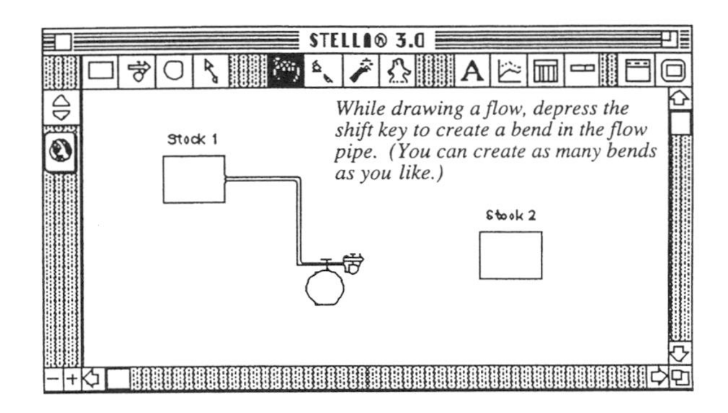

A3.6 Repositioning Flow Pipes

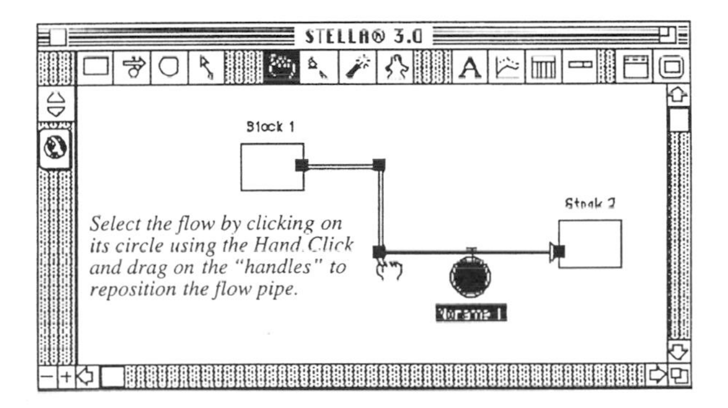

A3.7 Reversing Direction of a Flow

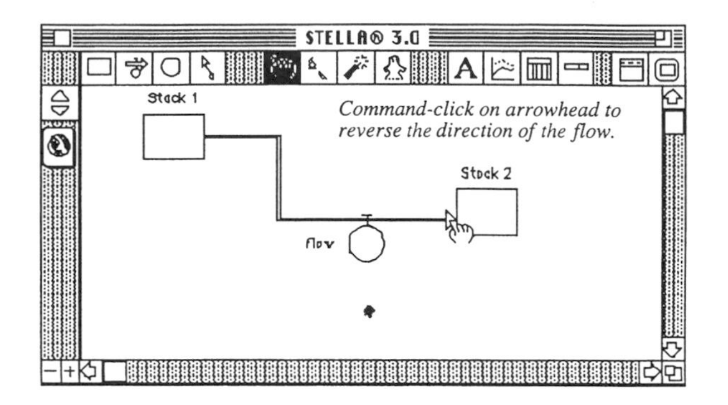

A3.8 Flow Define Dialog—Builtins

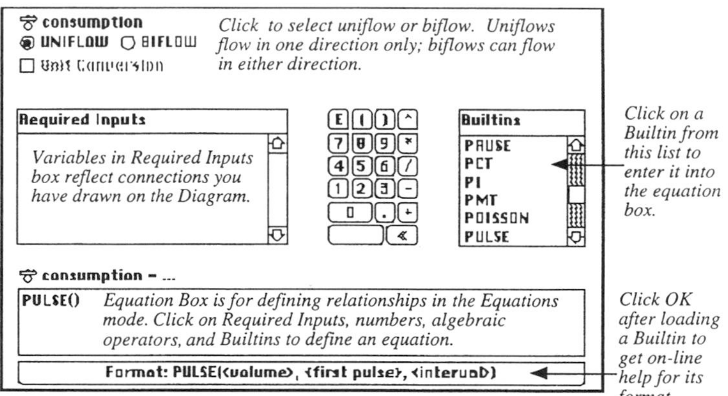

A3.9 Moving Variable Names

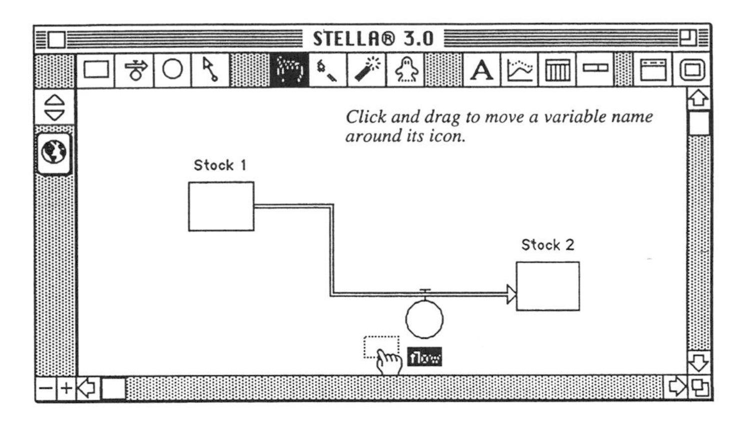

A3.10 Drawing Connectors

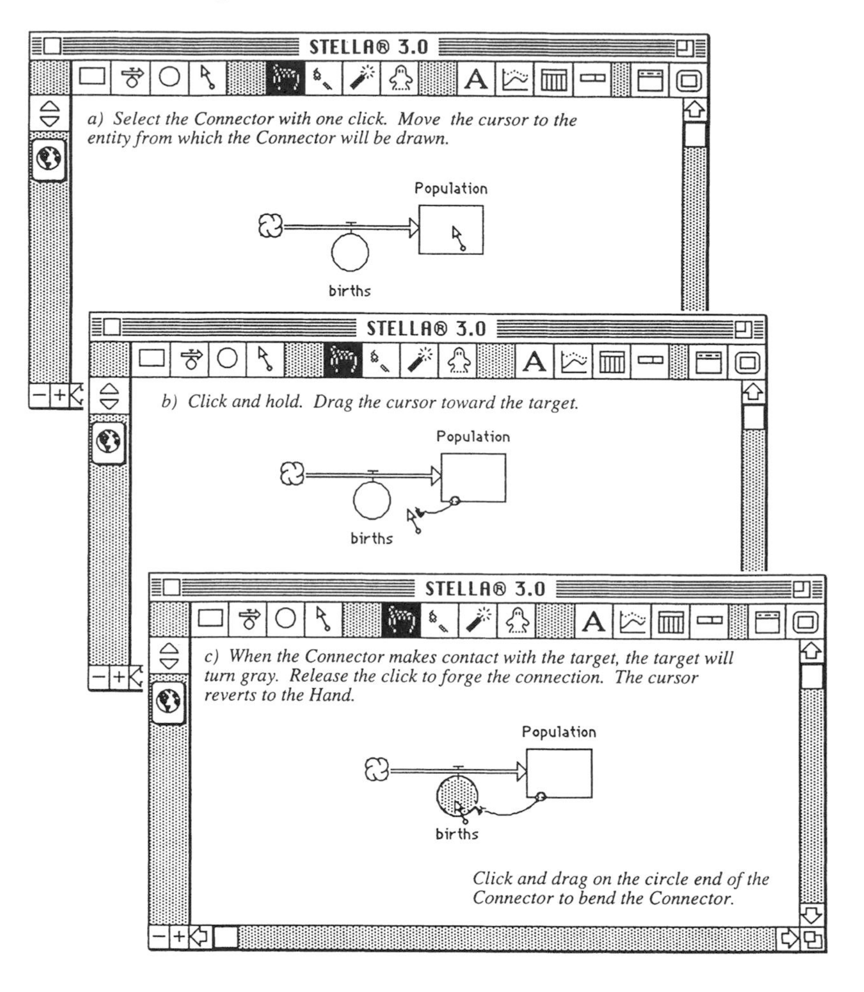

A3.11 Defining Graphs and Tables

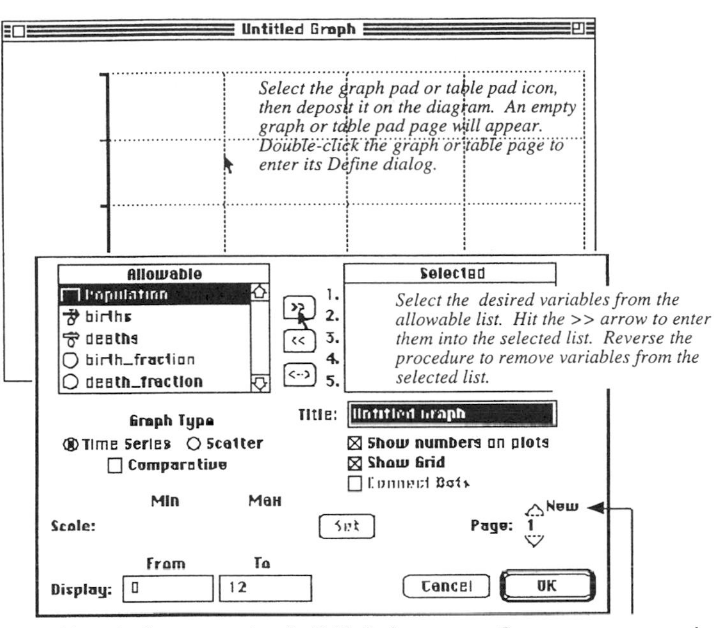

Note: Analogous operations for Table Pads.

Create as many new graph pad pages as you like.

A3.12 Dynamite Operations on Graphs and Tables

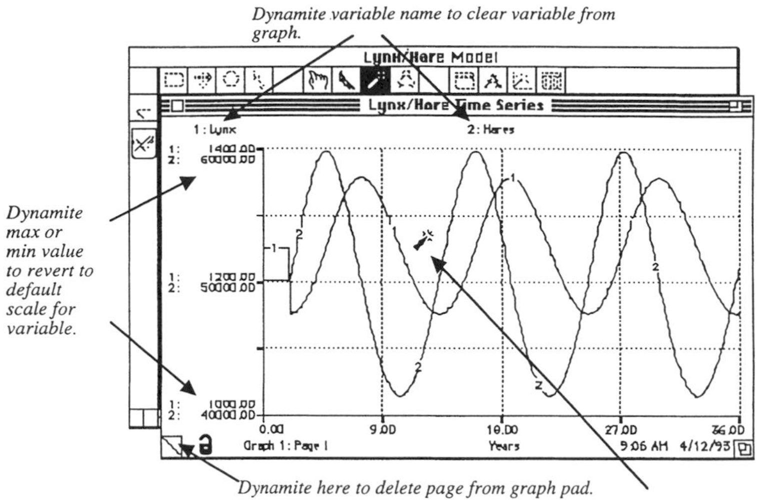

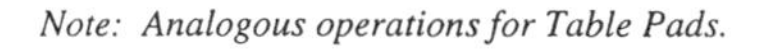

Note: Analogous operations for Table Pads.

MADONNA Quick Start

B1 Installation of MADONNA

If your machine has a math coprocessor (floating point unit), copy MADONNA-FPU to your hard disk. If your Mac does not have an FPU (e.g., PowerBook 500 series) use the Madonna-SANE version. On PowerPC machines, use the SANE version. Connectix Speed Doubler™ is highly recommended for PowerPCs.

B2 Example: Harmonic Oscillator

B2.1 Writing the Model Equations

Open MADONNA and bring up the equation window (Edit Equations in the Model menu, or -E). Type in the following model for the simple harmonic oscillator:

```
dX/dt = -Y
dY/dt = K*X
```

MADONNA doesn't care whether you use upper or lower case, or what order you type in the items, but it's best to organize your equation window in complicated problems. To import a model from STELLA™, open STELLA's Equation window and save the equations as Text. Then switch to MADONNA and open the equation file. With the exception of a few built-ins, MADONNA supports all STELLA syntax.

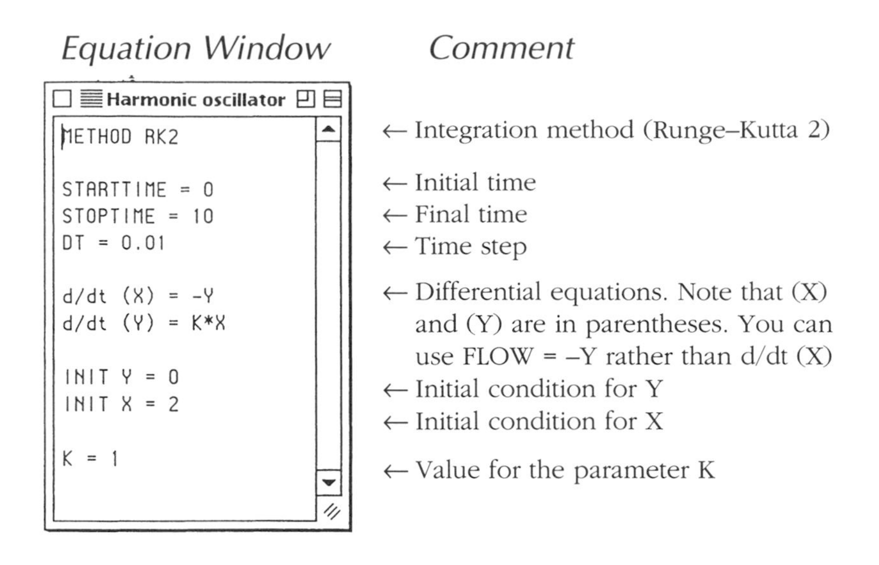

B.2.2 Compiling the Model

Choose Edit Equations in the Model Menu (or press -E) to compile the model; the parameter window will appear, listing all of the model variables. The drop-down menu shows the integration method you have chosen. You can select another method here if the Runge–Kutta method proves unsatisfactory. You can change any quantity in the parameter window by clicking on it and typing a new value in the window.

Parameter Window *Comments*

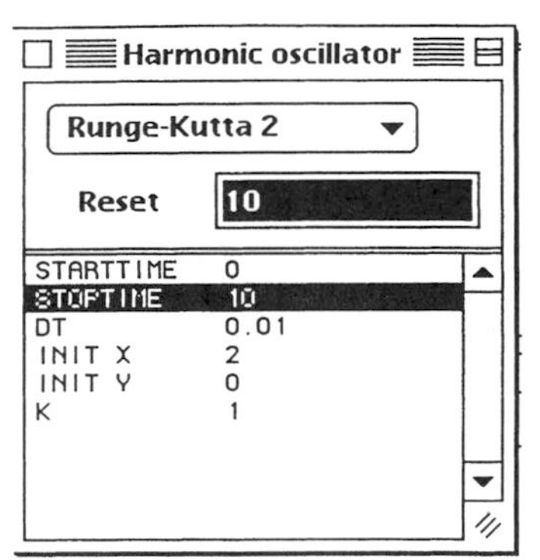

← Integration method

← Set parameter value window

← Highlight the parameters to change

B.2.3 Running the Model

Choose Run in the Model menu (or type -R). The first time you run the model, the New Graph dialog box will open to allow you to select the variables you wish to plot. Click OK and MADONNA will execute the model and the graph will appear.

B.2.4 The Graph Window

If you have a color monitor, the variables will be plotted in different colors. If you have a monochrome monitor, choose Use Dashed Lines in the Graph Menu. MADONNA will put a Parameter List and a Legend on the window. You can move them around with the mouse, or toggle them on and off in the Graph Menu. You can also put the number of steps and run time on the graph. Before printing a graph, choose 25% page reduction in the Page Setup menu for a higher resolution printout.

To change the x or y axes, simply double-click on them. To change the plotting variables, double-click anywhere inside the graph window. To zoom in on a portion of the graph, use the mouse to drag a bounding box. Use the Zoom Out command in the Graph Menu to decrease the magnification.

If you make more than one run, you can choose to overlay the plots, or to plot only the most recent run by choosing Overlay Last Plot or Discard Last Run from the Graph Menu.

B.2.5 Multiple Runs

You can run the model for several successive parameter values, or take the average of many runs by choosing Multiple Runs . . . or Average Runs . . . from the Model Menu.

References

Bak, P and K. Chen, Self-Organized Criticality, *Scientific American* January; 46–53, 1991.

Bak, P., K. Chen, and M. Creutz, Self-organized Criticality in the "Game of Life," *Nature* 342:780–782, 1989.

Bak, P., K. Chen, and C. Tang, A Forest Fire Model and Some Thoughts on Turbulence, *Physics Letters,* 147:297–300, 1990.

Beltrami, E., *Mathematics for Dynamic Modeling,* Boston: Academic Press, 1987.

Botkin, D., Life and Death in a Forest: The Computer as an Aid to Understanding, in: C. Hall and J. Day (eds.), *Ecosystem Modeling in Theory and Practice: An Introduction with Case Studies,* New York: John Wiley and Sons, 1977, p. 217.

Brown, D. and P. Rothery, *Models in Biology: Mathematics, Statistics and Computing,* New York: Wiley, 1993.

Câmara, A.S., F.C. Ferreira, J.E. Fialho, and E. Nobre, Pictorial Simulation Applied to Water Quality Modeling, *Water Science and Technology* 24:275–281, 1991.

Casti, J.L., *Alternate Realities: Mathematical Models of Nature and Man,* New York: John Wiley and Sons, 1989.

Cohen, D., Maximizing Final Yield When Growth Is Limited by Time or by Limited Resources, *Journal of Theoretical Biology* 33:299–307, 1971.

Decker, H., A Simple Mathematical Model of Rodent Population Cycles, *Journal of Mathematical Biology* 2:57–67, 1975.

Denning, P., Modeling Reality, *American Scientist* 78:495–498, 1990.

Edelstein-Keshet, L., *Mathematical Models in Biology,* New York: Random House, 1988.

Finger, R., J. Hughes, B.J. Meade, A.R. Pelletier, and C.T. Palmer, Age-Specific Incidence of Chickenpox, *Public Health Reports* Nov/Dec:750–755, 1994.

Garber, P. and B. Hannon, Modeling Monkeys: A Comparison of Computer Generated and Empirical Measures, *International Journal of Primatology* 14:827–852, 1993.

Gates, D.M., Biophysical Ecology, New York: Springer-Verlag, 1980.

Gertner, G. and B Guan, *Using an Error Budget to Evaluate the Importance of Component Models Within a Large-scale Simulation Model,* Proc. Conf. on Math. Modeling of Forest Ecosystems, Frankfurt am Main, Germany: J.D. Verlag, 62–74, 1991.

Gertner, G., X. Cao, and H. Zhu, A Quality Assessment of a Weibull Based Growth Projection System, *Forest Ecology and Management* 71:235–250, 1991.

Hannon, B., The Optimal Growth of *Helianthus annuus*, *Journal of Theoretical Biology* 165:523–531, 1993.

Hannon, B. and M. Ruth, *Dynamic Modeling,* New York: Springer-Verlag, 1994.

Hastings, A., Structured models of metapopulation dynamics, *Journal of the Biology Linnean Society* 42:57–71, 1991.

Hethcote, H.W., Qualitative Analyses of Communicable Disease Models, *Mathematical Biosciences* 28:335–356, 1976.

Iqbal, M., *An Introduction to Solar Radiation*, Academic Press, Toronto, New York, 1983.

Jenson, R.V., Classical Chaos, *American Scientist* 75:168–181, 1987.

Johnson, A., Spatiotemporal Hierarchies in Ecological Theory and Modeling, Second International Conf. on Integrating Geographic Inform. Systems and Environ. Mod., Sept 26–30, 1993, Breckenridge, CO.

Kamien M. and N. Schwartz, *Dynamic Optimization*, North Holland, 1983, pp. 186–192.

Kaufman, S., *The Origins of Order: Self Organization and Selection in Evolution*, New York: Oxford University Press, 1993.

Lavorel, S., Gardner, R., and R. O'Neill, Analysis of Patterns in Hierarchically Structured Landscapes, *OIKOS* 67:521–528, 1993.

Levin, S., The Problem of Pattern and Scale in Ecology, *Ecology* 73–6:1943–1967, 1992.

Lightman, A. and O. Gingerich, When do Anomalies Begin?, *Science* 255:690–695, 1991.

Loucks, O., M. Plumb-Mentjes, and D. Rodgers, Gap Processes and Large-Scale Disturbances in Sand Prairies, in: *The Ecology of Natural Disturbances and Patch Dynamics*, San Diego, CA: Academic Press, 1985, pp. 71–83.

Lovelock, J.E., *The Ages of Gaia,* New York: W.W. Norton & Co., 1988.

May, R.M. (ed.), *Theoretical Ecology*, 2nd ed., Oxford: Blackwell Scientific Publishers, 199_, pp. 5–29.

May, R.M., Parasitic Infections as Regulator of Animal Populations, *American Scientist* 71:36–45, 1983.

Myers, H.J.H. and C.J. Krebs, Genetic, Behavioral and Reproductive Attributes of Dispersing Field Voles *Microtus pennsylvanicus* and *Microtus ochragaster*, *Ecological Mongraphs* 41:53–78, 1971.

Nee, S. and R. May, Dynamics of Metapopulations: Habitat Destruction and Competitive Coexistence, *Journal of Animal Ecology* 61:37–40, 1992.

Orgel, L.E., The Origin of Life on the Earth, *Scientific American* October:77–83, 1994.

Penrose, R., *The Emperor's New Mind,* New York: Penguin Books, 1989.

Poundstone, W., *The Recursive Universe,* Chicago: Contemporary Books, 1985.

Preston, R., *The Hot Zone*, New York: Anchor Books, 1995.

Prigogine, I., From Being to Becoming: Time and Complexity in the Physical Sciences, New York: W. H. Freeman and Company, 1980.

Reader's Digest Association, *The American Medical Association Family Medical Guide,* New York: Random House, 1982.

Roughgarden, J., Models of Population Processes, *Lectures on Mathematics in the Life Sciences*, American Mathematical Society, Vol. 18, 1986, pp. 235–267.

Ruth, M. and F. Pieper, Modeling Spatial Dynamics of Sea Level Rise in a Coastal Area, *System Dynamics Review* 10:375–389, 1994.

Ruth, M. and B. Hannon, *Modeling Dynamic Economic Systems,* New York: Springer-Verlag, 1996.

Schrödinger, E., *What is Life?,* Cambridge: Cambridge University Press, 1944.

Spain, J.D., *Basic Microcomputer Models in Biology,* Reading, MA: Addison-Wesley Publishing, 1982.

Starfield, A.M. and A.L. Bleloch, *Building Models for Conservation and Wildlife,* New York: MacMillan Publishing, 1986.

Strogatz, S. and I. Stewart, Coupled Oscillators and Biological Synchronization, *Scientific American* —:102–109, 1993.

Tamarin, R.H., Demography of the Beach Vole (*Microtus breweri*) and the Meadow Vole (*Microtus pennsylvanicus*) in Southern Massachusetts, *Ecology* 58:1310–1321, 1977.

Toffoli, T. and N. Margolus, *Cellular Automata: A New Environment for Modeling,* Cambridge, MA: MIT Press, 1987.

von Neumann, J., Theory of Self-Reproducing Automata, Chicago: University of Illinois Press, 1966.

Westervelt, J. and B. Hannon, A Large-Scale, Dynamic Spatial Model of the Sage Grouse in a Desert Steppe Ecosystem, Mimeo, Department of Geography, University of Illinois, Urbana, Illinois, 1993; and http://ice.gis.uiuc.edu.

Wiley, M. J., P. P. Tazik, and S. T. Sobaski, *Controlling Aquatic Vegetation with Triploid Grass Carp,* Champaign, IL: Illinois Natural History Survey, Circular 57, 1987.

Index

QH 324.2 .H36 1997

Hannon, Bruce M.

Modeling dynamic biological systems

IMPORTANT: Read Before Opening CD-ROM Package

HIGH PERFORMANCE SYSTEMS, INC. LICENSE AGREEMENT

NOTE: This book is non-returnable if the CD-ROM package has been opened.

WINDOWS® and MACINTOSH® Version 0-387-94850-3